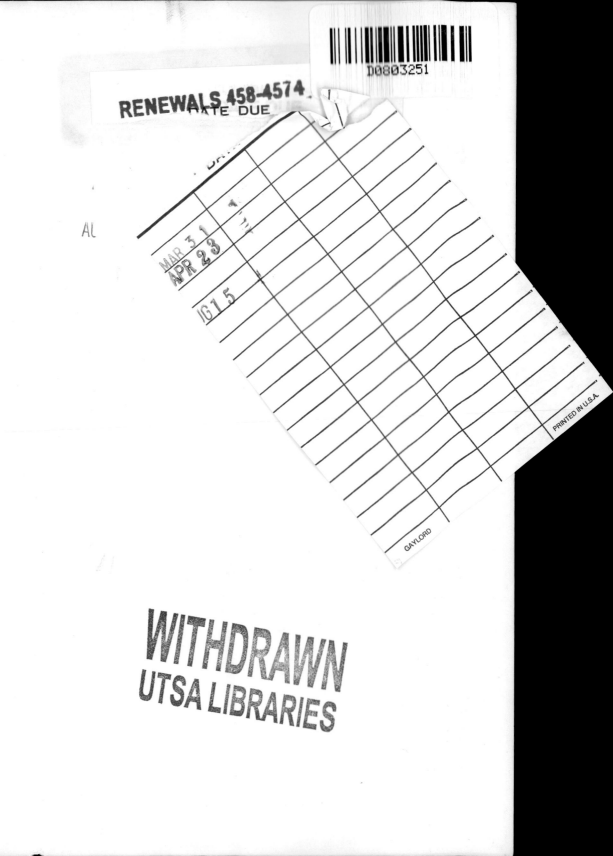

INTEGER PROGRAMMING

Theory, Applications, and Computations

OPERATIONS RESEARCH
AND INDUSTRIAL ENGINEERING

Consulting Editor: J. William Schmidt

CBM, Inc., Cleveland, Ohio

Applied Statistical Methods
I. W. Burr

Mathematical Foundations of Management Science
and Systems Analysis
J. William Schmidt

Urban Systems Models
Walter Helly

Introduction to Discrete Linear Controls: Theory and Application
Albert B. Bishop

Integer Programming: Theory, Applications, and Computations
Hamdy A. Taha

In preparation:

Transform Techniques for Probability Modeling
Walter C. Giffin

Analysis of Queueing Systems
J. A. White, J. W. Schmidt, and G. K. Bennett

INTEGER PROGRAMMING
Theory, Applications, and Computations

Hamdy A. Taha

Department of Industrial Engineering
University of Arkansas
Fayetteville, Arkansas

 1975

Academic Press   New York   San Francisco   London

A Subsidiary of Harcourt Brace Jovanovich, Publishers

ACADEMIC PRESS, INC.
111 Fifth Avenue, New York, New York 10003

United Kingdom Edition published by
ACADEMIC PRESS, INC. (LONDON) LTD.
24/28 Oval Road, London NW1

Library of Congress Cataloging in Publication Data

Taha, Hamdy A
  Integer programming—theory, applications, and
computations.

  (Operations research and industrial engineering)
  Bibliography: p.
  Includes index.
  1. Integer programming. I. Title.
T57.7.T33   519.7'7   74-10205
ISBN 0–12–682150–X

To my sons

TAREK and SHARIF

Contents

Preface   xi

Chapter 1   **Integer Optimization and Its Applications**

1.1   What Is Integer Optimization?   1
1.2   "Solving" the Integer Problem by Rounding the Continuous Optimum   4
1.3   Examples of the Applications of Integer Programming   8
1.4   Methods of Integer Programming   27
1.5   Organization of the Book   29
      Problems   30

Chapter 2   **Linear Programming**

2.1   Introduction   34
2.2   Definition of Linear Programming   35
2.3   The Simplex Method   37
2.4   The Revised Simplex Method   49
2.5   The Dual Problem   52
2.6   Bounded Variables   74
      Problems   79

Chapter 3   **Zero–One Implicit Enumeration**

3.1	Introduction	85
3.2	Zero–One Equivalence of the Integer Problem	86
3.3	Concept of Implicit Enumeration	87
3.4	Enumeration Scheme	89
3.5	Fathoming Tests	94
3.6	Nonlinear Zero–One Problem	116
3.7	Mixed Zero–One Problem	119
3.8	Concluding Remarks	133
	Problems	133

Chapter 4   **Branch-and-Bound Methods**

4.1	The Concept of Branch-and-Bound	139
4.2	Branch-and-Bound Principle	142
4.3	General (Mixed) Integer Linear Problem	144
4.4	Solution of Nonlinear Integer Programs by Branch-and-Bound	171
4.5	Concluding Remarks	172
	Problems	173

Chapter 5   **Cutting Methods**

5.1	Introduction	177
5.2	Dual Cutting Methods	179
5.3	Primal Cutting Methods	213
5.4	Comments on Computational Experience	224
5.5	Concluding Remarks	225
	Problems	225

Chapter 6   **The Asymptotic Integer Algorithm**

6.1	Introduction	230
6.2	The Idea of the Asymptotic Algorithm	231
6.3	Development of the Asymptotic Algorithm	234
6.4	Solution of the Group (Relaxed) Problem	241
6.5	Solution of Integer Programs by the Group Problem	253
6.6	Reducing the Number of Congruences	258
6.7	Faces of the Corner Polyhedron	259
6.8	Concluding Remarks	260
	Problems	260

Chapter 7   **Algorithms for Specialized Integer Models**

7.1	Introduction	263
7.2	Knapsack Problem	263
7.3	Fixed-Charge and Plant Location Problems	285
7.4	Traveling Salesman Problem	304

7.5 Set Covering Problem 316
7.6 Concluding Remarks 335
Problems 335

Chapter 8 **Computational Considerations in Integer Programming**

8.1 Introduction 342
8.2 Model Formulation in Integer Programming 343
8.3 A Composite Algorithm 346
8.4 "General" Approximate Methods for Integer Programming 347
8.5 Concluding Remarks 351

References 353

Index 375

Preface

Integer programming is a subject of tremendous potential applications. The past two decades have witnessed extensive theoretical research. The result is a vast collection of solution methods and algorithms. Progress in the computational aspects of integer programming has not been as impressive, however, in spite of tremendous developments in the power and accuracy of the digital computer.

This book is intended to give a balanced coverage of the theory, applications, and computations of integer programming. It is emphasized, however, that while much progress has been achieved on the theoretical front, computations with the devised techniques have been less satisfactory. The general orientation of the presentation is to expose the potential computational advantages of each technique in a manner that may assist the practitioner in choosing the most suitable technique (or combination of techniques) for *solving* problems.

Chapter 1 presents a general categorization of integer applications which also is intended to underscore the importance of integer programming in real-life applications. Because linear programming forms the basis for most of the developments in integer programming, Chapter 2 presents a self-contained treatment of the subject which assumes no prior knowledge of linear programming.

The three basic techniques of integer programming are covered in the subsequent four chapters. Zero–one implicit enumeration is treated in Chapter 3. Chapter 4 presents the branch-and-bound methods. The cutting-plane method and its closely related asymptotic problem are covered in Chapters 5 and 6.

Because of the computational difficulty that characterizes general integer techniques, attempts are made to develop special methods for integer problems with specialized structures. Chapter 7 presents different methods for each of the knapsack, fixed-charge, traveling salesman, and set covering problems.

The closing chapter (Chapter 8) summarizes a number of observations about the formulations and executions of integer programming models. The objective is to condition the models in a manner that will improve the chances of actually realizing a solution to the problem.

The encouragement and support provided by Dr. John L. Imhoff, head of the Industrial Engineering Department at the University of Arkansas, through all the stages of writing this book are greatly appreciated.

As always, I am gratefully indebted to my students who patiently provided comments and criticisms on all the drafts that were used as class notes since the Spring of 1971. Special recognition is given to my graduate students Major Frank R. Giordano (U.S. Army) and John E. Moore (NASA).

INTEGER PROGRAMMING

Theory, Applications, and Computations

Integer Optimization and Its Applications

1.1 What Is Integer Optimization?

Any decision problem (with an objective to be maximized or minimized) in which the (quantifiable) decision variables must assume nonfractional or discrete values may be classified as an integer optimization problem. In general, an integer problem may be constrained or unconstrained, and the functions representing the objective and constraints may be linear or nonlinear. In the strict sense, every integer problem should be regarded as nonlinear, since its functions are defined only at discrete values of the variables. However, this technical detail may be overlooked for the sake of establishing a classification that is more meaningful from the viewpoint of developing solution methods for the integer problems. Namely, an integer problem is classified as linear if, by relaxing the integer restriction on the variables, the resulting functions are strictly linear. Otherwise, the problem is nonlinear. It is shown later that this classification serves as an important basis for developing solution methods for the integer problem. Indeed, most of the developments in the area are concentrated on the linear problem, primarily because of its relative ease.

Integer optimization, in the sense defined above, is not a new mathematical subject, but until the applications of operations research became recognized in the late 1940s and early 1950s, most of the problems tackled were primarily of a purely mathematical nature. Examples include determination of the maximum number of parts into which n planes in general position divide three spaces and the coloring problem in which it is required to determine the minimum number of colors needed to color the regions of an arbitrary planar map so that no two regions that have a boundary *segment* in common have the same color. [See Saaty (1970) for more detailed discussion.] Unfortunately, efforts in this direction, unlike in continuous mathematics, have produced little unifying theory for integer optimization. Instead, only special cases were studied.

The importance of integer optimization in solving practical problems evolved as a result of the impressive developments in the field of operations research, particularly the subject of linear programming. It was then that both researchers and practitioners recognized the need for solving programming models in which some or all of the decision variables are integers. Although several important problems in various areas of application were formulated as integer models (see Section 1.3), it was only in 1958 that Gomory (1958) developed the first *finite* integer programming technique for solving linear integer problems. Since then, other specialized algorithms have been and are still being developed.

1.1.1   Mathematical Definition of the Integer Programming Problem

The general integer problem may be defined as:

maximize (or minimize)

$$z = g_0(x_1, x_2, \ldots, x_n)$$

subject to

$$g_i(x_1, x_2, \ldots, x_n) \begin{Bmatrix} \leqslant \\ = \\ \geqslant \end{Bmatrix} b_i, \qquad i \in M \equiv \{1, 2, \ldots, m\}$$

$$x_j \geqslant 0, \qquad j \in N \equiv \{1, 2, \ldots, n\}$$

$$x_j \text{ an integer}, \qquad j \in I \subseteq N$$

If $I = N$, that is, all the variables x_j are restricted to integer values, the problem is called a *pure* integer problem. Otherwise, if $I \subset N$, then one is dealing with a *mixed* problem. Sometimes the concept of pure and mixed programs is extended to include the slack variables associated with the constraints of the problem. This is particularly important when the algorithm itself necessitates that all the variables (slacks and otherwise) be restricted to integer values before it can be utilized. It is shown later that, in most cases that have thus far been treated, such a restriction presents no special inconvenience.

Most of the serious developments in the field of integer programming have been directed to the cases where the functions g_i, $i \in \{0\} \cup M$, are linear. For the sake of standardization, the linear problem is written as:

maximize (or minimize)

$$z = \sum_{j \in N} c_j x_j$$

subject to

$$\sum_{j \in N} a_{ij} x_j + S_i = b_i, \quad i \in M$$

$$S_i \geqslant 0, \quad i \in M$$

$$x_j \geqslant 0, \quad j \in N$$

$$x_j \text{ integer}, \quad j \in I \subset N$$

where S_i is a slack variable. If the constraint is originally in equation form, then the associated slack variable is nonexistent.

It must be noted that it has been traditional in some integer programming literature to write the objective and constraint functions in the forms $\sum_{j \in N} (-c_j)(-x_j)$ and $\sum_{j \in N} (-a_{ij})(-x_j)$. However, this (needlessly complicating) notation has no actual bearing on the integer programming developments, and there is no real advantage in following it.

In the absence of the integrality condition, the problem becomes an ordinary (continuous) linear or nonlinear program. In other words, integer programming methods seek the determination of the optimum point among all the discrete points included in the continuous feasible solution space.

On the surface, it may appear that the additional integrality condition should not present a serious problem. Indeed, the solution space is "better" defined now since one no longer needs to "search" among an infinite number of points as in the continuous case (assume, for simplicity, that the continuous space is bounded). Unfortunately, the above conclusion cannot be any more erroneous. Although the solution space of the integer problem is structurally better defined (in the sense given above) than in the continuous problem, it has proved to be *computationally* formidable. To date, in spite of two decades of continuous theoretical research, together with a tremendous increase in the speed and power of the digital computer, the developed integer algorithms have not yielded satisfactory computational results.

The fact of the matter is that the integrality condition often destroys the "nice" properties of the solution space. A typical illustration is the integer *linear* problem. In the absence of the integer condition, the solution space is convex. This is primarily the basic property that leads to the tremendously successful simplex method for solving linear programs (see Chapter 2).

Because of the successful accomplishments in the fields of linear and nonlinear *continuous* programs, it is no wonder that almost all integer algorithms have been developed mainly by converting the discrete space into an equivalent continuous one. This is generally achieved by modifying the original continuous solution space such that the desired best integer point(s) is singled out. Even in the situations where it may appear that the continuous space is not utilized (such as problems in which all the variables are binary), one can usually show that the method of solution can be traced to the continuous version (see Theorem 3-3, Section 3.5.1).

1.2 "Solving" the Integer Problem by Rounding the Continuous Optimum

The disappointment experienced with available integer programming codes has led most potential users to avoid them. Consequently, some believe that it may be better to solve the problem as a continuous model and then round the results "intelligently" to acquire a good, if not optimal, solution quickly. After all, the error in estimating the different parameters of the model should allow such flexibility in locating an "optimum" solution.

Unfortunately, researchers in the field have not addressed themselves seriously to challenging this viewpoint. As a result, its advocates appear to be gaining momentum, especially since the long theoretical research has not produced significant computational advantages.

The idea of rounding is not without merit, at least for the time being, and especially when one is confronted with having either "a" solution or no solution at all. However, one must not assume that every integer problem can be dealt with in this manner. In other words, one must be aware of the (sometimes severe) limitations of rounding.

To assist in establishing the limitations of rounding, a categorization of integer programming applications is introduced in the next section. It will be seen that the nature of the application as specified by this categorization has a great deal to do with the limitation of utilizing rounding. In order to assert the significance of this categorization, typical illustrative models will be introduced for each case. This also should acquaint the reader with the potential uses of integer programming. In the end, the primary purpose is to convince the reader that further research is still warranted in order to develop workable algorithms for this important class of problems.

1.2.1   Classifying Integer Applications Based on Rounding the Continuous Optimum

Integer programming problems are usually categorized according to their area of application. A different categorization will be followed here for the purpose of establishing the importance of integer programming as a useful tool. An application is categorized as a *direct* integer model if the decision involves variables that are *naturally* quantifiable but for which fractional values are inadmissible. Examples of such variables include the number of machines needed for a given job and the number of buses needed to provide passengers service on a given route.

It must be noted that there is a class of direct integer *linear* models for which the continuous solutions are automatically integer. These models, such as the well-known transportation problem, possess what is called the *unimodular property*, which guarantees that every basic solution is integer. [See Heller and Tompkins (1956).]

A *coded* integer model is one in which the decision variables describe (quantitatively) a qualitative aspect or relationship that implies a finite number of possibilities. A typical example describes the "yes–no" or "go–no go" type of decision. In this case, a binary variable can be used as a *code* representing this situation with the one-value implying the yes or go decision and the zero-value representing the only other possibility.

Finally, a *transformed* integer model is one whose formulation may not include any integer variables but whose method of solution may necessitate the use of some "artificial" integer variables in order for the model to be analytically tractable. Normally, these models include some logical relationships that cannot be handled directly by the available solution methods, but this difficulty can be rectified by utilizing artificial binary variables. For example, in regular mathematical models, all constraints are assumed to be satisfied simultaneously, but there are situations where, say, either one of two given constraints must be satisfied. A binary variable can be used to convert the problem in the apparent sense into two simultaneous constraints without actually disturbing the original logical relationship. As will be seen later, the binary variable in this case may be regarded as a code with the one-value implying that one of the constraints is active and the zero-value indicating that the remaining constraint is inactive.

1.2.2   Limitations of the Rounding Procedure

The above categorization relates to the idea of rounding as follows: Rounding implies a sense of *approximation*, that is, $x = 10.2$ may be approximated (rounded) to $x = 10$, while $x = 10.8$ may be approximated to $x = 11$. The rounding "error" in this case depends on how large the continuous value is. In terms of the above categorization, only direct integer models seem susceptible to applying approximation or rounding. For example, if a continuous solution indicates that the number of ships needed is 20.3 this can be taken as approximately 20 ships. However, in the coded and transformed integer models, a fractional value of the integer variable is meaningless. Primarily, one is interested in the value of the variable only in the sense that it represents an element of a (numerical) code. Indeed, in order for the value of a variable to make sense, it must be *decoded* first in terms of the decisions for the original problem. For example, for a binary variable, a zero-value may be decoded as a no-decision, in which case a one-value is taken as a yes-decision. In this respect, there is no logical foundation for applying approximation (or rounding) to these problems. Of course, it may be argued that in capital budgeting a variable may be interpreted meaningfully on the continuous 0–1 scale, so that if $x_j = \frac{1}{2}$ this means that 50% of the jth project must be completed in the current period. But if this is allowable, then the problem is actually continuous in nature and there is no need in this case to impose the integrality condition.

The above discussion is aimed at confining the possibility of using rounding to what is referred to as direct integer models. The objective now is to specify the limitations of utilizing rounding in this case. There are two points to be considered:

1. If a *feasible* solution is obtained by rounding, one should not be under the illusion that such a solution is optimal or even close to optimal. The fact that continuous optimization is used does not necessarily mean that it will lead to a "good" integer solution. The rounding procedure at best may be regarded as a *heuristic*. But even in this case, there may be other heuristics that would yield better solutions than when rounding is used.

2. Any integer model having an *original equality* constraint can never yield a feasible integer solution through rounding. This is based on the assumption that only basic variables can be rounded, if necessary, and that all the nonbasic variables remain at zero level. The assumption is not unreasonable since it is generally difficult to consider elevating a nonbasic variable above zero while maintaining feasibility.† This result was observed by Glover and Sommer (1972) and is illustrated by the following numerical example.

---

† Actually, the idea of rounding in which the nonbasic variables are allowed to assume positive (integer) values serves as the basis for developing a sophisticated, but still computationally difficult, method for solving integer linear programming problems (see Chapter 6).

▶**EXAMPLE 1.2-1**   The optimal continuous solution to the problem:

maximize

$$z = 20x_1 + 10x_2 + x_3$$

subject to

$$3x_1 + 2x_2 + 10x_3 = 10$$

$$2x_1 + 4x_2 + 20x_3 \leqslant 15$$

$$x_1, x_2, x_3 \geqslant 0 \quad \text{and} \quad \text{integer}$$

is given by $x_1 = \frac{5}{4}$, $x_2 = \frac{25}{8}$, and $x_3 = 0$, with x_3 the nonbasic variable. Under the assumption that x_3 is fixed at zero level, it is impossible that the rounded solution $(x_1 = 1, x_2 = 3)$ can be made to satisfy the first *equality* constraint. This follows since, by definition, the *basic* solution $(x_1 = \frac{5}{4}, x_2 = \frac{25}{8})$ is unique, and it is thus impossible to secure different values with x_1 and x_2 remaining as basic variables.

Now suppose that the (first) equation is changed to a ($\leqslant$) constraint. The slack variables S_1 and S_2 associated with the first and second constraints have the convenient property that each is associated with a unique constraint, and hence the adjustments resulting from rounding the basic variables can be "absorbed" by each slack variable independently (compare this with the difficulty that may result from adjusting x_3 in the above example). This is the reason these slack variables, even though they are nonbasic, are allowed to assume positive values. Thus for this situation, the rounded solution $x_1 = 1$, $x_2 = 3$, yields $S_1 = S_2 = 1$. This, incidentally, is not the optimum solution, since $x_1 = x_2 = 2$ obviously yields a better value of z.

It may appear that this last situation leads to contradiction, since the values of the basic variables x_1 and x_2 are no longer unique. This is not true since the uniqueness of the basic solution is dictated by the fact that the associated nonbasic variables remain at zero level. However, this restriction is no longer active when the nonbasic variables can be (conveniently) allowed to be positive, as in the case where the constraints are of the type ($\leqslant$). Actually, it is a perfectly valid theory in linear programming to have *positive nonbasic* variables, as illustrated by the case of upper bounded variables (see Section 2.6).◀

It must be stated that if the original equations are such that some of the original decision variables take the role of slack variables then rounding may be possible, provided these variables are nonbasic in the continuous optimum solution.

The main conclusion from the above dicussion is that the use of rounding is not a reliable procedure for dealing with the general integer problem. This emphasizes the importance of continued research in order to develop efficient algorithms for the integer problem.

1.3 Examples of the Applications of Integer Programming

In terms of the above categorization, there is little that distinguishes the formulation of a direct integer model from an ordinary (linear or nonlinear) programming model except that the variables are integers. However, interesting formulations exist for the cases of coded and transformed integer models. Consequently, all the examples in this section will be selected from these two areas. In addition to exposing the reader to various practical applications, emphasis is placed on how the assumptions and the choice of the decision variables can lead to different model formulations with varying degrees of complexity for the same problem.

1.3.1 Examples of "Coded" Integer Models

The examples cited in this section include applications to capital budgeting, sequencing, scheduling, and the knapsack, traveling salesman, and set covering problems. The list is not meant to be exhaustive of all the applications under this category, but it should cover all the well-known cases. Further examples are also included in the problems at the end of the chapter.

▶EXAMPLE 1.3-1 (CAPITAL BUDGETING)   Capital budgeting is concerned with the availability of a variety of investment projects, and the objective is to select the most promising set of projects under the condition of a fixed capital budget. The problem may be considered for a single period or a multiperiod horizon.

There is a large variety of formulations for the capital budgeting problem. Perhaps the main source of variation originates from the objective function. In general, the capital budgeting problem is not a deterministic case. Some researchers argue that the use of *expected* return is not sufficient as an objective criterion since it ignores the dispersion around the mean and hence the concept of risk. Others insist that higher-order moments of the distribution of return must also be accounted for. In this brief summary, a number of formulations will be presented. It is shown how the different arguments can lead to drastically different integer models, with each formulation requiring an entirely different solution method as a result. The bulk of this presentation is based on an article by Peterson and Laughhunn (1971).

Let the following symbols represent the characteristics of the investment problem: c_j a random variable representing the net present value of the jth investment project ($j = 1, 2, \ldots, N$), a_{tj} a random variable representing the net cost of the jth investment project in the tth period ($t = 0, 1, 2, \ldots, T$), and b_t a random variable representing the net cash available in the tth period for allocation to the investment projects. It is assumed that the density functions of the random variables are available or that at least enough information can be secured to determine their first three moments.

In the simplest case, the model can be formulated under *assumed certainty*, that is, each random variable is replaced by its expected value. This formulation was first introduced by Weingartner (1963). If $x_j = 0$ denotes rejection and $x_j = 1$ denotes acceptance of project j, the resulting deterministic model becomes:

maximize

$$E\{z\} = \sum_{j=1}^{N} E\{c_j\}x_j$$

subject to

$$\sum_{j=1}^{N} E\{a_{tj}\}x_j \leqslant E\{b_t\}, \qquad t = 0, 1, 2, \ldots, T$$

$$x_j = (0, 1), \qquad j = 1, 2, \ldots, N$$

The operator E signifies the expected value of the random variable.

The above formulaton is based on the assumption that the decision maker's utility function is linear in z. This assumption may not be satisfactory, in general, since it does not account for the risk factor, which can often be expressed in terms of the dispersion of z. Thus another formulation of the problem may be given as:

maximize

$$y = E\{z\} - KV_z$$

subject to

$$\sum_{j=1}^{N} E\{a_{tj}\}x_j \leqslant E\{b_t\}, \qquad t = 0, 1, 2, \ldots, T$$

$$x_j = (0, 1), \qquad j = 1, 2, \ldots, N$$

where V_z = variance of $z = \sum_{i,j} x_i \sigma_{ij} x_j$, $(i, j) = 1, 2, \ldots, N$, σ_{ij} is the covariance of (c_i, c_j), and K is a weighting factor to be specified by the decision maker. The proposed objective function seeks to maximize $E\{z\}$ and simultaneously minimize V_z. The weighting factor K may be interpreted as a coefficient of "risk aversion." Thus large values of K reflect the decision maker's conservative attitude, since he is taking into account the possibility of loss. A nonconservative decision maker may thus select small values of K. The new formulation yields a quadratic objective function as compared with a linear function in the previous case.

Although the new objective function has improved the decision criterion, there is still the difficulty of using the expected values in the constraints. Hillier (1967) considers this problem by expressing the constraints in a chance-constrained form. That is,

$$\text{Prob}\left\{ \sum_{j=1}^{N} a_{tj} x_j \leq b_t \right\} \geq \alpha_t$$

where α_t is a constant whose value lies between zero and one. The basic assumption here is that a_{tj} and b_t may be arbitrarily distributed but that $\sum a_{tj} x_j - b_t$ is approximately normal. This assumption is based on some version of the central limit theorem, which can be shown to hold under rather general conditions, especially if the random variables are independent. Thus if $F(\cdot)$ denotes the standard normal cumulative distribution function of $(\sum_{j=1}^{N} a_{tj} x_j - b_t)$ and K_β is defined such that

$$F(K_\beta) = \beta, \quad 0 \leq \beta \leq 1$$

the above stochastic constraint may be converted to the following equivalent deterministic constraint:

$$E\left\{ \sum_{j=1}^{N} a_{tj} x_j - b_t \right\} + K_{\alpha_t}\left(\text{Var}\left\{ \sum_{j=1}^{n} a_{tj} x_j - b_t \right\} \right)^{\frac{1}{2}} \leq 0$$

or

$$\sum_{j=1}^{N} E\{a_{tj}\} x_j + K_{\alpha_t}\left(\text{Var}\left\{ \sum_{j=1}^{N} a_{tj} x_j - b_t \right\} \right)^{\frac{1}{2}} \leq E\{b_t\}$$

From the knowledge of the covariance matrix of a_{tj} and b_t, it is possible to determine the variance expression, which in general should include nonlinear terms in x_j. However, Hillier (1967) shows how linear approximations can be utilized to replace the resulting constraints.◀

▶**EXAMPLE 1.3-2** (SCHEDULING) Suppose in a given school it is desired to schedule C classes over a finite horizon of H (class) periods. (If the horizon is represented by a one-week period, then period 1 is the first period on Monday, while period H is the last period on Friday.) It is assumed that the total number of periods which teacher t can allocate to class c is equal to p_{tc}.

Define the binary variables

$$
x_{tch} = \begin{cases} 1, & \text{if teacher } t \text{ is assigned} \\ & \text{to class } c \text{ in period } h \\ 0, & \text{otherwise} \end{cases}
$$

The constraints of the model are specified as follows:

(i) In a general period of time, only one teacher is assigned to a given class and no one teacher is shared between two or more classes. Thus

$$
\sum_{t=1}^{T} x_{tch} = 1, \qquad \text{for all} \quad c \text{ and } h
$$

$$
\sum_{c=1}^{C} x_{tch} = 1, \qquad \text{for all} \quad t \text{ and } h.
$$

The above $(C + T)H$ restrictions may be replaced by one constraint yielding exactly the same result, namely,

$$
\sum_{\substack{t_1, t_2, = 1 \\ t_1 \neq t_2}}^{T} \sum_{h=1}^{H} \sum_{c=1}^{C} x_{t_1 ch} x_{t_1 ch} = 0
$$

The reduction in the number of constraints is accomplished at the expense of creating nonlinearities, which is known to worsen the computational speed of the problem. Also, the number of crossproducts $(x_{t_1 ch} x_{t_2 ch})$ increases exponentially relative to the original number of variables. It must be asserted here that typically in integer programming, unlike continuous programming problems, the computational efficiency is worsened by the increase in the number of variables (crossproducts) and that, indeed, the increase in the number of constraints may be computationally advantageous. This point should always be kept in mind when formulating integer programming models.

(ii) Each teacher t must work exactly p_{tc} periods in class c, which is expressed as

$$
\sum_{h=1}^{H} x_{tch} = p_{tc}, \qquad \text{for all} \quad t \text{ and } c
$$

Other restrictions relating to the real situation may also be handled in a similar fashion.

An objective criterion of the scheduling problem is to minimize the number of periods during which classes may be idle, expressed as

minimize

$$z = \sum_{t_1, t_2, t_3, = 1}^{H} x_{t_1, c, h-1}(1 - x_{t_2, c, h})x_{t_3, c, h+1}$$

This expression indicates that, if class c is occupied during periods $h - 1$ and $h + 1$, the associated term in the objective function is equal to zero (minimum) only if period h is used; otherwise, if h is not occupied, the term is equal to one. Although this objective criterion minimizes the number of one-period gaps, it says nothing about gaps with two periods or more. A more complex objective function is required to account for this point. ◀

▶EXAMPLE 1.3-3 (SEQUENCING) Sequencing is concerned with determining the order in which a number of jobs are processed in a shop so that a given objective criterion is optimized. Assuming that each job has a prespecified due date, this criterion may be expressed in a variety of ways:

(i)   Minimize total throughput time (time in the shop) for all the jobs;
(ii) minimize makespan, that is, the time by which all jobs are completed; and
(iii) minimize the total lateness or lateness penalty for all projects.

The constraints of the model generally include:

(i)   precedence relationship between jobs, if any;
(ii) due dates;
(iii) job splitting possibilities;
(iv) limited resources;
(v)   concurrent and nonconcurrent job performance requirements;
(vi) substitution of job performance requirements.

Pioneering work in sequencing models was introduced by E. H. Bowman (1959), H. M. Wagner (1959), and Manne (1960). [See also Conway *et al.* (1967) for more detailed discussion on the problem.] However, the work of Pritsker *et al.* (1969) provides more general models for the sequencing problem and consequently will be presented here. One of the objectives of the presentation is to show how a change in the definition of the variables can lead to equivalent models but with drastically varied sizes.

Consider first the situation in which N *independent* jobs are sequenced on a single facility so that the lateness penalty is minimized. It is assumed that the processing time for job k is a_k, while the due date is d_k with the penalty per late period p_k. The problem can be formulated as a zero–one model as follows. Let i represent the job number while j represents its position in the sequencing of the N jobs. Define

$$x_{ij} = \begin{cases} 1, & \text{if job } i \text{ is processed in position } j \\ 0, & \text{otherwise} \end{cases}$$

As in the assignment problem, only one job can occupy a position and only one position can be assigned to a job. Thus

$$\sum_{j=1}^{N} x_{ij} = 1, \qquad \text{for all } i$$

$$\sum_{i=1}^{N} x_{ij} = 1, \qquad \text{for all } j$$

These are the only direct constraints for this simplified model. It is implicitly assumed here that once a job is started it must be completed.

In order to establish the objective criterion, it is necessary to compute the completion date for each job. The completion date for job j is given by its processing time a_j plus the sum of the processing times of all the jobs that precede it. To establish if job k precedes job i, consider the following: If job k is processed in position j and if job k precedes job i, then by definition $x_{kj} = 1$ and $x_{i1} = x_{i2} = \cdots = x_{ij} = 0$. These conditions are both necessary and sufficient. The result can be presented in a compact expression: Given that job k is processed in position j, then k precedes i if and only if

$$x_{kj}\left(1 - \sum_{r=1}^{j} x_{ir}\right) = 1$$

Since the exact position j in which job k is processed is not known a priori, the above expression reduces to

$$\sum_{j=1}^{N} x_{kj}\left(1 - \sum_{r=1}^{j} x_{ir}\right) = 1$$

if job k precedes job i regardless of its position in the sequence. Thus the completion date for job i is

$$c_i = a_i + \sum_{k=1}^{N} a_k \sum_{j=1}^{N} x_{kj}\left(1 - \sum_{r=1}^{j} x_{ir}\right)$$

The objective function then becomes:

minimize

$$z = \sum_{i=1}^{N} p_i \max\{0, (c_i - d_i)\}$$

This function is not readily suitable for applying the available solution techniques. However, the following trick can be employed to simplify the expression. Let

$$y_i = \max\{0, (c_i - d_i)\}.$$

The objective function expression becomes equivalent to

minimize

$$z = \sum_{i=1}^{N} p_i y_i$$

subject to

$$\left\{ a_i + \sum_{k=1}^{N} a_k \sum_{j=1}^{N} x_{kj} \left(1 - \sum_{r=1}^{j} x_{ir} \right) - d_i \right\} \leqslant y_i, \qquad i = 1, \ldots, N$$

$$y_i \geqslant 0 \quad \text{and integer}$$

It is noted that the integrality of y_i is dictated by its definition, so that the converted problem now includes both ordinary and binary integer variables.

The above formulation results in a problem with $(N + N^2)$ integer variables and $3N$ constraints. Moreover, the nonlinear terms in the first N constraints should generally result in a computationally difficult problem. It is shown now that by changing the definition of the decision variables, the size and complexity of the problem are improved and the nonlinearity is eliminated.

Assume that the makespan (time span during which all jobs are completed) is divided into equal time periods and define

$$x_{it} = \begin{cases} 1, & \text{if job } i \text{ is completed} \\ & by \text{ the beginning of period } t \\ 0, & \text{otherwise} \end{cases}$$

Now given that a_i is the processing time (in number of periods) for job i, it is seen that

$$x_{i,t+a_i} - x_{it} = \begin{cases} 1, & \text{if job } i \text{ is being} \\ & \text{processed in period } t \\ 0, & \text{otherwise} \end{cases}$$

The logic behind the last expression is that if job i is begun by period t then it will be completed by period $t + a_i$, in which case $x_{it} = 0$, while $x_{i,t+a_i} = 1$. Otherwise, both x_{it} and $x_{i,t+a_i} = 0$.

Since there is only one processing facility, then only one job may be processed in any period t. This is expressed as

$$\sum_{i=1}^{N} (x_{i,t+a_i} - x_{it}) = 1, \qquad t = 1, 2, \ldots, T \equiv \sum_{i=1}^{N} a_i$$

In order to ensure that once a job is started it must be completed, the following restriction is imposed:

$$x_{it} - x_{i,t+1} \leqslant 0, \qquad \text{for all } i \text{ and } t$$

which shows that if job i is completed by period t, then $x_{ik} = 1$ for $k > t$ in order to ensure that the same job is not rescheduled later.

The formulation of the objective function is straightforward. The number of delay periods for job i is equal to $\sum_{t=d_i+1}^{T} (1 - x_{it})$, so that the objective function becomes

minimize

$$z = \sum_{i=1}^{N} p_i \sum_{t=d_i+1}^{T} (1 - x_{it})$$

or, equivalently,

maximize

$$z' = \sum_{i=1}^{N} p_i \sum_{t=d_i+1}^{T} x_{it}$$

The new formulation is completely linear in the new variable x_{it}. Also, there is no need to introduce auxiliary variables (such as y_i in the first formulation) so that the problem is a pure zero–one model. It is noticed, however, that the number of variables in this case is equal to NT, where $T = \sum_{i=1}^{N} a_i$, thus depending directly on how large T is. It is possible to reduce the number of variables by scaling down T as far as practically possible. For example, instead of using one-hour periods, one may use two-hour periods. Notice, however, that the present formulation may prove superior to the preceding one, since y_i, being general integer variables, should generally require larger computation time than a problem including the same number of zero–one variables.

A more generalized sequencing problem is now considered. It is assumed that there are M projects with each project consisting of a number of jobs. These jobs are interdependent in the sense that a predetermined precedence relationship must be satisfied. In addition, the sequencing is determined under the restriction of limited resources.

Define

i project number, $i = 1, 2, \ldots, M$;

j job number within project i, $j = 1, 2, \ldots, N_i$;

t time period, $t = 1, 2, \ldots, \max D_i$, where D_i is the absolute due date for project i. If D_i is not specified, then it is set equal to the last period in the scheduling horizon;

d_i desired due date for project i. No delay penalty is incurred if project i is completed on or before d_i;

E_i earliest possible period by which project i is completed.

a_{ij} arrival period of job j, project i. Arrivals occur at the beginning of periods;

n_{ij} (fixed) number of periods required to complete job j of project i;

e_{ij} earliest possible period for completing job j of project i;

l_{ij} latest possible period for completing job j of project i;

k resource facility number, $k = 1, 2, \ldots, K$;

r_{ijk} amount of resource k required for job j of project i;

R_{kt} amount of resource k available during period t;

$$x_{ijt} = \begin{cases} 1, & \text{if job } j \text{ of project } i \text{ is} \\ & \text{completed } in \text{ period } t \\ 0, & \text{otherwise} \end{cases}$$

$$x_{it} = \begin{cases} 1, & \text{if project } i \text{ is completed} \\ & by \text{ period } t \\ 0, & \text{otherwise} \end{cases}$$

Notice that, by definition, $x_{it} \leqslant x_{i,\,t+1}$, so that if $x_{iq} = 1$, then $x_{it} = 1$ for all $t > q$. On the other hand, $x_{ijt} = 1$ for *one* value of t only.

The constraints of the model are given as follows:

(i) *Precedence constraints* Suppose that in project i, job v must precede job w. If T_{iv} and T_{iw} denote the completion periods of jobs m and n, respectively, then

$$T_{iv} + n_{iw} \leqslant T_{iw}$$

But since

$$T_{iv} = \sum_{t=e_{iv}}^{l_{iv}} t x_{ivt} \quad \text{and} \quad T_{iw} = \sum_{t=e_{iw}}^{l_{iw}} t x_{iwt}$$

then job v precedes job w if

$$\sum_{t=e_{iv}}^{l_{iv}} t x_{ivt} + n_{iw} \leqslant \sum_{t=e_{iw}}^{l_{iw}} t x_{iwt}$$

(ii) *Job completion*   Each job has exactly one completion period. Thus

$$\sum_{t=e_{ij}}^{l_{ij}} x_{ijt} = 1, \qquad i = 1, 2, \ldots, M, \quad j = 1, 2, \ldots, N_i$$

(iii) *Project completion*   Project i cannot be completed by period t until $\sum_{q=e_{ij}}^{t-1} x_{ijq} = 1$ for all N_i jobs of project i. This is expressed as

$$x_{it} \leq \frac{1}{N_i} \sum_{j=1}^{N_i} \sum_{q=e_{ij}}^{t-1} x_{ijq}, \qquad i = 1, 2, \ldots, M, \quad t = E_i, \ldots, D_i$$

(iv) *Resources constraints*   It is assumed that a job requires the same types and amounts of resources throughout its processing duration. Otherwise the job must be broken down into appropriate "subjobs." It is still possible to require that these jobs be performed contiguously by imposing an *equality* restriction on the appropriate precedence constraints.

   A job is being processed in period t if the job is completed in period q, where $t \leq q \leq t + n_{ij} - 1$. Thus the constraint for resource k is given as

$$\sum_{i=1}^{N} \sum_{j=1}^{N_i} \sum_{q=t}^{t+n_{ij}-1} r_{ijk} x_{ijq} \leq R_{kt}, \qquad \begin{aligned} & t = \min n_{ij}, \ldots, \max D_i; \\ & k = 1, 2, \ldots, K \end{aligned}$$

(v) *Concurrency and nonconcurrency of jobs*   If jobs v and w are to be performed concurrently, then this is satisfied by requiring $x_{ivt} = x_{iwt}$.

   The nonconcurrency of v and w implies that they cannot be performed simultaneously. Any job j is performed in period t if $\sum_{q=t}^{t+n_{ij}-1} x_{ijq} = 1$. Thus the nonconcurrency constraint is

$$\sum_{q=t}^{t+n_{iv}-1} x_{ivq} + \sum_{q=t}^{t+n_{iw}-1} x_{iwq} \leq 1,$$

$$t = \max\{e_{iv}, e_{iw}\}, \ldots, \min\{l_{iv}, l_{iw}\}$$

   Other constraints concerning job splitting and resources substitutability can be developed in a similar fashion.

The sequencing problem has several acceptable objective criteria. These are illustrated for the given problem as follows:

(i) *Minimization of total project throughput time*   Throughput time is defined as the elapsed time between the arrival and completion of a project. If a_i is the arrival period of project i, its throughput time is

$$d_i - \sum_{t=E_i}^{D_i} x_{it} + 1 - a_i$$

and the minimization of the sum of throughput times for all projects becomes equivalent to

maximize

$$z = \sum_{i=1}^{M} \sum_{t=E_i}^{D_i} x_{it}$$

(ii) *Minimization of makespan* Makespan is equal to the time by which *all* projects are completed. If x_t is defined such that

$$x_t = \begin{cases} 1, & \text{if all projects are completed in } t \\ 0, & \text{otherwise} \end{cases}$$

then the objective criterion becomes

minimize

$$z = \sum_{t=\max E_i}^{\max D_i} x_t$$

In this case the project completion constraint given above must be changed to

$$x_t \leqslant \left(1 \Big/ \sum_{i=1}^{M} N_i \right) \sum_{i=1}^{M} \sum_{j=1}^{N_i} \sum_{q=e_{ij}}^{t-1} x_{ijq}, \qquad t = \max E_i, \ldots, \max D_i$$

The lateness penalty criterion is treated in the same manner used in the simplified sequencing example given above.◀

▶**EXAMPLE 1.3-4** (KNAPSACK PROBLEM)   Consider the situation in which a hiker must decide on the various items he could take under the limitation that he cannot carry more than a specified weight. His objective is to maximize the total value of the items he takes.

Let a_j be the weight of the jth item and c_j its value as decided by the hiker. If x_j is a binary variable with $x_j = 1(0)$ implying that the jth item is selected (not selected), then the problem is formulated as

maximize

$$z = \sum_{j=1}^{n} c_j x_j$$

subject to

$$\sum_{j=1}^{n} a_j x_j \leqslant b$$

$$x_j = (0, 1), \qquad j = 1, 2, \ldots, n$$

where n is the number of items from which the selection is made and b the weight limitation.

The knapsack problem is sometimes referred to as the *cargo loading* problem or the *flyaway kit* problem. These names stem from similar applications of loading a vessel having a limited capacity and of packing a kit with valuable items for use on an airplane.

The knapsack problem has been receiving considerable attention in the literature mainly because it can be used in developing a decomposition algorithm for the well-known trim loss (or cutting stock) problem (see Section 7.2.1). Recent research indicates the possibility of converting general integer linear problems into knapsack-type models. The results of this research have motivated further work on this problem in the hope that it may assist in developing efficient solution methods for the general integer problem (see Section 7.2.2 for details).

The above formulation is referred to as a one-dimensional knapsack problem. A multidimensional knapsack problem includes more than one constraint. For example, a volume restriction may also be imposed in addition to the weight constraint.◄

►**EXAMPLE 1.3-5** (TRAVELING SALESMAN PROBLEM)   Consider the situation of sequencing n jobs on a single facility. A setup cost is incurred every time a new job is scheduled, but this setup cost depends on the immediately preceding job that was processed on the facility. For example, in a paint factory, the setup cost for producing white paint after red paint is higher than when the order is reversed, that is, when white paint precedes red paint. The setup costs in this case are said to be *sequence dependent*. The objective is to determine the sequence that minimizes the total sum of setup costs.

This type of problem was given the classical name "traveling salesman," because each job may be thought of as a "city." The "distances" between the cities take the place of the setup costs. The salesman is to visit each city once, and the *shortest* tour by which he can visit all the cities must be determined.

The traveling salesman problem can be formulated mathematically as follows: A tour including n cities is said to consist of n *directed* arcs [an arc represents a (one-way) route between two consecutive cities]. These arcs will be designated according to their order in the tour, so that the kth directed arc is associated with the kth leg of the itinerary.

Define

$$x_{ijk} = \begin{cases} 1, & \text{if the } k\text{th directed arc is from city } i \text{ to city } j \\ 0, & \text{otherwise} \end{cases}$$

The constraints of the problem may be classified under four types:

(i)   Only one other city may be reached from a specific city i:

$$\sum_j \sum_k x_{ijk} = 1, \qquad i = 1, 2, \ldots, n$$

(ii)   Only one route may be assigned to a specific k:

$$\sum_i \sum_j x_{ijk} = 1, \qquad k = 1, 2, \ldots, n$$

(iii)   Only one other town may initiate a directed arc to a specific city j:

$$\sum_i \sum_k x_{ijk} = 1, \qquad j = 1, 2, \ldots, n$$

(iv)   Given that the kth directed arc ends at some specific city j, the $(k + 1)$st arc must start at the same city j. This ensures that the trip consists of "connected" segments:

$$\sum_{\substack{i \\ i \neq j}} x_{ijk} = \sum_{\substack{r \\ r \neq j}} x_{jr(k+1)}, \qquad \text{for all } j \text{ and } k$$

The objective function is then given as:

minimize

$$z = \sum_i \sum_j \sum_k d_{ij} x_{ijk}, \qquad i \neq j$$

where d_{ij} is the shortest (one-way) distance from city i to city j.

The above formulation, although relatively simple, leads to problems of monumental sizes even for small n. For example, for $n = 5$ there are 125 variables and 40 constraints, while for $n = 10$ there are 1000 variables and 130 constraints. The feasibility of solving problems of a practical size using the above formulation is thus highly questionable.

It is shown in Section 7.4 that the special structure of the problem may be exploited to develop an efficient specialized algorithm. Namely, one notices that if constraints (iv) are ignored temporarily, the problem can be reduced to an ordinary two-dimensional assignment problem. This is the basis for developing a new algorithm, but special provisions must be made to ensure that the obtained solution constitutes a tour (of connected itinerary legs).◀

▶**EXAMPLE 1.3-6**   (COVERING PROBLEM)   Consider the problem of delivering orders from a warehouse to m different destinations, each of which receives its order in one delivery. A carrier may combine at most k orders to be delivered simultaneously. Determination of a feasible combination depends on the route assigned to each carrier. Because a destination may fall on more than one route, alternatives exist as to which carrier should deliver the order.

Define an *activity* to be a single feasible combination of one, two, ..., or k orders and let n be the number of such activities. Let c_j be the cost associated with the jth activity. The variable x_j is defined equal to one if the jth activity is carried out and zero otherwise. The delivery problem then becomes:

minimize

$$z = \sum_{j=1}^{n} c_j x_j$$

subject to

$$\sum_{j=1}^{n} \alpha_j x_j = \alpha_0$$

$$x_j = (0, 1)$$

where α_j is a column vector having m entries, with the ith entry equal to one if activity j delivers order i and zero otherwise. The column vector α_0 has m 1's. Thus the ith constraint implies that only one order is delivered to a given destination.

There are other practical problems that can be formulated in a similar manner. In the *airline crew scheduling* problem one is concerned with assigning crews to different time periods. Thus, each flight "leg" corresponds to a constraint that requires that one of the available crews be assigned to that flight leg. Each activity (column) corresponds to a possible period assignment for a specific crew, so that the entry of the column is equal to 1 if the flight leg can be handled by such an assignment and is zero otherwise. Naturally, $x_j = 1(0)$ implies that the jth assignment is utilized (not utilized). The objective is to minimize the total cost of all assignments. A similar model can be constructed for the design of a combinatorial circuit with a prespecified switching function [see Hohn (1955)].

The name "covering problem" was acquired because the above model generalizes a case that occurs in graph theory. Given a graph $G = (N, E)$, with N nodes and E edges, each edge joining certain pairs of nodes, it is required to find a "cover" with the minimum number of edges, where a cover is defined to be a subset of edges such that each of the N nodes is incident to some edge of the subset. The formulation of the problem is very similar to that of the delivery problem, except that each $\alpha_j = (a_{1j}, \ldots,$ $a_{mj})^T$ has exactly two entries of 1 with the remaining entries equal to zero. In this case, α_j corresponds to the edge of G joining nodes p and q if $a_{pj} = a_{qj} = 1$. Actually $A = (\alpha_1, \ldots, \alpha_n)$ represents the *incidence matrix* of the graph with one column for each edge and one row for each node [see Ford and Fulkerson (1962)]. The variable x_j is thus equal to one if the jth edge is included in the cover; otherwise, it is equal to zero. By definition, the objective function is to minimize $x_1 + x_2 + \cdots + x_n$.

The covering problem has received considerable attention in the literature because, in addition to its substantial practical applications, it possesses special properties that lead to very efficient computational results. This point is discussed further in Section 7.5.◄

1.3.2   Examples of "Transformed" Integer Models

►EXAMPLE 1.3-7   (FIXED-CHARGE PROBLEM)   Suppose it is required to decide on the locations of production plants among m existing sites. All plants produce homogeneous products. Plant i has a capacity of a_i units and necessitates a fixed investment f_i. The products are shipped to n customers, with the jth customer demanding b_j units. If c_{ij} is the cost of producing a unit at plant i and shipping it to customer j, it is required to determine the operating capacities of the plants so that the total production and investment costs are minimized.

Let x_{ij} be the amount manufactured at plant i for customer j; then the model becomes

minimize

$$z = \sum_{i=1}^{m} \sum_{j=1}^{n} F_{ij}(x_{ij})$$

subject to

$$\sum_{i=1}^{m} x_{ij} \geqslant b_j, \qquad j = 1, 2, \ldots, n$$

$$\sum_{j=1}^{n} x_{ij} \leqslant a_i, \qquad i = 1, 2, \ldots, m$$

where

$$F_{ij}(x_{ij}) = \begin{cases} c_{ij} x_{ij} + f_i, & \text{if } \sum_{j=1}^{n} x_{ij} > 0 \\ 0, & \text{otherwise} \end{cases} \qquad x_{ij} \geqslant 0$$

The above model is very similar to an ordinary linear program except for the fact that z is a nonlinear function. However, there is no direct algorithm that can handle the model as it is stated.

The model can be converted into a mixed (zero–one) integer problem using a convenient substitution. Let

$$y_i = \begin{cases} 1, & \text{if } \sum_{j=1}^{n} x_{ij} > 0 \\ 0, & \text{otherwise} \end{cases}$$

The model then becomes:

minimize

$$z = \sum_{i=1}^{m} \left(\sum_{j=1}^{n} c_{ij} x_{ij} + f_i y_i \right)$$

subject to

$$\sum_{i=1}^{m} x_{ij} \geq b_j, \qquad j = 1, 2, \ldots, n$$

$$\sum_{j=1}^{n} x_{ij} \leq a_i y_i, \qquad i = 1, 2, \ldots, m$$

$$x_{ij} \geq 0$$

$$y_i = (0, 1), \qquad \text{for all } i.$$

If for a given i, $x_{ij} > 0$, then y_i is forced to be equal to 1 and the corresponding capacity constraint remains unchanged. If, on the other hand, $x_{ij} = 0$ for all i, then $y_i = 0$ since, with the capacity constraint being redundant, the minimum of z can only be achieved with $y_i = 0$.

Notice that, unlike the coded integer models, the auxiliary variables y_i are introduced primarily for analytic convenience and as such only yield *redundant* information about the solution to the problem. Although $y_i = 1(0)$ can be interpreted as plant i being (not being) constructed, the same information is readily secured by observing whether $\Sigma_j x_{ij}$ is positive or zero.

The above model can be generalized to include production cost function with breaks; that is, instead of assuming that each unit is produced at the same cost, the per unit cost assumes decreasing marginal values with the level of production. Moreover, the change from a given marginal cost to a smaller one is accomplished at the expense of incurring a given fixed charge.

For plant i and level of production k, define a_{ik} = break point representing level of production; w_{ik} = number of units produced, $0 \leq w_{ik} \leq a_{ik} - a_{i,k-1}$;

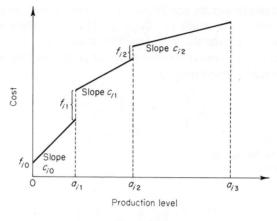

Figure 1-1

c_{ik} = marginal cost; f_{ik} = fixed cost. Further, let d_{ij} be the per unit transportation cost from plant i to customer j. A typical production cost function is illustrated in Fig. 1-1. The model may be formulated as:

minimize

$$z = \sum_i \left(\sum_j d_{ij} x_{ij} + \sum_k c_{ik} w_{ik} + \sum_k f_{ik} y_{ik} \right)$$

subject to

$$\sum_i x_{ij} \geqslant b_j, \qquad \text{for all} \quad j$$

$$\sum_j x_{ij} = \sum_k w_{ik}, \qquad \text{for all} \quad i$$

$$\frac{w_{i,k-1}}{a_{ik} - a_{i,k-1}} \geqslant y_{ik} \geqslant \frac{w_{ik}}{a_{i,k+1} - a_{ik}}$$

$$x_{ij}, w_{ik} \geqslant 0$$

$$y_{ik} = 0, 1$$

It is seen that if $w_{ik} > 0$, $y_{ik} = 1$ since it must satisfy $0 < y_{ik} \leqslant 1$. If $w_{ik} = 0$, then $0 \leqslant y_{ik} \leqslant 1$ and $y_{ik} = 0$, as per the minimization of the objective function. Notice that $w_{i,k-1} = a_{ik} - a_{i,k-1}$ if $w_{ik} > 0$, because y_{ik} is a zero–one variable, for otherwise $0 < y_{ik} < 1$ would produce an infeasible solution.◀

▶**EXAMPLE 1.3-8** (SEPARABLE PROGRAMMING)   Consider the nonlinear problem maximize (or minimize)

$$z = \sum_{i=1}^{m} f_i(x_i)$$

subject to

$$\sum_{i=1}^{m} g_{ij}(x_i) \leq b_j, \qquad j = 1, 2, \ldots, n$$

$$x_j \geq 0, \qquad j = 1, 2, \ldots, n$$

This problem is separable because all its functions can be separated, each with a single variable.

An approximate solution can be obtained for this problem by linearizing each of the functions f_i and g_{ij} over its permissible domain. An illustration of a typical linear approximation in shown in Fig. 1-2. The breaking points a_0, a_1, ... are chosen so that the linear approximation closely represents the true function.

Let the number of breaking points for the ith variable x_i be equal to K_i and define a_{ik} as its kth breaking point. Let t_{ik} be a weight associated with the kth breaking point of the ith variable and define y_{ik} as an associated binary variable. The problem can be approximated (Dantzig, 1960a) as a mixed zero-one integer model as follows:

maximize (or minimize)

$$z = \sum_{i=1}^{m} \sum_{k=1}^{K_i} f_i(a_{ik}) t_{ik}$$

subject to

$$\sum_{i=1}^{m} \sum_{k=1}^{K_i} g_{ij}(a_{ik}) t_{ik} \leq b_j, \qquad j = 1, 2, \ldots, n$$

$$0 \leq t_{i1} \leq y_{i1}$$
$$0 \leq t_{i2} \leq y_{i1} + y_{i2}$$
$$0 \leq t_{i3} \leq y_{i2} + y_{i3}$$
$$\vdots$$

$$0 \leq t_{i, K_i - 1} \leq y_{i, K_i - 2} + y_{i, K_i - 1}$$
$$0 \leq t_{i, K_i} \leq y_{i, K_i - 1}$$
$$y_{i1} + y_{i2} + \cdots + y_{i, K_i - 1} = 1, \qquad \text{all } i$$
$$t_{i1} + t_{i2} + \cdots + t_{i, K_i} = 1, \qquad \text{all } i$$
$$y_{ik} = (0, 1), \qquad \text{all } i \text{ and } k$$

The role played by the binary variables y_{ik} is to allow no more than two *adjacent* weights ($t_{i, k}$ and $t_{i, k+1}$) to assume positive values; otherwise, the linear approximation is invalid. The additional constraints relating the

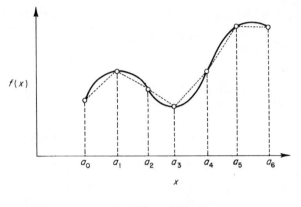

$f(x)$

$a_0 \quad a_1 \quad a_2 \quad a_3 \quad a_4 \quad a_5 \quad a_6$

x

Figure 1-2

variables t and y guarantee that this will always occur. (Other formulations are given in Problems 1-12 and 1-13.)

The above formulation guarantees attaining a global optimum (for the approximating problem), particularly for problems that possess several *local* optima. This can occur when the functions f_i and g_{ij} do not follow specific regularity conditions, e.g., a nonconvex solution space.◄

►**EXAMPLE 1.3-9** (DICHOTOMIES) An important class of programming problems occurs where the solution space is *dichotomized* in such a way that the problem loses the special properties that otherwise would allow it to be solved by available algorithms. These problems are illustrated here by several examples

(i) *Either–or constraints* (Manne, 1960) In a single machine scheduling problem where several jobs are to be sequenced, a constraint must be added to ensure that no two jobs are processed simultaneously. If x_i and x_j are the starting times for jobs i and j and if a_i and a_j are their respective processing times, noninterference between the two jobs can be expressed as

$$\text{either} \quad x_i + a_i \le x_j \quad \text{or} \quad x_j + a_j \le x_i$$

The either–or constraint cannot be implemented directly in a programming model. To overcome this difficulty, define y_{ij} as a binary variable and let M be a very large positive value. The either–or constraint may then be written as

$$x_i + a_i \le x_j + My_{ij} \quad \text{and} \quad x_j + a_j \le x_i + M(1 - y_{ij})$$

If $y_{ij} = 0$, the first constraint becomes active (that is, i precedes j), while the second becomes redundant. An opposite condition prevails if $y_{ij} = 1$.

(ii) *k-fold alternatives* (Dantzig, 1960a) In some problems, it is specified that only k out of m $(k < m)$ constraints must hold. If the m constraints are given by

$$\sum_j a_{ij} x_j \leqslant b_i, \qquad i = 1, 2, \ldots, m$$

this condition is satisfied by replacing the m constraints by

$$\sum_j a_{ij} x_j \leqslant b_i + My_i, \qquad i = 1, 2, \ldots, m$$

$$\sum_{i=1}^{m} y_i = m - k$$

$$y_i = (0, 1), \qquad i = 1, 2, \ldots, m$$

This shows that exactly $(m - k)$ of the y_i-variables will be equal to 1.

A special case of the above problem occurs when the right-hand side of

$$\sum_j a_j x_j = b$$

assumes only one of the exclusive values $b_1, b_2, \ldots, b_r$. This is expressed by

$$\sum_j a_j x_j = \sum_{k=1}^{n} b_k y_k, \qquad \sum_{k=1}^{n} y_k = 1$$

where $y_k = (0, 1)$ for $k = 1, 2, \ldots, r.$◀

1.4 Methods of Integer Programming

Integer programming techniques are generally categorized into two broad types: (1) search methods and (2) cutting methods. The first type is motivated by the fact that the integer solution space can be regarded as consisting of a *finite* number of points.† In its simplest form, search methods seek enumerating "all" such points. This would be equivalent to simple exhaustive enumeration. What makes search methods more promising than simple exhaustive enumeration, however, is that techniques can be developed

---

† In the obvious sense, this statement may seem ambiguous, if not erroneous, when applied to the mixed integer problem. Technically, however, it is shown later that, even in the mixed case, the search methods are primarily controlled by the integer variables.

to enumerate only a portion (hopefully small) of all the candidate solutions while *automatically* discarding the remaining points as nonpromising. Clearly, the efficiency of the resulting "search" algorithm depends on the power of the techniques that are developed to discard the nonpromising solution points.

Search methods primarily include implicit enumeration techniques and branch-and-bound techniques. The first type is mostly suited for the zero–one problem, and may actually be considered as a special case of the branch-and-bound methods.

Cutting methods are developed primarily for the (mixed or pure) integer linear problem. These methods are motivated by the fact that the simplex solution to a linear program must occur at an extreme point. The idea then is to add specially developed secondary constraints that are violated by the current noninteger solution but never by any feasible (integer) point. The successive application of such a procedure should eventually result in a new (convex) solution space with its optimum extreme point properly satisfying the integrality condition. The name "cutting" methods is suggested by the fact that the secondary constraints "cut" off infeasible parts of the continuous solution space.

Another method for the integer problem is inspired by the cutting techniques and calls for identifying the *convex hull* of all the feasible integer points by constructing a set of proper linear inequalities. Once this is done, the application of the regular simplex method would clearly produce the desired result. The underlying development is usually referred to as the *asymptotic* problem.

There are several specialized methods for solving structured problems such as the knapsack problem, the set covering problem, the traveling salesman problem, and the fixed-charge problem. These methods, although designed specifically to exploit the special structure of the problem, are actually in the spirit of either the search methods or the cutting methods (or a combination of both).

A common property that characterizes almost all the cutting methods is that no feasible (integer) solution is available until the very end. In other words, there is no available information about the integer solution if the calculations are stopped prematurely. Such methods are usually known as dual techniques and their property represents a major disadvantage. There has been some effort to develop "primal" algorithms (that is, with integer feasibility maintained at all times), but from the computational viewpoint, these algorithms have been less satisfactory.

The search methods are sometimes referred to as "almost dual" techniques. Although primal feasibility is not guaranteed at all times, occasionally the algorithms may produce a good feasible solution. This is an important advantage as compared with the cutting methods. However, the search methods (especially the branch-and-bound type) usually result in severe taxation of the computer memory, a difficulty not present in the cutting methods. The reason for this will be made evident after each method is explained.

Interestingly enough, the cutting and search methods possess mutually exclusive properties which, if combined, may result in a superior method of solution, namely, an algorithm that yields feasible solutions and in the meantime does not require a large computer memory. This idea is the basis for some interesting developments that will be presented in Chapters 4 and 5.

1.5 Organization of the Book

Linear programming forms an important foundation for most of the developments in integer programming. For this reason, Chapter 2 is devoted to presenting the basic theory of linear programming with a slant toward its use in integer programming. The chapter assumes no prior knowledge of the subject.

Although cutting methods were the first to be developed (1958) and implicit enumeration was the last to be introduced (1965), implicit enumeration will be presented first (Chapter 3). This is followed by the branch-and-bound techniques (Chapter 4) and then the cutting techniques (Chapters 5 and 6). The reasoning behind this is that the implicit enumeration methods appear to be the most natural (and certainly the simplest) way to tackle the integer problem. Why these methods did not evolve earlier is somewhat peculiar, but one suspects that researchers in the field were heavily biased by the developments (and success) in the area of linear programming, which motivated them to think along these lines in search for a solution method to the integer problem. The nonchronological presentation of the solution methods should not cause any inconvenience in understanding the material. Indeed, it will be seen that this organization may lead to better appreciation of the relationship between the three methods.

Specialized algorithms for the knapsack problem, the fixed-charge problem, the traveling salesman problem, and the covering problem are presented in Chapter 7.

Although discussion of the computational performance of the different integer methods is presented in their respective chapters, Chapter 8 is devoted to drawing general guidelines for the utilization of these methods. The chapter also discusses the use of heuristics to find "good" solutions to the integer linear problem.

Problems

1-1 *Quadratic Assignment Problem* (Koopmans and Beckmann, 1957). Consider the problem of locating n plants in n locations with each location accomodating exactly one plant. The cost of shipping fixed volumes from plant i to plant j is $c_{ij}{}^{pq}$ when plant $i(j)$ is assigned to location $p(q)$. Formulate the problem as a zero–one problem so that the total transportation costs are minimized.

1-2 *Four-Color Problem* (Fortet, 1959). It is required to color a map of various countries by using four colors only so that every two neighboring countries have different colors. Let each country represent a node on a graph. An arc joins two nodes only if the represented countries have common borders. Let $x_i = 1$ if node i has color 1 or color 2 and $x_i = 0$ otherwise. Also, let $y_i = 1$ if node i has color 1 or color 3 and $y_i = 0$ otherwise. Derive an equation such that any of its feasible solutions will provide a legitimate coloring of the map.

1-3 *Four-Color Problem* [Gomory, cf. Dantzig (1963, p. 548)]. Another formulation of the four-color problem may be obtained by defining new values as follows. Let $t_i = 0, 1, 2,$ or 3 be a variable associated with node i, where each value represents one of the four colors. Thus for every two nodes i and j joined by an arc, it is necessary that $t_i \neq t_j$. Formulate the inequalities whose solutions solve the coloring problem. (*Hint*: use either–or constraints.) Note that this is the first example in which a coded variable assumes values other than zero or one.

1-4 Show how the following step function can be represented as a zero–one expression:

$$f(x) = b_i, \qquad a_{i-1} \leqslant x \leqslant a_i, \quad i = 1, 2, \ldots, n$$

where $b_i > b_{i-1}$ for all $i = 1, 2, \ldots, n$. In particular, show how the zero–one variables relate to the variable x by specifying the appropriate constraints.

1-5 *Sequencing* (Taha, 1971a). In the sequencing problem described in part (i) of Example 1.3-9, suppose a setup cost S_{ij} is incurred when job j immediately follows job i. This is called a *sequence-dependent* setup cost. Formulate a model to minimize the total setup cost of sequencing the n jobs and give the associated constraints. (*Hint*: In Example 1.3-9, the variable y_{ij} may be used to represent the "frequency" by which job i precedes job j, that is, $y_{ij} = 1$ if job i succeeds job j and $y_{ij} = 0$ if job j succeeds job i.)

1-6 Consider the constraints $a_i \leqslant f_i(x) \leqslant b_i$, $i = 1, 2, 3$ and a_i and b_i are given constants for all i. Show how the following conditional constraints can be expressed as manageable forms by using auxiliary zero–one variables:

(i) $f_1(x) > 0 \Rightarrow f_2(x) \geqslant 0$

(ii) $f_1(x) > 0 \Rightarrow f_2(x) \geqslant 0$ and $f_1(x) < 0 \Rightarrow f_3(x) \geqslant 0$

1-7 Express the L-shaped region in the accompanying figure by using auxiliary zero–one variables.

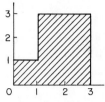

1-8 *Political Districting* (Garfinkel and Nemhauser, 1970). Suppose a certain state has m counties. The population of the ith county is p_i. Each county is to be included in a political district as an indivisible unit. Depending on the population of the state, a total of k representatives are assigned to it. The problem is to assign each county to a district such that the population differences among the districts are minimized. Suppose that $n \, (> k)$ districts can be formed, and formulate the problem as a zero–one model so as to select the most appropriate district representation. (*Hint*: This problem is similar to the set covering model of Example 1.3-6.)

1-9 Modify the objective function in Problem 1-8 so that the worst district is as acceptable as possible.

1-10 Convert the following constraint into a manageable form by simulating the effect of the absolute value using auxiliary zero–one variables:

$$\left| \sum_{j=1}^{n} a_{ij} x_j \right| \geqslant b > 0$$

1-11 *Minimal Cuts in Network Flows* (Ford and Fulkerson, 1962). In the theory of network flows, two nodes n_0 and n_k are linked by a collection of directed arcs via nodes $n_1, n_2, \ldots, n_{k-1}$. Associated with each arc is a nonnegative integer value representing its capacity. There are no arcs coming into n_0 (the source node) or departing from n_k (the sink node). Define the sets N_1 and N_2 such that $N_1 \cap N_2 = \varnothing$, $n_0 \in N_1$, $n_k \in N_2$, and $N_1 \cup N_2 = \{n_0, n_1, \ldots, n_k\}$. Thus N_1 and N_2 are said to define a *cut* of the network. The capacity of the cut is the sum of the capacities of all the arcs joining the nodes in N_1 and N_2. It can be proved that the maximal flow between n_0 and n_k is determined by the cut having the minimal capacity. Formulate the problem as a zero–one model in order to determine the minimal cut.

1-12 *Linearization of a Curve* (Healy, 1964).   Let $f(x)$ be the function to be linearized and let its domain be divided into the intervals $[a_i, b_i]$, where $a_{i+1} = b_i$. Suppose that $f(x)$ is approximated in the ith interval by $m_i x + c_i$ and define $y_1, y_2, \ldots, y_n$ as binary variables. Then

$$f(x) \simeq v = \sum_{i=1}^{n} y_i(m_i x + c_i)$$

This means that depending on whether y_i equals zero or one, x will fall inside or outside the interval $[a_i, b_i]$. Show that the above approximation is equivalent to the following linear form:

$$v = \sum y_i(c_i + m_i a_i) + \sum m_i w_i$$
$$x = \sum w_i + \sum a_i y_i, \qquad w_i = y_i(x - a_i)$$
$$\sum y_i = 1, \qquad\qquad y_i = (0, 1), \quad i = 1, 2, \ldots, n$$
$$w_i \leqslant y_i(b_i - a_i)$$

1-13 *Linearization of a Curve* (Markowitz and Manne, 1957).   Show that the same approximation in Problem 1-12 (and Example 1.3-8) can be effected by using the condition:

$$\text{either} \quad u_i - (b_i - a_i) \geqslant 0 \quad \text{or} \quad u_{i+1} \geqslant 0 \qquad \text{(or both)}$$

where $0 \leqslant u_i \leqslant (b_i - a_i)$ and $x = u_1 + u_2 + \cdots + u_n$.

1-14 *Global Minimization of a Concave Function* (Dantzig, 1963, p. 543). Suppose that a concave function $f(x_1, x_2, \ldots, x_n)$ is reasonably approximated over its feasible domain by a set of m tangent hyperplanes of the form

$$\sum_{j=1}^{n} a_{ij} x_j - b_i = 0, \qquad i = 1, 2, \ldots, m$$

Then at any point $(x_1, x_2, \ldots, x_n)$, the value of f can be assumed bounded from below by *at least* one of the m tangent hyperplanes. Assuming that the solution space is a convex polyhedron Q, express the problem as a mixed zero–one integer linear model.

1-15 *Ratio Problem* (Healy 1964).   Consider the following ratios:

$$\frac{x_1}{x_1 + x_2} = \frac{y_1}{y_1 + y_2} = k$$

where k assumes one of the values $k_1, k_2, \ldots, k_n$. If $x_1, x_2, y_1,$ and y_2 are nonnegative variables, show how these variables can be converted to linear constraints by using auxiliary binary variables.

1-16 *Information Retrieval* (Day, 1965). There are n different files from which to retrieve information. The length of the jth file is $d_j, j = 1, 2, \ldots, n$. It is required to fill m requests of information, each of which, considered as a unit, is stored in at least one file. Formulate the model for determining the files to be searched so that the total searched length is minimized. (*Hint*: This is an application of the covering problem.)

CHAPTER 2

Linear Programming

2.1 Introduction

The purpose of this chapter is to present a foundation of linear programming with emphasis on the topics that contribute to the development of integer programming theory. In addition, the chapter presents other related specialized topics including the so-called Benders approach for partitioning programming problems having two different types of variables into two subproblems each having homogeneous variables. The Benders approach is especially useful in solving mixed integer problems.

The material in this chapter is comprehensive and self-contained. Students with no background in linear programming, in addition to acquiring an adequate knowledge of the subject, will find it suitable and sufficient for providing the necessary foundation for understanding integer programming. Those already familiar with linear programming are encouraged to review the chapter in order to maintain continuity and uniformity throughout the remainder of the book.

2.2   Definition of Linear Programming

A linear programming problem is defined as:

maximize (or minimize)

$$z = \sum_{j=1}^{p} c_j x_j$$

subject to

$$\sum_{j=1}^{p} a_{ij} x_j (\leqslant, =, \text{ or } \geqslant) b_i, \qquad i = 1, 2, \ldots, m$$

$$x_j \geqslant 0, \qquad j = 1, 2, \ldots, p$$

where c_j, a_{ij}, and b_i are known constants for all i and j, and x_j are nonnegative variables.

The constraints of the problem can be converted into equations by adding a (nonnegative) slack variable x_{p+i} if the ith inequality is of the type $\leqslant$ and subtracting a (nonnegative) surplus variable x_{p+k} if the kth inequality is of the type $\geqslant$. Assuming that the augmentation of the slack and surplus variables will result in a total of n variables, the problem can be put into matrix form as:

maximize (or minimize)

$$z = CX$$

subject to

$$AX = b$$
$$X \geqslant 0$$

where C is an n-row vector, A an $m \times n$ matrix, and b an m-column vector. The elements of C corresponding to the slack and surplus variables are all zero.

It can always be asumed that $m < n$, since if $m > n$, at least $m - n$ equations must be redundant, and $m = n$ results in a trivial situation in which $AX = b$ has a unique solution point. When $m < n$, the system $AX = b$ generally has an infinite number of solutions. (The case where there is no solution is considered later.) However, the following theorem shows that only a finite number of solution points are of interest in determining the optimum of the linear program.

▶**THEOREM 2-1**   The optimum solution to the above linear program, when finite, must be associated with a basic solution of the system $AX = b$.◀

A basic solution to $AX = b$ is obtained by setting $n - m$ variables equal to zero and then solving for the remaining variables, provided the solution exists and is unique. In this case, the variables set equal to zero are called *nonbasic*, while the remaining ones are called *basic*.

Mathematically, a basic solution is identified as follows: Let the elements of X be rearranged so that its first m elements are the basic variables. Thus X can be partitioned to X_B and X_N, where X_B is called the basic vector. The equations $AX = b$ can then be written as

$$(B, N)\begin{pmatrix} X_B \\ X_N \end{pmatrix} = b$$

where B and N are obtained by arranging the columns of A to match the elements of X_B and X_N. The matrix B is square of size m.

Since $X_N = 0$ by definition, it follows that $BX_B = b$. Thus X_B is a basic solution if and only if B is nonsingular (det $B \neq 0$). This can be satisfied so long as the vectors comprising B are independent. In this case, the matrix B is said to form a *basis*. Naturally, the nonsingularity of B does not guarantee that $X_B \geqslant 0$. Thus the solution of $BX_B = b$ is of interest only if X_B is a (basic) feasible solution.

Geometrically, a basic solution X_B defines an *extreme* (or a corner) *point* of the solution space $AX = b$. In general, an extreme point is uniquely defined by a basic solution except when such a point is overdetermined; that is, in an n-dimensional problem, there are more than n hyperplanes passing through the same corner point. In this case, the same extreme point is identified by more than one basic solution and the solution is said to be *degenerate*. It is impossible, however, to identify more than one extreme point with the same basic solution. A typical two-dimensional solution space is shown in Fig. 2-1. Points A, B, C, and D are the extreme

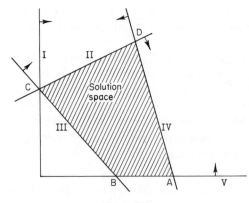

Figure 2-1

points of the solution space, but point C is degenerate because it is defined by three lines. The three basic solutions associated with C are obtained as the intersections of lines 1 and 2, lines 1 and 3, and lines 2 and 3.

By definition, the number of basic solutions associated with the solution space $AX = b$ is limited by

$$C_m^{\,n} = \frac{n!}{m!\,(n-m)!}$$

This suggests that the optimum solution can be obtained by enumerating all possible basic solutions. The optimum is associated with the (basic) *feasible* solution yielding the largest (smallest) objective value z, assuming the problem is of the maximization (minimization) type. This procedure is inefficient since, in addition to enumerating a possibly very large number of solutions, some of these solutions may be nonexistent ($BX_B = b$ is inconsistent) or infeasible ($X_B \not\geq 0$). An efficient method was thus developed that will judiciously "enumerate" a relatively small number of the extreme points, but it never encounters any inconsistent or infeasible solutions. This is the simplex method, which is introduced in the next section.

2.3   The Simplex Method

The selective search of the simplex method (among all the basic solutions) depends on two fundamental conditions: (1) The feasibility condition, given an initial basic feasible solution to the problem, guarantees generating new basic solutions that are always feasible. This is accomplished by replacing *one* (current) basic vector in B by an associated nonbasic vector chosen from N. (2) The optimality condition ensures that the nonbasic vector selected by the feasibility condition will improve the objective value or otherwise indicate that no further improvements can be achieved by generating new basic solutions. At this point the optimum has been reached.

2.3.1   Initial Basic Feasible Solution

If all the constraints of the problem are originally of the type $\leq$ and provided the right-hand side vector b is nonnegative, then by converting the inequalities into equations, the slack variables will provide an obvious starting basic feasible solution. This follows since the constraints put in equation form can be written as

$$AX = (D,\ I)\begin{pmatrix} X_d \\ X_s \end{pmatrix} = b$$

where X_d are the original decision variables and X_s the slack variables. The vectors associated with X_s form an identity matrix I. Thus, by regarding X_d as nonbasic variables at zero level, the constraints reduce to $IX_s = b$ or $X_s = b$. In this case, the matrix $B = I$ obviously forms a basis, and hence $X_s = b$ is a basic solution. Because $b \geqslant 0$, by definition X_s must also be feasible.

For the case where a constraint is of the type $=$, no slack variable can be used to provide a starting solution. Also, the surplus variables associated with $\geqslant$ constraints will yield a negative, and hence infeasible, solution. Thus for constraints of the types $=$ and $\geqslant$, starting basic variables are secured by using the so-called *artificial variables*. Let the ith equation representing the $=$ or $\geqslant$ constraint after the surplus is subtracted be given by

$$\sum_j a_{ij} x_j = b_i$$

and let x_{Ri} be an (nonnegative) artificial variable. Then the ith equation may be written as

$$\sum_j a_{ij} x_j + x_{Ri} = b_i$$

Now, $x_{Ri} = b_i$ can be used to provide a starting basic variable. However, this variable is artificial, and if the optimal solution is to be feasible for the original problem, then a condition must be imposed to guarantee x_{Ri} be at zero level in the final solution. This can be achieved by assigning a very high per unit penalty for x_{Ri} in the objective function, provided of course that the problem has a feasible solution.

To illustrate the artificial variables technique, consider the following example:

maximize

$$z = 2x_1 + 3x_2$$

subject to

$$x_1 + 2x_2 = 10$$
$$2x_1 + 3x_2 \geqslant 8$$
$$3x_1 + 4x_2 \leqslant 20$$
$$x_1, x_2 \geqslant 0$$

The form yielding the starting solution may be written as:

maximize

$$z' = 2x_1 + 3x_2 - Mx_{R1} - Mx_{R2}$$

subject to

$$x_1 + 2x_2 \quad\quad + x_{R1} \quad\quad\quad\quad\quad = 10$$
$$2x_1 + 3x_2 - x_3 \quad\quad + x_{R2} \quad\quad = 8$$
$$3x_1 + 4x_2 \quad\quad\quad\quad\quad\quad + x_4 = 20$$
$$x_1, x_2, x_3, x_4, x_{R1}, x_{R2} \geqslant 0$$

where x_3 is a surplus variable and x_4 a slack variable, while x_{R1} and x_{R2} are artificial variables. The penalty $-M$ $(M > 0)$ guarantees that x_{R1} and x_{R2} will be at zero level at the optimum, assuming the solution space is feasible. This necessitates that M be selected large enough so that the artificial variables remain "unattractive" in the optimal solution. The above linear program yields the starting basic solution $x_{R1} = 10$, $x_{R2} = 8$, and $x_4 = 20$, with all the remaining variables being nonbasic at zero level.

Another procedure for finding a starting basic solution is the *two-phase* method, which employs artificial variables in the same manner given above. However, rather than modifying the objective function by assigning high penalties to the aritificial variables, a new objective function is formed, which *minimizes* the sum of the artificial variables. This yields the problem:

minimize

$$w = \sum_i x_{Ri}$$

subject to

$$AX = b$$
$$X \geqslant 0 \quad\quad\quad\quad\quad\quad (Phase\ I\ problem)$$

where X includes the slack, surplus, and artificial variables.

The above problem initiates Phase I of the two-phase procedure. It is a regular linear program whose optimum solution can be obtained by the simplex algorithm (Section 2.3.4). However, as indicated in Section 2.2, the simplex algorithm produces optimum solutions that are necessarily basic. Thus, by minimizing the sum of the artificial variables subject to the constraints of the solution space, the resulting optimum *basic solution* provides a starting basis for the original problem provided it is feasible. This means that Phase II is initiated by using the optimal (feasible) basis from Phase I together with the original objective function. From that point on, the simplex algorithm is applied to secure the optimum of the original problem.

The rationale behind the above procedure is as follows. The artificial variable x_{Ri} may be regarded as a "measure of infeasibility" of the ith

constraint equation. Thus, by minimizing $\sum_i x_{Ri}$ subject to the solution
space $AX = b$, $X \geqslant 0$, two cases will result:

(1) If the solution space is infeasible (e.g., $x_1 \geqslant 2$ and $x_1 \leqslant 1$), then at
least one artificial variable must be positive in the optimum solution to
the Phase I problem. In this case, $\min \sum_i x_{Ri} > 0$, which indicates that the
original problem has no feasible solution and hence the computations of
Phase II will not be necessary.

To illustrate this situation consider the simple problem minimize $z = x_1$
subject to $x_1 \geqslant 2$, and $x_1 \leqslant 1$. Augmenting the slack, surplus, and artificial
variables, the Phase I problem is given by:

minimize

$$w = x_{R1}$$

subject to

$$x_1 - x_{s1} + x_{R1} \qquad = 2$$
$$x_1 \qquad\qquad + x_{s2} = 1$$
$$x_1, x_{s1}, x_{s2}, x_{R1} \geqslant 0$$

Simple inspection yields the optimum solution $x_{R1} = 1$, $x_1 = 1$, which shows
that the problem is infeasible. Actually, $x_{R1} = 1$ is interpreted as: $x_1 \geqslant 2$
must be converted to $x_1 \leqslant 2$ in order for a feasible solution to exist.

(2) If $\min \sum_i x_{Ri} = 0$, then the solution space is feasible. Thus the optimum
basic solution of the Phase I problem can be used as a starting basis for
Phase II, the original problem. In general, one expects all the artificial
variables at the end of Phase I to be nonbasic at zero level, in which case
they are deleted all together in Phase II calculation. However, it is possible
that an artificial variable be basic at zero level. In this case, this variable
will be part of the initial basic variables for Phase II. Because no penalty
is assigned to such a variable in the objective function, special provisions
must be made in Phase II so that this variable is never allowed to assume
a positive value. Eventually, the artificial variable *may* become nonbasic,
in which case its column is dropped from the computations. (See Problem 2-9
for details of this point.)

The advantage of the two phase method has to do with elimination of
machine round-off error that could result from the introduction of the high
penalty M in the objective function.

2.3.2   The Feasibility Condition

Using the notation in Section 2.2, the basic and nonbasic vectors X_B
and X_N are related by

$$BX_B + NX_N = b$$

It is assumed that for $X_N = 0$, $BX_B = b$ yields $X_B \geqslant 0$. Thus, B defines the current basis. The generation of a new basis from B requires replacing one of the (current) basic variables in X_B (called the *leaving variable*) by an associated nonbasic variable in X_N (called the *entering variable*). The selection of the entering variable is based on the optimality condition (next section) and requires that such a variable have the potential to improve the objective value when its value increases above its current zero level. The leaving variable is determined by the feasibility condition, which ensures that the new solution not only is basic, but also is feasible (none of the variables is negative).

Given B^{-1}, the inverse of B, then

$$X_B = B^{-1}b - B^{-1}NX_N$$

where $X_N = 0$ by definition and $X_B = B^{-1}b$ is a nonnegative vector. Suppose x_j is selected from X_N as the entering variable, and let $\theta \geqslant 0$ be the value at which x_j enters the new (basic) solution. Thus if P_j is the vector of N associated with x_j, then

$$X_B = B^{-1}b - B^{-1}P_j x_j = B^{-1}b - \theta B^{-1}P_j$$

The value of $x_j = \theta$ is selected such that: (1) none of the elements of X_B becomes negative, and (2) one of the elements of X_B must assume a zero value, signifying that the corresponding "old" basic variable now becomes nonbasic (at zero level).

Consider the ith element of X_B, $(X_B)_i$ given by

$$(X_B)_i = (B^{-1}b)_i - \theta(B^{-1}P_j)_i$$

Thus $(X_B)_i \geqslant 0$ is satisfied if θ is selected such that

$$(B^{-1}b)_i - \theta(B^{-1}P_j)_i \geqslant 0, \qquad \text{for all} \quad i = 1, 2, \ldots, m$$

The left-hand side can become negative only if $(B^{-1}P_j)_i > 0$. Hence, feasibility is maintained so long as θ satisfies

$$\theta \leqslant (B^{-1}b)_i/(B^{-1}P_j)_i, \qquad \text{for all} \quad i \quad \text{such that} \quad (B^{-1}P_j)_i > 0$$

If $(B^{-1}P_j)_i \leqslant 0$ for all i, then θ has no upper bound, which means that the value of x_j can be increased to infinity without violating feasibility. In this case, the solution space is unbounded.

Assuming that $(B^{-1}P_j)_i > 0$ for at least one i, then (at least) one element of X_B is driven to zero by selecting θ such that

$$\theta = \min_i \{(B^{-1}b)_i/(B^{-1}P_j)_i, (B^{-1}P_j)_i > 0\} = (B^{-1}b)_r/\alpha_{rj}$$

where α_{rj} (>0) is the rth element of $\alpha_j \equiv B^{-1}P_j$. This means the basic variable x_r is the leaving variable. If θ is associated with more than one minimum ratio, the tie may be broken arbitrarily. The new basis is then obtained by replacing the rth column of B with P_j.

2.3.3   The Optimality Condition

The selection of the entering variable x_j is made so that the new basis will improve the objective value. This condition is formalized mathematically as follows. Let C_B and C_N be the vectors representing the basic and nonbasic coefficients of the objective function associated with the current basis. Thus the objective function can be written as

$$z = C_B X_B + C_N X_N$$

Since

$$X_B = B^{-1}b - B^{-1}NX_N$$

the objective function can be expressed in terms of the nonbasic vector X_N as follows:

$$z = C_B B^{-1}b - (C_B B^{-1}N - C_N)X_N$$

Let P_j be the jth vector of N and c_j the jth element of C_N. Define

$$z_j = C_B B^{-1}P_j$$

then

$$(C_B B^{-1}N - C_N)X_N = \sum_{j \in NB} (z_j - c_j)x_j$$

where NB represents the set of nonbasic subscripts.

At the current basis, $X_N = 0$ and the corresponding objective value is $z' = C_B B^{-1}b$. Let z'' be the objective value that results when $x_j = \theta$, $j \in NB$, is the entering variable. Since all the remaining nonbasic variables are at zero level, it follows that

$$z'' = z' - \theta(z_j - c_j), \qquad \text{or} \qquad z'' - z' = -\theta(z_j - c_j)$$

If the problem is maximization, the objective value increases ($z'' - z' > 0$) by making x_j ($=\theta$) the entering variable only if $z_j - c_j < 0$. Conversely, in the minimization case, $z'' < z'$ is satisfied only if $z_j - c_j > 0$ is satisfied.

The optimality condition thus selects the entering variable $x_j, j \in NB$, such that:

(i)   *Maximization case*

$$z_j - c_j = \min_{k \in NB}\{z_k - c_k \,|\, z_k - c_k < 0\}$$

If $z_k - c_k \geqslant 0$ for all $k \in NB$, the current basis is optimum.

(ii)   *Minimization case*

$$z_j - c_j = \max_{k \in NB} \{z_k - c_k | z_k - c_k > 0\}$$

If $z_k - c_k \leqslant 0$ for all $k \in NB$, the current basis is optimum.

The above conditions rely on the rule of thumb that the largest *per unit* improvement in the objective value, that is, $z_j - c_j$, may result in the largest improvement in z. However, this is not true in general since the total improvement $\theta(z_j - c_j)$ depends also on the value of θ, which changes with the selection of $x_j, j \in NB$. Computationally, the comparison of the values of $\theta(z_j - c_j)$ is costly, and experimental results have shown that the selection of x_j based on $z_j - c_j$ only is generally more efficient from the standpoint of computational time.

2.3.4   The Simplex Algorithm

The steps of the algorithm can be distilled from the above information as follows.

Step 0   Find a starting basic feasible solution (Section 2.3.1). If none exists, terminate. Otherwise, go to step 1.

Step 1   Compute $z_k - c_k, k \in NB$, for all the nonbasic variables associated with the current basis B. If none of the nonbasic variable can improve solution, stop; current basis is optimum. Otherwise, select the entering variable x_j (Section 2.3.3). Go to step 2.

Step 2   If $\alpha_j \leqslant 0$, stop; the solution is unbounded. Otherwise, select the leaving variable x_r (Section 2.3.2). Form the new basic vector by replacing x_r by x_j in X_B. Accordingly, form the new basic matrix by replacing P_r in B by P_j. Go to step 1.

From the computational standpoint, let X_2 be the vector defining the starting basic solution and define C_2 as its associated vector of objective coefficients. Then X_1 will define the remaining variables. The initial linear program can be arranged in equation form as

$$z - C_1 X_1 - C_2 X_2 = 0, \qquad A_1 X_1 + A_2 X_2 = b$$

where $X_1, X_2 \geqslant 0$. Because X_2 is a starting basic vector, then according to Section 2.3.1, $B = A_2 = I$ and $C_B = C_2$. Expressing the z-equation in terms of X_1 only, one gets

$$z + (C_2 A_1 - C_1)X_1 + 0X_2 = C_2 b$$

The new coefficients of the objective function directly yield the values of $z_j - c_j$ for all elements of nonbasic X_1.

It is customary to put this information in the following tabular form:

Basic vector	X_1^T	X_2^T	Solution
z	$C_2 A_1 - C_1$	0	$C_2 b$
X_2	A_1	I	b

(*Initial tableau*)

By putting $X_1 = 0$, the tableau gives directly the initial solution $z = C_2 b$ and $X_2 = b$.

In general, if B and X_B represent the current basic solution, the associated tableau, expressed in terms of the nonbasic variables, appears as:

Basic vector	X_1^T	X_2^T	Solution
z	$C_B B^{-1} A_1 - C_1$	$C_B B^{-1} - C_2$	$C_B B^{-1} b$
X_B	$B^{-1} A_1$	$B^{-1} I$	$B^{-1} b$

(*General tableau*)

The basic vector X_B consists of m elements taken from the vectors X_1 and X_2. The remaining elements are nonbasic. The solution is given directly as $z = C_B B^{-1} b$ and $X_B = B^{-1} b$ with all the remaining nonbasic variables equal to zero.

The initial and general tableaus express the objective and constraint equations in terms of the nonbasic variables. This automatically allows the determination of the entering variable x_j by inspecting the objective equation, and the leaving variable by taking the ratios of the elements of the right-hand side $B^{-1} b$ to the positive elements of the constraint coefficients $B^{-1} P_j$ under the entering variable.

The process of introducing the entering variable and eliminating the leaving variable, which produces the new basic solution, can be achieved by applying the so-called Gauss–Jordan elimination method to the current tableau. This will automatically result in a new tableau to which the optimality and feasibility conditions can be applied. The method is explained by a numerical example.

▶**EXAMPLE 2.3-1**   Maximize

$$z = -6x_1 + 5x_2$$

subject to

$$-x_1 + 2x_2 \leqslant 4$$
$$x_1 \qquad\quad \geqslant 2$$
$$x_1, x_2 \geqslant 0$$

Adding the slack and artificial variables, this gives:

maximize

$$z = -6x_1 + 5x_2 - Mx_{R1}$$

subject to

$$
\begin{aligned}
-x_1 + 2x_2 &+ x_{s1} && = 4 \\
x_1 &&- x_{s2} &+ x_{R1} = 2 \\
x_1, x_2, x_{s1}, x_{s2}, x_{R1} &\geqslant 0
\end{aligned}
$$

Writing the objective equation as

$$z + 6x_1 - 5x_2 + Mx_{R1} = 0$$

the problem can be put in the following tableau form:

Basic Vector	x_1	x_2	x_{s2}	x_{s1}	x_{R1}	
z	6	-5	0	0	M	0
x_{s1}	-1	2	0	1	0	4
x_{R1}	1	0	-1	0	1	2

The above tableau must be conditioned so that the objective equation is expressed in terms of the nonbasic variables x_1, x_2, and x_{s2} only. This is equivalent to making the coefficients under the starting variables x_{s1} and x_{R1} equal to zero. (Such a step is unnecessary when the starting solution variables are all slacks.) The desired result is achieved by adding or subtracting proper multiples of the constraint equations to or from the objective equation. In the above tableau, the x_{R1}-row is multiplied by $-M$ and added to the z-row. This yields:

Basic vector	X_1^{T}			X_2^{T}			
	x_1	x_2	x_{s2}	x_{s1}	x_{R1}	Solution	
z	$6-M$	-5	M	0	0	$-2M$	
$X_B\begin{cases} x_{s1} \\ x_{R1} \end{cases}$	-1	2	0	1	0	4	*(Initial*
	$\boxed{1}$	0	-1	0	1	2	*tableau)*

Note that the use of row operations produces the same result obtained by utilizing the matrix manipulations given above.

From the objective equation, x_1 is the entering variable since it has the most negative objective coefficient, that is,

$$z_1 - c_1 = \min\{6 - M, -5\} = 6 - M$$

From the constraint coefficients under x_1, one has

$$\alpha_1 = \begin{pmatrix} -1 \\ 1 \end{pmatrix} \quad (= B^{-1}P_1)$$

Also,

$$X_B = \begin{pmatrix} x_{s1} \\ x_{R1} \end{pmatrix} = \begin{pmatrix} 4 \\ 2 \end{pmatrix} \quad (= B^{-1}b)$$

Thus

$$\theta = \min\{-, 2/1\} = 2$$

which shows that x_{R1} must leave the solution.

The new tableau is obtained by introducing x_1 into the basic solution to replace x_{R1}. But according to the above theory, the new tableau is ready for carrying out further iterations only if its equations are expressed in terms of its associated nonbasic variables, namely, x_2, x_{s2}, and x_{R1}. This can be achieved by creating a unit vector under x_1 with the 1-element being at the x_{R1}-row of the x_1-column. This is called the *pivot element* and the x_{R1}-row the *pivot row*.

The Gauss–Jordan elimination technique is used to create the unit vector under the entering variable. It is based again on adding proper multiples of the "normalized" pivot row to all other rows. Normalization means that the entire row is divided by its pivot element, and this will create the 1-element of the unit vector under x_1. In the above tableau, the pivot element is already equal to 1 so that this step is unnecessary. Next, to create zero elements in all other rows, multiply the normalized pivot row by the appropriate element and add it to the corresponding row. Thus the normalized row is multiplied by $-(6 - M)$ when added to the z-row, and by $+1$ when added to the x_{s1}-row. This leads to the new tableau:

Basic vector	x_1	x_2	x_{s2}	x_{s1}	x_{R1}	Solution
z	0	-5	6	0	$-6 + M$	-12
$X_B \begin{cases} x_{s1} \\ x_1 \end{cases}$	0	$\boxed{2}$	-1	1	1	6
	1	0	-1	0	1	2

(*First tableau*)

(The x_{R1}-column may be dropped at this point, since x_{R1} will never become basic again.)

The first tableau shows that x_2 is an entering variable. Taking the ratios to determine θ, it follows that x_{s1} is the (only eligible) variable to leave the solution. The pivot element is equal to 2 and the Gauss–Jordan elimination method yields the following tableau:

Basic vector	x_1	x_2	x_{s2}	x_{s1}	x_{R1}	Solution
z	0	0	7/2	5/2	$M - 7/2$	3
$X_B \quad x_2$	0	1	$-1/2$	1/2	1/2	3
$\qquad\; x_1$	1	0	-1	0	1	2

(Optimum tableau)

Since all $z_k - c_k$ are nonnegative, the last tableau is optimal. This yields $z = 3$, $x_1 = 2$, and $x_2 = 3$ with all the remaining variables being at zero level.◀

In the preceding tableaus, the columns under the current basic variables X_B always constitute an identity matrix in the constraints with all zero coefficients in the objective row. This special structure allows the use of a more compact simplex tableau. Namely, the initial tableau is given as:

	Solution	X_2^{T}
z	$C_2 b$	$C_2 A_1 - C_1$
X_1	b	A_1

(Initial tableau)

(The solution column is moved to the left of the tableau to conform with the standard notation.) The given tableau provides all necessary information to start the simplex iterations. When the entering and leaving variables are determined, then in the next iteration the X_1 and X_2 vectors are changed by exchanging the positions of the entering and leaving variables in the two vectors. The columns of the tableau are changed by using the Gauss–Jordan row operations while imagining that the unit vector associated with the leaving variable has been augmented to the table. After carrying out the row operations, this additional column will give the elements under the leaving variable, now a nonbasic variable. Notice that the unit vector associated with any basic variable is determined by the position of that variable in the basic vector X_B.

▶**EXAMPLE 2.3-2**   The tableaus of Example 2.3-1 are reproduced by using the new format:

	x_1	x_2	x_{s2}	
z	$-2M$	$6-M$	-5	M
x_{s1}	4	-1	2	0
x_{R1}	2	1	0	-1

(Initial)

	x_{R1}	x_2	x_{s2}	
z	-12	$-6+M$	-5	6
x_{s1}	6	1	2	-1
x_1	2	1	0	-1

→

	x_{s1}	x_{s2}
z	3	5/2	7/2
x_2	3	1/2	$-1/2$
x_1	2	0	-1

(Optimum)◀

2.3.5   Special Cases of the Simplex Applications

A.   Degeneracy and Cycling

If the feasibility condition of the simplex method produces the same minimum ratio θ for more than one basic variable, then the next iteration must yield at least one *zero basic* variable. In this case, the basic solution is said to be *degenerate*. Degeneracy here means that there is a chance that at least one α_{ij} associated with a basic $(X_B)_i = 0$ is positive. Thus, in the following iteration, $\theta = 0$, and hence the objective value will not improve. [If for *every* $(X_B)_i = 0$ the associated $\alpha_{ij} \leqslant 0$, then $\theta > 0$ and degeneracy will not occur in the following iteration.]

The difficulty resulting from degeneracy is that *cycling* may occur, that is, the same sequence of iterations may repeat itself without ever terminating the simplex algorithm. This special case rarely occurs in practice, however, and the cases where cycling was shown to occur are theoretical fabrications [see Hoffman (1953), Beale (1955)]. There are methods for ensuring that cycling will not occur (Charnes, 1952). However, this involves excessive additional computations, and because of the rare occurrence of cycling, special provisions in the simplex algorithm are not warranted generally.

B.   *Alternative Optima*

The optimum solution of a maximization (minimization) problem is achieved when $z_j - c_j \geqslant 0$ $(z_j - c_j \leqslant 0)$ for all nonbasic $j \in NB$. If at the optimum, $z_j - c_j = 0$ for any $j \in NB$, then an alternative solution exists, which yields the same optimum objective value but for different values of the variables. If, according to the feasibility condition, x_j can be made basic, then the alternative optimum is basic. Otherwise, it is nonbasic. In general, $z_j - c_j = 0$ signifies the existence of an infinite number of alternative *nonbasic* solutions. For example, if two basic alternative optima exist, then an infinite number of nonbasic alternative optima is defined by all the points on the line segment joining the two basic optimum points.

C.   *Unbounded Solutions*

If in a maximization (minimization) problem, $z_j - c_j < 0$ $(z_j - c_j > 0)$ but $\alpha_{ij} \leqslant 0$ for all i, then x_j, although it can improve the solution, has no effect on the problem feasibility. This means that x_j can be increased indefinitely, which is equivalent to improving the objective value indefinitely. In this case, the optimum solution is said to be unbounded. When this situation occurs in a real-life problem, it means that some of the relevant constraints of the problem have not been accounted for.

2.4   The Revised Simplex Method

The steps, and indeed the basic theory, of the revised simplex method are precisely the same as in the regular simplex method given in Section 2.3. The only difference occurs in computational details. Namely, one observes from the theory developed in Section 2.3 that the entire simplex tableau can be generated from the knowledge of the associated basis B. Furthermore, if B_t is the basis at the tth iteration, then B_{t+1} in the immediately succeeding iteration differs from B_t in exactly one column. This is the idea that led to the development of the revised simplex method.

A well-known theorem in matrix algebra shows that given B_t and its inverse B_t^{-1}, if B_{t+1} is obtained from B_t by replacing its rth column by the column vector P_j, then the inverse of B_{t+1} can be derived from B_t^{-1} directly by using the following formula:

$$B_{t+1}^{-1} = E_t B_t^{-1}$$

where E_t is an m-identity matrix (m is the size of B) whose rth column is replaced by

$$h_r = \begin{pmatrix} -\alpha_{1j}/\alpha_{rj} \\ -\alpha_{2j}/\alpha_{rj} \\ \vdots \\ +1/\alpha_{rj} \\ \vdots \\ -\alpha_{mj}/\alpha_{rj} \end{pmatrix} \quad \leftarrow r\text{th place}$$

where $\alpha_j = (\alpha_{1j}, \ldots, \alpha_{rj}, \ldots, \alpha_{mj})^{\mathrm{T}} = B_t^{-1} P_j$. This requires that $\alpha_{rj} \neq 0$, which is guaranteed by the simplex method feasibility condition.

The use of the above formula shows that it is never necessary to invert any matrices in the course of the simplex computations. Since the initial basis B_0 is always an identity matrix, $B_0^{-1} = I$ is automatically available. Thus, $B_1^{-1} = E_0 B_0^{-1} = E_0$, $B_2^{-1} = E_1 B_1^{-1} = E_1 E_0$, and, generally, $B_t^{-1} = \prod_{i=0}^{t-1} E_i$.

The computational advantage of the revised simplex method is that instead of carrying the entire tableau and applying the Gauss–Jordan technique to it, one need only compute the new inverse B_{t+1}^{-1} from B_t^{-1}. This is then used to determine the entering and leaving variables by using the formulas in Sections 2.3.2 and 2.3.3. The immediate advantage is the savings in computer storage, since it is necessary to carry B_{t+1}^{-1} only rather than the entire tableau. In addition, depending on the density of the matrix A as well as the relative numbers of constraints and variables, the revised simplex method could lead to computational savings [see Dantzig (1963, p. 217)]. Perhaps the important advantage from the viewpoint of integer programming is that the nature of computations in the revised method reduces the (damaging) effect of machine round-off error. In the tableau-type computations, the round-off error accumulates from one tableau to the next. In the revised method, the only error is associated with the computation of B_t^{-1} from B_{t-1}^{-1}, and there are numerical methods for controlling such errors.

▶**EXAMPLE 2.4-1**   Example 2.3-1 is solved by the revised method. The reader is urged to compare the two methods to convince himself that their steps are exactly the same.

Initial iteration

$$X_B = (x_{s1}, x_{R1})^{\mathrm{T}}, \qquad\qquad C_B = (0, -M)$$

$$B_0 = \begin{pmatrix} 1 & 0 \\ 0 & 1 \end{pmatrix} = B_0^{-1}, \qquad C_B B_0^{-1} = (0, -M)$$

To determine the entering variables, compute $z_k - c_k$ for the nonbasic variables x_1, x_2, and x_{s2}. Thus

$$(z_1 - c_1, z_2 - c_2, z_{s2} - c_{s2}) = C_B B_0^{-1} (P_1 P_2 P_{s2}) - (c_1, c_2, c_{s2})$$

$$= (0, -M) \begin{pmatrix} -1 & 2 & 0 \\ 1 & 0 & -1 \end{pmatrix} - (-6, 5, 0)$$

$$= (6 - M, -5, M)$$

(Compare with initial tableau in Example 2.3-1.)

Thus x_1 enters the new basis. The leaving variable is determined as follows:

$$X_B = B_0^{-1} b = \begin{pmatrix} 4 \\ 2 \end{pmatrix}, \qquad \alpha_1 = B^{-1} P_1 = \begin{pmatrix} -1 \\ 1 \end{pmatrix}$$

Hence $\theta = 2/1 = 2$ and x_{R1} leaves the solution. This means that the second column of B_0 is replaced by P_1, which yields

$$B_1^{-1} = E_0 = \begin{pmatrix} 1 & -1/1 \\ 0 & 1/1 \end{pmatrix} = \begin{pmatrix} 1 & 1 \\ 0 & 1 \end{pmatrix}$$

First iteration

$$X_B = \begin{pmatrix} x_{s1} \\ x_1 \end{pmatrix} = B_1^{-1} b = \begin{pmatrix} 6 \\ 2 \end{pmatrix}$$

$$C_B = (0, -6), \qquad C_B B_1^{-1} = (0, -6)$$

$$(z_2 - c_2, z_{s2} - c_{s2}) = (0, -6) \begin{pmatrix} 2 & 0 \\ 0 & -1 \end{pmatrix} - (5, 0) = (-5, 6)$$

This shows that x_2 enters the solution.

For the leaving variable,

$$B_1^{-1} P_2 = \begin{pmatrix} 1 & 1 \\ 0 & 1 \end{pmatrix} \begin{pmatrix} 2 \\ 0 \end{pmatrix} = \begin{pmatrix} 2 \\ 0 \end{pmatrix}$$

$$\theta = \frac{6}{2} = 3$$

corresponding to x_{s1}. Thus x_{s1} leaves the solution:

$$B_2^{-1} = E_1 B_1^{-1} = \begin{pmatrix} +1/2 & 0 \\ -0/2 & 1 \end{pmatrix} \begin{pmatrix} 1 & 1 \\ 0 & 1 \end{pmatrix} = \begin{pmatrix} 1/2 & 1/2 \\ 0 & 1 \end{pmatrix}$$

Second iteration

$$X_B = \begin{pmatrix} x_2 \\ x_1 \end{pmatrix} = B_2^{-1}b = \begin{pmatrix} 3 \\ 2 \end{pmatrix}$$

$$C_B = (5, -6), \qquad C_B B_2^{-1} = (5/2, -7/2)$$

$$(z_{s1} - c_{s1}, z_{s2} - c_{s2}) = (5/2, -7/2)\begin{pmatrix} 1 & 0 \\ 0 & -1 \end{pmatrix} - (0, 0) = (5/2, 7/2)$$

This shows that the last solution is optimal.◀

2.5 The Dual Problem

Consider the linear program:

maximize

$$z = CX$$

subject to

$$AX \leqslant b \qquad (Primal\ problem)$$
$$X \geqslant 0$$

which is arbitrarily referred to as the *primal* (given) problem. There exists another problem called the *dual*, whose properties are closely related to the primal. This problem is defined as:

minimize

$$w = b^T Y$$

subject to

$$A^T Y \geqslant C^T \qquad (Dual\ problem)$$
$$Y \geqslant 0$$

The dual problem is derived from its primal by using the following conditions:

(1) The primal objective is maximization, and the dual objective is minimization;
(2) The number of variables in the dual is equal to the number of constraints in the primal;
(3) The number of constraints in the dual is equal to the number of variables in the primal;

(4)   The objective coefficients in the primal form the right-hand side of the dual.

(5)   The right-hand side of the primal forms the objective of the dual;

(6)   All variables are nonnegative in both problems.

One notices that, if the above rules are reversed so that the dual takes the place of the primal, then the above (modified) rules, when applied to the dual, will yield the primal. This means that the dual of the dual is the primal.

▶**EXAMPLE 2.5-1**   Consider the primal problem:

maximize

$$z = 2x_1 - 3x_2$$

subject to

$$x_1 + 5x_2 \leqslant 11$$
$$7x_1 - 6x_2 \leqslant 22$$
$$9x_1 + 4x_2 \leqslant 33$$
$$x_1, x_2 \geqslant 0$$

Its dual is given by:

minimize

$$w = 11y_1 + 22y_2 + 33y_3$$

subject to

$$y_1 + 7y_2 + 9y_3 \geqslant 2$$
$$5y_1 - 6y_2 + 4y_3 \geqslant -3$$
$$y_1, y_2, y_3 \geqslant 0 ◀$$

As will be shown below, the optimum solution of *one* of the two problems immediately yields the optimum solution to the remaining problem. Since in linear programming, the efficiency of computations is reduced by the increase in the number of constraints (rather than the number of variables), it is advantageous to solve the dual when the number of constraints in the primal exceeds the number of variables. This situation is illustrated by the above example.

The above primal–dual definitions do not account for the cases where equality constraints and/or unrestricted variables exist in the primal problem. The provisions for these cases must be consistent with the above definitions, however.

Consider an equality constraint of the form

$$\sum_{j=1}^{n} a_{ij}x_j = b_i$$

which can be written as

$$\sum_{j=1}^{n} a_{ij}x_j \le b_i, \qquad \sum_{j=1}^{n}(-a_{ij})x_j \le -b_i$$

Thus, in writing the dual, let y_i^+ and y_i^- be the associated dual variables. Since both constraints are inequalities of the type $\le$, y_i^+ and y_i^- are defined as *nonnegative* variables. The terms associated with y_i^+ and y_i^- in the dual constraints then appear in the n respective constraints as

$$a_{i1}(y_i^+ - y_i^-), \quad a_{i2}(y_i^+ - y_i^-), \ldots, a_{in}(y_i^+ - y_i^-)$$

Hence, by letting $y_i = y_i^+ - y_i^-$, this is equivalent to saying that the original equation constraint corresponds to an *unrestricted* variable y_i in the dual.

Next, if the variable x_k is unrestricted in the primal, then it can be substituted by $x_k = x_k^+ - x_k^-$, where $x_k^+, x_k^- \ge 0$. By following a procedure similar to the one above but in the reverse order, one concludes that the dual constraint associated with x_k must be an equality.

To summarize: (1) an equality constraint in the primal yields an associated *unrestricted* variable in the dual, and (2) an unrestricted primal variable yields an associated *equality* constraint in the dual. These properties are equally applicable when applied to the dual for the purpose of determining the primal.

▶**EXAMPLE 2.5-2**   Suppose in Example 2.5-1 that the first constraint is an equation and x_2 is unrestricted. The dual becomes:

minimize

$$w = 11y_1 + 22y_2 + 33y_3$$

subject to

$$y_1 + 7y_2 + 9y_3 \ge 2$$
$$5y_1 - 6y_2 + 4y_3 = -3$$
$$y_1 \text{ unrestricted}$$
$$y_2, y_3 \ge 0.◀$$

2.5.1   Relationships between the Solutions of Primal and Dual

This section shows how the solution of one problem (primal or dual) gives directly the solution to the other problem (provided solutions exist and are finite). It also reveals interesting relationships that lead to the discovery of the so-called *dual simplex* algorithm which is basic to the development of cutting-plane methods in integer programming.

▶**THEOREM 2-2**   In the primal $\max\{CX \,|\, AX \leqslant b, X \geqslant 0\}$ and the dual $\min\{b^T Y \,|\, A^T Y \geqslant C^T, Y \geqslant 0\}$, if X_0 and Y_0 are *feasible* solutions to the primal and dual problems respectively, then

$$z_0 = CX_0 \leqslant b^T Y_0 = w_0$$

Moreover, if $CX_0 = b^T Y_0$, then X_0 and Y_0 are *optimal* solutions to the primal and dual problems, respectively.

   Proof   From the primal $AX_0 \leqslant b$, but since $Y_0 \geqslant 0$, it follows that $Y_0^T AX_0 \leqslant Y_0^T b$. Similarly, from the dual, $A^T Y_0 \geqslant C^T$, but because $X_0 \geqslant 0$, hence $X_0^T A^T Y_0 \geqslant X_0^T C^T$, or $Y_0^T AX_0 \geqslant CX_0$. From the two inequalities, it follows that

$$z_0 = CX_0 \leqslant Y_0^T AX_0 \leqslant Y_0^T b = b^T Y_0 = w_0$$

This means that for any *feasible* pair (X, Y) the objective value of the minimization problem acts as an upper bound on the objective value of the maximization problem.

   The proof of the second part is as follows. By assumption, $CX_0 = b^T Y_0$. Let X be any feasible primal solution. Then from the result of the first part of the theorem $CX \leqslant b^T Y_0$. But since $b^T Y_0 = CX_0$ by assumption, hence $CX \leqslant CX_0$ and X_0 must be optimum. Similarly, if Y is any feasible solution, then

$$b^T Y \geqslant CX_0 = b^T Y_0$$

and Y_0 must be optimum.◀

   The above results are equally applicable when some (or all) of the constraints are equations and/or some (or all) of the variables are unrestricted. The reader is asked to verify this result in Problem 2-15.

▶**EXAMPLE 2.5-3**   Consider Example 2.3-1. Since the problem includes both $\geqslant$ and $\leqslant$ constraints, the most convenient way to develop the dual is to convert the primal into equations by augmenting the appropriate slacks, which yields:

   maximize
$$z = -6x_1 + 5x_2$$
subject to

$$
\begin{aligned}
-x_1 + 2x_2 + x_{s1} \quad\quad\quad &= 4 \\
x_1 \quad\quad\quad\quad - x_{s2} &= 2 \\
x_1, x_2, x_{s1}, x_{s2} &\geqslant 0
\end{aligned}
$$

This "all-equation" form is particularly suitable since it is the one used in solving the primal by the simplex. Hence, in deriving the dual solution from its optimal (primal) tableau, there will be no confusion as to the signs of the dual variables.

The associated dual is:

minimize

$$w = 4y_1 + 2y_2$$

subject to

$$-y_1 + y_2 \geqslant -6$$
$$2y_1 \quad\quad \geqslant \quad 5$$
$$\quad\quad\quad\quad\quad (y_1, y_2 \text{ are unrestricted})$$
$$y_1 \quad\quad \geqslant \quad 0$$
$$-y_2 \geqslant \quad 0$$

The two dual constraints $y_1 \geqslant 0$ and $-y_2 \geqslant 0$ correspond to the two primal variables x_{s1} and x_{s2}, respectively. Notice that these constraints override the conditions that y_1 and y_2 be unrestricted, which is created by the fact the primal constraints are equations. (The reader should notice what is meant by "controlling the signs of the dual," since the condition $y_2 \leqslant 0$ is not immediately obvious from the original problem.)

A primal feasible solution is $x_1 = 2$, $x_2 = 3$, which is obtained by inspection. Similarly, a dual feasible solution is $y_1 = \frac{5}{2}$, $y_2 = 0$. This yields $z = 3$ and $w = 10$, which shows that $z < w$ so that $3 \leqslant \max z = \min w \leqslant 10$. Actually, the results of Example 2.3-1 shows max $z = 3$, which means that the optimum could coincide with one of the boundary values.◀

▶**Theorem 2-3**   If B is the *optimal primal* basis and C_B is its associated vector in the objective function, then the *optimal* solution of the associated *dual* problem is $Y = C_B B^{-1}$.

Proof   According to Theorem 2-2, $Y = C_B B^{-1}$ is optimal if (i) it satisfies the dual constraints $A^T Y \geqslant C^T$, where $Y \geqslant 0$; and (ii) it yields $b^T Y = CX$, where X is the optimal primal vector associated with B.

By the optimality condition (Section 2.3.3), X is optimal if and only if all the objective row coefficients are nonnegative (maximization), that is, if and only if

$$C_B B^{-1} A - C \geqslant 0, \quad \text{and} \quad C_B B^{-1} \geqslant 0$$

By putting $Y = C_B B^{-1}$, it follows immediately that $Y = C_B B^{-1}$ is a feasible dual solution.

The second point is proved by observing that

$$\max z = C_B X_B = C_B B^{-1} b = Y^\mathrm{T} b = \min w. \blacktriangleleft$$

Theorem 2-3 applies equally to primal and dual problems in which constraints are equations and variables are unrestricted. The proof is very similar to the one given above.

By inspecting the general (matrix) tableau (Section 2.3.4) one finds that the objective coefficients under the starting variables, $X_2{}^\mathrm{T}$, are given by $C_B B^{-1} - C_2$, where C_2 gives the coefficients of X_2 in the original objective function. If X_2 is all slack, $C_2 = 0$ and the indicated coefficients in the objective row yield the optimal dual solution directly. Otherwise, C_2 must be added to $C_B B^{-1} - C_2$. The reader should not find it difficult to obtain the optimal dual values if some other tableau form, such as the one in Example 2.3-2, is used.

In the nonoptimal iterations of the simplex tableaus, the associated values $Y = C_B B^{-1}$ are usually referred to as the "simplex multipliers" or "shadow prices." This standard definition will be used in later developments.

▶**EXAMPLE 2.5-4**   The dual of Example 2.3-1 is given in Example 2.5-3. From Example 2.3-1, the objective coefficients under the starting variables are $C_B B^{-1} - C_2 = (5/2, M - 7/2)$. Since $C_2 = (0, -M)$, it follows that $(y_1, y_2) = C_B B^{-1} = (5/2, M - 7/2) + (0, -M) = (5/2, -7/2)$. This solution yields $\min w = 4(5/2) + 2(-7/2) = 3$, which is equal to max z.◀

▶**THEOREM 2-4**   (a) When the primal (dual) is unbounded, the dual (primal) remains infeasible.

(b) When the primal (dual) is infeasible the dual (primal) is either unbounded or infeasible.

Proof   (a) The primal is unbounded means that the entering variable x_j has $z_j - c_j < 0$, but its $\alpha_j = B^{-1} P_j \leqslant 0$. This means that x_j can never be made basic and, without a change of basis, the condition $z_j - c_j < 0$ will persist. Let $Y^\mathrm{T} = C_B B^{-1}$. Since $z_j - c_j = C_B B^{-1} P_j - c_j$, it follows that if the primal is unbounded, then $Y^\mathrm{T} P_j - c_j < 0$ or $Y^\mathrm{T} P_j < c_j$, which means that the dual constraint is not satisfied. The same result can be proved by another procedure. From Theorem 2-2,

$$\max z = \max CX \leqslant b^\mathrm{T} Y$$

provided Y is feasible. Because max $CX \to \infty$, $b^\mathrm{T} Y$ cannot form a finite upper bound on max z for any feasible Y. Thus Y must be infeasible.

(b) If primal is infeasible, then CX forms no legitimate lower bound on min $b^\mathrm{T} Y$, and the latter may be decreased indefinitely. To show that when primal is infeasible, the dual also could be infeasible the following example is constructed.

Primal   maximize

$$z = 4x_1 + 3x_2$$

subject to

$$x_1 - x_2 \leqslant 3/5$$
$$x_1 - x_2 \geqslant 2$$
$$x_1, x_2 \geqslant 0$$

Dual   minimize

$$w = \tfrac{3}{5}y_1 - 2y_2$$

subject to

$$y_1 - y_2 \geqslant 4$$
$$-y_1 + y_2 \geqslant 3$$
$$y_1, y_2 \geqslant 0$$

(The dual is obtained after converting all primal constraints to $\leqslant$.) Clearly, the constraints of both primal and dual cannot be satisfied simultaneously. ◀

▶**THEOREM 2-5**   (COMPLEMENTARY SLACKNESS)   Let $(s_1, s_2, \ldots, s_i, \ldots, s_m)$ be the vector of slack variables associated with the primal and let $(y_1, y_2, \ldots, y_i, \ldots, y_m)$ be the variables of the dual. Then for any pair of the primal and dual optimal solutions,

$$y_i s_i = 0, \qquad i = 1, 2, \ldots, m$$

Proof   The primal constraints are given by

$$(A, I)\binom{X}{S} = b, \qquad X, S \geqslant 0$$

where $S = (s_1, s_2, \ldots, s_m)^{\mathrm{T}}$ is the slack vector. Premultiplying both sides by $Y = (y_1, y_2, \ldots, y_m)$, where Y is a feasible dual solution, then

$$YAX + YS = Yb$$

At the optimal solution (X^*, Y^*),

$$CX^* = Y^*AX^* = Y^*b$$

(See Theorem 2-2, Section 2.5.1.) Thus

$$Y^*S^* = 0, \qquad \text{or} \qquad y_i^* s_i^* = 0, \qquad i = 1, 2, \ldots, m ◀$$

This result implies that if the ith primal constraint is inactive $(s_i > 0)$, then the associated dual variable y_i must be zero. On the other hand, if $y_i > 0$, then the ith primal constraint must be active $(s_i = 0)$.

▶**THEOREM 2-6**   While the primal problem is seeking optimality, the dual problem is automatically seeking feasibility.

Proof   The proof is actually implied by the above theorem. The primal problem starts feasible and nonoptimal. However, nonoptimality of the primal means that at least one of the objective row coefficients is negative, that is, at least one $z_j - c_j$, $j \in NB$, is negative. By putting $Y^T = C_B B^{-1}$, $z_j - c_j = C_B B^{-1} P_j - c_j = Y^T P_j - c_j < 0$ implies that the associated dual solution is infeasible. This continues to be the case for every iteration until, at the optimum, $z_j - c_j \geqslant 0$, for all $j \in NB$, which means that only at the optimum iteration does the dual solution $Y^T = C_B B^{-1}$ become feasible.◀

The implication of the above theorem is that, while the (primal) simplex method starts feasible and nonoptimal and maintains feasibility until the optimal is attained, a similar procedure may be developed for problems that start infeasible and optimal. In this case, successive iterations will maintain optimality but will move toward feasibility until a feasible optimum is achieved. This idea is the basis for the so-called dual simplex method. The dual algorithm is particularly important because of its direct use in integer cutting-plane algorithms.

For notational convenience, it is customary to refer to linear programs that start optimal as "dual feasible," since this implies that the associated dual constraints are automatically feasible. "Primal feasible" programs are those solutions satisfying all primal constraints.

2.5.2   Dual Simplex Algorithm†

In the regular simplex method, the feasibility condition is used to maintain the feasiblity of the basic solution by using a ratio test. This same idea can be utilized to maintain the optimality of the objective row.

To be specific, consider the linear program of the form:

minimize

$$z = CX$$

subject to

$$AX \leqslant b$$
$$X \geqslant 0$$

† Lemke (1954).

Let $X_s \geq 0$ be the slacks associated with the constraints. Then the problem can be expressed by the following equations:

$$z - CX = 0, \qquad AX + IX_s = b$$

Thus the starting tableau may be written as:

Basic vector	X^T	X_s^T	Solution	
z	$-C$	0	0	*(Initial tableau)*
X_s	A	I	b	

The starting basic vector is $X_s = b$, which is infeasible (otherwise, the problem is of the primal type).

A necessary condition for the application of the dual algorithm is that the objective row must be dual feasible (optimal). This means that the vector C must be nonnegative, so that $\{z_j - c_j\} = -C \leq 0$ satisfies the optimality condition of the minimization problem. In general, this may not be satisfied. In this case an *artificial constraint* of the form

$$1X = x_1 + x_2 + \cdots + x_n \leq M$$

must be augmented to the constraints, where 1 is an n-row vector with all its components equal to 1, and M is selected sufficiently large so that the constraint is redundant with respect to the original constraints $AX \leq b$. Let c_k be the kth component of the vector C and define

$$-c_j = \max\{-c_k| -c_k > 0, k = 1, 2, \ldots, n\}$$

Then x_j is selected as an entering variable while the slack variable associated with the artificial constraint leaves the solution, thus making the artificial equation the pivot row. A change of basis under this condition should render an optimal starting tableau.

At any iteration of the dual simplex method, the entering variable is determined by the optimality condition while the leaving variable is obtained from the feasibility condition. Once this is achieved, a change of basis is made by following the exact Gauss–Jordan elimination method, which was introduced with the primal simplex algorithm (Section 2.3.4).

Suppose the basis at the current iteration is B and that the columns of B are a proper subset of the columns of (A, I). Define $\alpha_j = B^{-1}P_j$ where, again P_j is a column of (A, I). Thus the current basic solution is $X_B = B^{-1}b$.

Feasibility condition   Let $(X_B)_i = (B^{-1}b)_i$ be the ith element of $(B^{-1}b)$ and define

$$(B^{-1}b)_r = \min_i \{(B^{-1}b)_i \,|\, (B^{-1}b)_i < 0, i = 1, 2, \ldots, m\}$$

Then the basic variable x_r is the leaving variable. This means that the most infeasible (most negative) basic variable is dropped from the basic vector. If r is undefined the solution is feasible.

Optimality condition   Let $z_k - c_k$ be the kth element of the objective row associated with the current basis B. Consider α_{rk}, $k \in NB$, that is, the rth elements of $B^{-1}P_k$ for all nonbasic P_k, where r is the index of the row of the leaving variable. Determine

$$|(z_j - c_j)/\alpha_{rj}| = \min_{k \in NB} \{|(z_k - c_k)/\alpha_{rk}|, \alpha_{rk} < 0, k \in NB\}$$

Then the nonbasic variable x_j enters the solution. If all $\alpha_{rk} \geq 0$, the problem has no feasible solution. (With some reflection, the reader will find that this criterion is based on the same minimum ratio idea used with the feasibility condition of the primal simplex method). Notice that by using the absolute values, the above condition is applicable to both maximization and minimization problems.

▶**EXAMPLE 2.5-5**   Consider Example 2.3-1. This can be written in tableau form as:

Basic vector	x_1	x_2	x_{s1}	x_{s2}	Solution
z	6	-5	0	0	0
x_{s1}	-1	2	1	0	4
x_{s2}	-1	0	0	1	-2

The tableau is neither primal nor dual feasible. In Example 2.3-1, it was made primal feasible by adding an artificial *variable* to the second constraint. In this example, the tableau is made dual feasible by adding the artificial constraint $x_1 + x_2 \leq M$, which gives:

Basic vector	x_1	x_2	x_{s1}	x_{s2}	x_{s3}	Solution
z	6	-5	0	0	0	0
x_{s1}	-1	2	1	0	0	4
x_{s2}	-1	0	0	1	0	-2
x_{s3}	1	1	0	0	1	M

To obtain dual feasibility, x_2 enters the basic vector and x_{s3} leaves. This yields the following tableau:

Basic vector	x_1	x_2	x_{s1}	x_{s2}	$\downarrow$ x_{s3}	Solution
z	11	0	0	0	5	$5M$
$\leftarrow$ x_{s1}	-3	0	1	0	$\boxed{-2}$	$4 - 2M$
x_{s2}	-1	0	0	1	0	-2
x_2	1	1	0	0	1	M

The succeeding tableaus are given below:

Basic vector	$\downarrow$ x_1	x_2	x_{s1}	x_{s2}	x_{s3}	Solution
z	7/2	0	5/2	0	0	10
x_{s3}	3/2	0	$-1/2$	0	1	$-2 + M$
$\leftarrow$ x_{s2}	$\boxed{-1}$	0	0	1	0	-2
x_2	$-1/2$	1	1/2	0	0	2

Basic vector	x_1	x_2	x_{s1}	x_{s2}	x_{s3}	Solution	
z	0	0	5/2	7/2	0	3	
x_{s3}	0	0	$-1/2$	3/2	1	$-5 + M$	
x_1	1	0	0	-1	0	2	
x_2	0	1	1/2	1/2	0	3	(*Optimum*)

This yields the same solution as in Example 2.3-1. It is interesting to notice that the dual solution of the given problem can be secured from the objective row coefficient of x_{s2} and x_{s2}, namely, $y_1 = 5/2$, and $y_2 = 7/2$. Observe, however, that the dual must be constructed from the initial primal tableau of the example, which gives the dual objective as $w = 4y_1 - 2y_2$ (cf. Example 2.5-4). ◄

As in the primal simplex method (Example 2.3-2), the dual simplex tableau can be presented in a compact form. This is achieved by deleting the unit vectors associated with the current basic variables.

2.5.3 Treatment of Additional Constraints by Dual Simplex

A typical solution method in integer programming is first to solve the problem as a linear program. If the resulting solution does not satisfy the integer conditions, then specially designed constraints are added to the optimum linear programming tableau in order to force the solution toward integrality. This is a typical situation in which the dual simplex method is used. The optimum linear program is dual feasible, but the addition of the new constraint will make it primal infeasible.

▶**EXAMPLE 2.5-6**   Suppose the constraint $2x_1 + x_2 \leqslant 6$ is added to the optimum tableau in Example 2.5-5. The optimum solution $(x_1 = 2, x_2 = 3)$ violates this constraint. Let x_{s4} be the slack variable associated with the new constraint. The first step is to express the additional constraint in terms of the nonbasic variables of the optimum tableau in Example 2.5-5. This yields the following tableau:

Basic vector	x_1	x_2	x_{s1}	x_{s2}	x_{s3}	x_{s4}	Solution	
z	0	0	5/2	7/2	0	0	3	
x_{s3}	0	0	$-1/2$	3/2	1	0	$-5 + M$	
x_1	1	0	0	-1	0	0	2	
x_2	0	1	1/2	$-1/2$	0	0	3	
x_{s4}	0	0	1/2	5/2	0	1	-1	←additional constraint

The variable x_{s4} leaves the solution. But since all its constraint coefficients are nonnegative, the new constraint cannot result in a feasible solution.◀

2.5.4   Column Simplex Method†

As a result of the strong relationships that exist between the primal and dual problems (mainly, the optimum solution to one problem yields optimum solution to the other), a new tableau form was developed which can result in efficient computations. Namely, if the number of constraints is smaller than the number of variables, it is more efficient to solve the primal, while in the opposite case, the dual is more efficient. One observes that the application of the Gauss–Jordan elimination method to the *rows* of the dual is equivalent to applying the same method to the *columns* of the primal. Thus, after the starting tableau is set up, the decision as to whether the Gauss–Jordan elimination is applied to the rows or to the columns depends on the relative values of the number of constraints and number of variables. In essence, the column tableau does not really add much new except for the convenience of not having to formulate the dual, should it be necessary to solve it.

The column tableau form is applicable to both the primal and dual simplex methods. The feasibility and optimality conditions are applied in

† Beale (1954).

the exact manner given previously for the primal and the dual simplex methods. Perhaps the only (minor) difference occurs in changing the basis. In row-tableau form, the *pivot row* is divided by the pivot element α_{rj}, but in the column-tableau, the *pivot column* is divided by the negative of the pivot element $(-\alpha_{rj})$. This follows since, in the primal problem, the starting solution is dual infeasible, that is, $z_j - c_j < 0$ for at least one j. From the *primal* feasibility condition, α_{rj} must be positive. Thus, by dividing the jth column by $(-\alpha_{rj})$, the objective row would be moving toward dual feasibility. A similar reason can be given to problems which start dual feasible since $\alpha_{rj} < 0$, and dividing by $(-\alpha_{rj})$ would maintain dual feasibility.

The linear program is usually put in the form:

maximize (or minimize)

$$z = CX$$

subject to

$$AX \leqslant b$$
$$X \geqslant 0$$

and it may be either primal or dual feasible. The starting tableau is obtained by augmenting the trivial constraints

$$IX - IX = 0$$

Letting X_s be the (nonnegative) slacks associated with $AX \leqslant b$, the starting tableau is usually written as:

$$X^{\mathrm{T}}$$

z	0	$-C$
X	0	$-I$
X_s	b	A

(*Initial tableau*)

The starting basic vector is $X = 0$ and $X_s = b$.

▶**EXAMPLE 2.5-7**   Consider the following dual feasible problem:

minimize

$$z = 4x_1 + 8x_2$$

subject to

$$-x_1 + 2x_2 \leqslant -2$$
$$-4x_1 - 2x_2 \leqslant -4$$
$$x_1 - 4x_2 \leqslant -8$$
$$x_1, x_2 \leqslant \quad 0$$

The following successive tableaus lead to the (optimum) feasible solution:

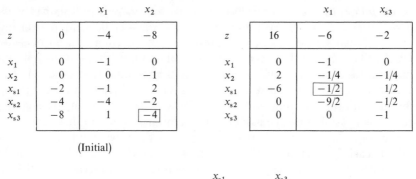

		x_1	x_2
z	0	-4	-8
x_1	0	-1	0
x_2	0	0	-1
x_{s1}	-2	-1	2
x_{s2}	-4	-4	-2
x_{s3}	-8	1	$\boxed{-4}$

(Initial)

		x_1	x_{s3}
z	16	-6	-2
x_1	0	-1	0
x_2	2	$-1/4$	$-1/4$
x_{s1}	-6	$\boxed{-1/2}$	$1/2$
x_{s2}	0	$-9/2$	$-1/2$
x_{s3}	0	0	-1

		x_{s1}	x_{s3}
z	88	-12	-8
x_1	12	-2	-1
x_2	5	$-1/2$	$-1/2$
x_{s1}	0	-1	0
x_{s2}	54	-9	-5
x_{s3}	0	0	-1

(Optimal) ◀

As in Example 2.3-2, it is possible to eliminate the rows associated with the negative unit vectors $(-I)$ without losing information. This will yield a more compact tableau. However, the (complete) column simplex tableau is convenient to use with integer cutting-plane methods because: (i) the basic vector never changes, and (ii) the cut (or additional constraint) can be deleted after pivoting. (The second point is explained in Chapter 5 in connection with the cutting methods.) In addition, this tableau form could prove useful in providing a proper starting tableau for the so-called lexicographic dual simplex method presented in the next section.

2.5.5 Lexicographic Dual Simplex Method

The lexicographic dual (or l-dual) simplex method is essential in proving the convergence of most cutting-plane methods in integer programming. The method was primarily developed to prevent cycling in the dual algorithm, which could occur when $z_j - c_j = 0$ for the entering variable x_j (cf. cycling in the primal algorithm, Section 2.3.5). Before introducing the method, the following definition is introduced.

A vector α is said to be lexicographically positive (*l*-positive) if its first nonzero element is positive. This is usually expressed as $\alpha \succ 0$. A vector β is *l*-negative ($\beta \prec 0$) if $-\beta$ is *l*-positive. The vector α is *l*-greater than the vector β ($\alpha \succ \beta$) if $\alpha - \beta \succ 0$. For example, $\alpha = (0, 0, 1, -3, 0) \succ 0$, while $\beta = (-1, 0, 2, 3, 1) \prec 0$. Also, $\alpha \succ \beta$ because $\alpha - \beta = (1, 0, -1, -6, -1) \succ 0$.

The *l*-dual simplex method requires that the starting tableau be *l-dual feasible*. This means that every nonbasic column must be *l*-positive for the maximization problem and *l*-negative for the minimization problem. This condition is always satisfied if $z_j - c_j, j \in NB$ is *strictly* positive (negative) for maximization (minimization) problems. However, if any $z_j - c_j = 0$, then it is possible that the associated column may not be *l*-dual feasible.

In some cases, a simple rearrangement of the rows would result in an *l*-dual feasible starting solution. But, in general, this condition can be satisfied as follows: If the problem is of the minimization type, then the use of the column starting tableau (Section 2.5.4) always guarantees an *l*-dual feasible solution because the $(-I)$ matrix is situated immediately below the objective row. In the maximization one can append the artificial constraint $\sum_{j \in NB} x_j \le M, M > 0$ and sufficiently large, immediately below the objective row. Actually, the artificial constraint can include only those nonbasic variables whose $z_j - c_j = 0$.

Because the column simplex method of Section 2.5.4 may be useful in providing a starting *l*-dual feasible solution, the presentation below assumes the use of column, rather than row, pivoting. Actually, this is only a matter of convenience, since row pivoting is also equally applicable.

The feasibility condition for the *l*-dual method is exactly the same as for the regular dual method (Section 2.5.2). The only difference occurs in the optimality condition used to select the entering variable.

Optimality Condition for the l-Dual Method

At a current iteration, let x_r be the leaving variable and define

$$\beta_k \equiv \left(\frac{z_k - c_k}{\alpha_k}\right) = \begin{pmatrix} z_k - c_k \\ \alpha_{1k} \\ \vdots \\ \alpha_{rk} \\ \vdots \\ \alpha_{nk} \end{pmatrix}$$

Let $Q = \{k \,|\, \alpha_{rk} < 0\}$ and compute

$$\beta_k = \beta_k / (-\alpha_{rk}), \qquad k \in Q$$

Then x_j is the entering variable if in case of maximization (minimization)

$\underline{\beta_j}$ is the *lexicographically* largest (smallest) among all $\underline{\beta_k}$ $k \in Q$, that is, x_j enters the solution if

$$\underline{\beta_j} = \begin{cases} l\text{-}\min\{\underline{\beta_k}|k \in Q\} & \text{(minimization)} \\ l\text{-}\max\{\underline{\beta_k}|k \in Q\} & \text{(maximization)} \end{cases}$$

Following the determination of the entering and leaving variables, column pivoting is carried out as in Section 2.5.4. The selection of x_j as above guarantees that column pivoting will always produce l-dual feasible tableaus. Under this condition, cycling cannot occur. This follows because the solution vector is always l-decreasing (l-increasing) in case of maximization (minimization). To prove this point, consider the *maximization* case. Let α_0 be the solution column (values of the basic variables) at the current iteration. Given that α_j is the current pivot column, define $\hat{\alpha}_0$ and $\hat{\alpha}_j$ as the resulting columns after pivoting. Further, let ρ_0 and $\hat{\rho}_0$ be the first r elements (x_r is the leaving variable) of the solution column before and after pivoting. The proof is complete if one shows that $\alpha_0 \succ \hat{\alpha}_0$ (maximization). This is equivalent to proving that $\rho_0 \succ \hat{\rho}_0$. Now, if ρ_j represents the first r elements of α_j, then

$$\hat{\rho}_0 = \rho_0 + \alpha_{r0}(\rho_j/-\alpha_{rj}) = \rho_0 + \alpha_{r0}\,\hat{\rho}_j$$

Since the selection of x_r implies $\alpha_{r0} < 0$, and since $\hat{\rho}_j \succ 0$ (because $\hat{\alpha}_j \succ 0$), it follows that $\alpha_{r0}\,\hat{\rho}_j \prec 0$ and $\rho_0 \succ \hat{\rho}_0$ as is to be proved.

As seen from the above development, the l-dual method primarily eliminates cycling and hence guarantees convergence of the algorithm. In real practice, the occurence of cycling (as in the primal simplex method) is rare. Consequently, in a practical sense, one need not be concerned about cycling in the cases where $z_k - c_k = 0$ for at least one $k \in NB$. Nevertheless, the convergence proofs of most cutting-plane algorithms are based on the assumption that the l-dual method is used to solve the problem. For this reason, the l-dual (column) tableau will be used in connection with most integer algorithms throughout this book. The reader should keep in mind, however, that any other tableau form is suitable so long as cycling does not occur. This point is mentioned here because there are books that introduce the integer algorithm without any reference to the l-dual method. Again, the authors are relying on the rare occurrence of cycling in practice.

2.5.6   Benders' Partitioning Approach

An important application of duality theory was developed by Benders (1962) in connection with the following problem:

maximize

$$z = CX + f(Y)$$

subject to

$$AX + g(Y) \leqslant b$$
$$X \geqslant 0, \qquad Y \geqslant 0 \qquad \text{(P)}$$

where C is an n-vector, A an $m \times n$ matrix, and b an m-vector.

Benders' procedure solves (P) by partitioning the variables so that, at any iteration of the algorithm, problems containing either X or Y (but not both) are solved. The advantage is that problem (P) in (X, Y) may be complex computationally, while those in either X or Y may be simpler. For example, if Y is an integer vector, the partitioning will result in a linear program in X and an integer program in Y. Another possibility is that $f(Y)$ and $g(Y)$ may have a special structure (such as transportation model), which cannot be taken advantage of computationally unless partitioning is effected.

The details of Benders' method are first presented. This is followed by a proof of convergence.

Suppose the vector Y is fixed at given values. Then problem (P) reduces to the form:

maximize

$$v = CX + f(Y)$$

subject to

$$AX \leqslant b - g(Y)$$
$$X \geqslant 0 \qquad \text{(PX)}$$

Let u be the dual vector; then the dual of (PX) is

minimize

$$w = u(b - g(Y))$$

subject to

$$uA \geqslant C$$
$$u \geqslant 0 \qquad \text{(D1)}$$

Problem (D1) has two important properties: (i) its feasible region $R = \{u \,|\, uA \geqslant C, u \geqslant 0\}$ is independent of the choice of Y, and (ii) its minimum occurs at an extreme point u^k of the feasible region R. (If R is unbounded, augment the regularity constraint $\sum_{i=1}^{m} u_i \leqslant M$, where $M > 0$ is very large.)

Properties (i) and (ii) indicate that (D1) can be replaced by:

$$\underset{u^k}{\text{minimize}} \ \{u^k(b - g(Y)) \,|\, u^k \geqslant 0, k = 1, \ldots, K\} \qquad \text{(D2)}$$

where u^k is the vector representing the kth extreme point of R.

One is interested only in the case where (D1) has a feasible solution, since if (D1) is infeasible, then (PX) is either infeasible or has no finite optimum. Both cases make (P) of no interest. Thus in the following discussion (D1) is assumed to possess a finite optimum.

Using (D2), then (P) may be written as

maximize

$$z = f(Y) + \min_{u^k}\{u^k(b - g(Y))\}$$

subject to

$$u^k \geqslant 0, \qquad k = 1, \ldots, K$$
$$Y \geqslant 0 \qquad\qquad\qquad\qquad\qquad (P^*)$$

Now consider the problem:

maximize z

subject to

$$z \leqslant f(Y) + u^k(b - g(Y)), \qquad k = 1, \ldots, r, \quad 1 \leqslant r \leqslant K$$
$$u^k \geqslant 0, \qquad Y \geqslant 0 \qquad\qquad\qquad\qquad (P^r)$$

When $r = K$, then (P^r) is the same as (P^*) and hence solves (P). If $u^k, k = 1, 2, \ldots, r$, are known, then (P^r) is a problem in the variables Y (and z) only. Thus the solution of (P^K) yields optimal Y for (P). In turn, optimal X is determined by substituting optimal Y in (PX) and solving as a linear program.

The above discussion may indicate that it is necessary to determine *all* the extreme points u^k of R before (P^K) yields optimal Y. But this is not practical, especially if the number of vertices of R is large. The following iterative procedure is used instead to determine the optimum solution.

The general idea is to establish legitimate upper and lower bounds on the true optimum value z^* of (P), that is, $\underline{z} \leqslant z^* \leqslant \bar{z}$. The upper bound $\bar{z}$ is changed successively such that each new value coincides with a *new* extreme point of R. Termination occurs at the extreme point having $\underline{z} = \bar{z}$, which then gives the optimum solution. The procedure must thus terminate after enumerating at most K extreme points provided the enumeration is nonredundant.

It is first shown how $\underline{z}$ and $\bar{z}$ can be determined. Let u^0 be a feasible point in R (not necessarily an extreme point). Thus in (D2),

$$u^0(b - g(Y)) \geqslant \min_{u^k}\{u^k(b - f(Y))\,|\,u^k \geqslant 0, \text{ all } k\}$$

Utilizing this result, the constraint

$$z \leqslant f(Y) + u^0(b - g(Y))$$

is legitimate with respect to (P^1). $[r \equiv 1$ in this case to show that (P^r) has one constraint.] Let the optimal solution to (P^1) be (z^1, Y^1). Then the associated upper bound is given by $\bar{z} = z^1$, since the constraints of (P^1) are only a subset of those of (P^K).

To determine the associated lower bound, substitute Y^1 in $(\mathrm{P}X)$ and solve to obtain optimal $X = X^1$. Since $(\mathrm{P}X)$ is the same as (P) except that the use of $Y = Y^1$ results in a more restrictive solution space, the optimal solution to $(\mathrm{P}X)$ must provide the required lower bound $\underline{z} = f(Y^1) + CX^1$. Also, because $(\mathrm{P}X)$ and $(\mathrm{D}1)$ are duals, if u^1 is the optimal solution to $(\mathrm{D}1)$ given Y^1, then it must be true that

$$\underline{z} = f(Y^1) + u^1(b - g(Y^1))$$

Thus $(\mathrm{D}1)$ can also be used to determine the lower bound of $\underline{z}$.

Suppose now that $\underline{z} < \bar{z} = z^1$, the search must continue for a better extreme point of R. In order *not* to encounter the point (z^1, Y^1, X^1), and hence u^0, again the constraint

$$z \leqslant f(Y) + u^1(b - g(Y))$$

which violates (z^1, X^1, Y^1), must be augmented to (P^1) to form (P^2). Now (P^2) is solved to yield a new upper bound $\bar{z} = z^2$ and new $Y = Y^2$. Then a new $\underline{z}$ is determined by using Y^2 in $(\mathrm{P}X)$, and the same procedure above is repeated. [For the sake of computational efficiency, if the slack variable associated with any constraint in (P^r) becomes positive, then such a constraint can be discarded as nonredundant.]

The termination of the procedure occurs if Y^r and z^r produce $\underline{z} = \bar{z}$, which gives the optimum solution as (Y^r, X^r, z^r).

The steps of the algorithm can be summarized as follows.

Step 0   Determine u^0, any feasible solution to $(\mathrm{D}1)$. If u^0 does not exist, stop; (P) has no feasible solution. Otherwise, set $r = 1$ and go to step 1.

Step 1   Solve (P^r) in z and Y, that is,

maximize z

subject to

$$z \leqslant f(Y) + u^k(b - g(Y)), \qquad k = 0, 1, \ldots, r - 1$$
$$Y \geqslant 0$$

Let (z^r, Y^r) be the optimum solution. Then set $\bar{z} = z^r$. Go to step 2.

Step 2   Solve the linear program $(\mathrm{D}1)$ given $Y = Y^r$, that is,

minimize

$$w = u(b - g(Y^r))$$

subject to

$$uA \geqslant C$$
$$u \geqslant 0$$

(If $uA \geqslant C$ is unbounded, add the regularity constraint $\sum_{i=1}^{m} u_i \leqslant M$.) Let u^r be the optimal solution. Then

$$\underline{z} = f(Y^r) + u^r(b - g(Y^r))$$

Go to step 3.

Step 3   If $\underline{z} = \bar{z}$, Y^r is optimum, go to step 4 to determine optimum X. Otherwise, set $r = r + 1$ and go to step 1.

Step 4   Let $Y^* = Y^r$, then solve (PX) given by

maximize

$$v = CX + f(Y^*)$$

subject to

$$AX \leqslant b - g(Y^*)$$
$$X \geqslant 0$$

Let X^* be its optimum solution; then (X^*, Y^*) gives the optimum solution to (P).

The convergence of the above algorithm is ensured by the fact that R has a finite number of extreme points. But it is necessary to show that the procedure enumerates these points nonredundantly. This is proved as follows:

Consider two excessive extreme points u^{i-1} and u^i. As shown above,

$$\bar{z} = z^i = f(Y^i) + u^{i-1}(b - g(Y^i)) \geqslant z^*$$

Again, from (PX) and (D1), we have

$$\underline{z} = f(Y^i) + CX^i = f(Y^i) + u^i(b - g(Y^i)) \leqslant z^*$$

Using the fact that $\underline{z} < z^* < \bar{z}$ for any nonoptimal solution, then

$$u^i(b - g(Y^i)) < u^{i-1}(b - g(Y^i))$$

This shows that the optimal objective value of (D1) given u^i is smaller than its equivalence given u^{i-1}. Hence u^i must be different from u^{i-1}.

The constraints of (P^r) are usually referred to as Benders' cuts. The name follows from the fact that each cut eliminates (cuts off) an extreme point that has been encountered previously.

▶**EXAMPLE 2.5-8**

   Maximize
$$z = 2x_1 + x_2 - 6y^2 \tag{P}$$

subject to

$$4x_1 + x_2 - 2y \leqslant 8$$
$$4x_1 + 3x_2 - y^2 \leqslant 12$$
$$x_1, x_2, y \geqslant 0$$

Thus (PX), $(D1)$, and (P') are given by

   maximize
$$v = 2x_1 + x_2 - 6y^2 \tag{PX}$$

subject to

$$4x_1 + x_2 \leqslant 8 + 2y$$
$$4x_1 + 3x_2 \leqslant 12 + y^2$$
$$x_1, x_2 \geqslant 0$$

   minimize
$$w = (8 + 2y)u_1 + (12 + y^2)u_2 \tag{D1}$$

subject to

$$4u_1 + 4u_2 \geqslant 2$$
$$u_1 + 3u_2 \geqslant 1$$
$$u_1, u_2 \geqslant 0$$

   maximize z $\tag{P'}$

subject to

$$z \leqslant -6y^2 + (8 + 2y)u_1^{\,k} + (12 + y^2)u_2^{\,k}, \qquad k = 0, 1, \ldots, r - 1$$
$$(u_1^{\,k}, u_2^{\,k}) \geqslant 0, \quad \text{for all } k$$
$$y \geqslant 0.$$

   Because the solution space of $(D1)$ is unbounded, the regularity constraint $u_1 + u_2 \leqslant M$ is added. The solution space of $(D1)$, after augmenting the regularity condition, is shown in Fig. 2-2. This is only provided to study the effect of Benders' cuts. Let $u^0 = (u_1^{\,0}, u_2^{\,0}) = (1, 1)$. This gives:

   maximize z $\tag{P^1}$

subject to

$$z \leqslant -5y^2 + 2y + 20$$
$$y \geqslant 0$$

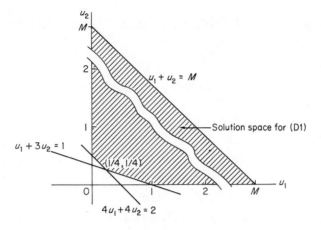

Figure 2-2

The right-hand side of the constraint assumes its maximum at $y = y^1 = 1/5$, with $\bar{z} = z^1 = 20.2$. Thus:

minimize

$$w = 8.4u_1 + 12.04u_2$$

subject to

$$4u_1 + 4u_2 \geqslant 2$$
$$u_1 + 3u_2 \geqslant 1$$
$$u_1 + u_2 \leqslant M$$
$$u_1, u_2 \geqslant 0$$

(D1)

This gives $u^1 = (u_1{}^1, u_2{}^1) = (1/4, 1/4)$ and $\underline{z} = 4.87$. Since $\underline{z} < \bar{z}$, (P²) is solved:

maximize z

subject to

$$z \leqslant -5y^2 + 2y + 20$$
$$z \leqslant -5.75y^2 + 0.5y + 5$$
$$y \geqslant 0$$

(P²)

Since the second constraint uniformly dominates the first, the optimum solution to (P²) is $y = y^2 = 1/23$ and $\bar{z} = z^2 = 5.0108$. [In this example, inspection is used to solve (P^r), but in general a finite algorithm may be needed.] Now:

minimize

$$w = 8.0869u_1 + 12.0019u_2$$

subject to

$$4u_1 + 4u_2 \geqslant 2$$
$$u_1 + 3u_2 \geqslant 1$$
$$u_1 + u_2 \leqslant M$$
$$u_1 + u_2 \geqslant 0 \qquad \text{(D1)}$$

The optimum solution is $(u_1{}^2, u_2{}^2) = (1/4, 1/4)$ with $\underline{z} = 5.0108$. Thus $y^2 = 1/23$ is optimum. To obtain optimum (x_1, x_2) consider

maximize

$$v = 2x_1 + x_2 - 6(1/23)^2$$

subject to

$$4x_1 + x_2 \leqslant 8.0869$$
$$4x_1 + 3x_2 \leqslant 12.0019$$
$$x_1, x_2 \geqslant 0 \qquad \text{(P}X)$$

The optimal solution is $(x_1, x_2) = (1.532, 1.9575)$.

Notice that u^1 and u^2 are exactly the same because this extreme point is the optimum. If this point was not optimum, this condition could not happen according to the convergence proof.◀

2.6 Bounded Variables

In branch and bound algorithms of integer programming (Chapter 4), typical linear programs are solved in which one or more variables are restricted by upper and/or lower bounds, that is, $x_j \leqslant u_j < \infty$ and $x_j \geqslant l_j > 0$. Although these bounding constraints can be treated as any other constraints, it is possible to account for their effect implicitly by using substitutions or by modifying the simplex method conditions, thus avoiding enlarging the size of the problem.

The lower bound constraints $x_j \geqslant l_j$ can be accounted for by using the substitution $x_j = x_j' + l_j$, where $x_j' \geqslant 0$. Thus the new problem will have x_j' (instead of x_j) for which the lower bound is zero. Observe that after the problem is solved in terms of x_j', $x_j = l_j + x_j'$ is feasible.

The upper bound constraint $x_j \leqslant u_j$ cannot be accounted for by using the substitution $x_j = u_j - x_j'$, where $x_j' \geqslant 0$. This follows since there is no guarantee x_j will be nonnegative in an optimal solution.

In this section, it is shown that both the (regular) primal and dual simplex methods can be equipped with mechanisms to account for upper bounds. The details for these cases are given below.

2.6.1   Primal Simplex Method for Bounded Variables†

Suppose the problem is given as:

maximize (or minimize)

$$z = CX$$

subject to

$$AX \leqslant b$$
$$0 \leqslant X \leqslant U$$

If x_j, the jth element of X, has no upper bound, then its upper bound u_j is taken equal to ∞.

The problem is primal feasible provided $b \geqslant 0$. Otherwise, artificial variables may be used to secure a starting basic feasible solution. The initial (primal-feasible) tableau generally appears as

Basic vector	X^{T}	X_s^{T}	
z	$-C$	0	0
X_s	A	I	b

The upper bound constraints $X \leqslant U$ do not appear explicitly in the above tableau.

The steps of the algorithm are similar to the regular primal simplex method except the feasibility condition is changed to account for the upper bounds. Suppose x_j is the entering variable (which is selected by using the same optimality condition of the regular primal simplex method); then the ith basic variable can be written as

$$(X_B)_i = \beta_i - \alpha_{ij} x_j$$

Thus any new value of x_j must be selected such that

$$0 \leqslant (X_B)_i \leqslant u_i, \qquad \text{and} \qquad 0 \leqslant x_j \leqslant u_j$$

† Dantzig (1955).

Now $(X_B)_i \geqslant 0$ implies that

$$x_j \leqslant \theta_1 \equiv \min_i \left\{ \frac{\beta_i}{\alpha_{ij}} \,\middle|\, \alpha_{ij} > 0 \right\}$$

which is the same as in the regular simplex method. On the other hand, $(X_B)_i \leqslant u_i$ implies that

$$x_j \leqslant \theta_2 \equiv \min_i \left\{ \frac{u_i - \beta_i}{-\alpha_{ij}} \,\middle|\, \alpha_{ij} < 0 \right\}$$

This means that feasibility is satisfied (on all fronts) by taking

$$x_j = \theta \equiv \min\{\theta_1, \theta_2, u_j\}$$

The above feasibility condition implies the following rules for changing the basis. Let $(X_B)_r$ be the variable associated with θ:

(1) If $\theta = \theta_1$, apply the regular row operations to effect a change in basis by introducing x_j and dropping $(X_B)_r$.
(2) If $\theta = \theta_2$, again change the basis by introducing x_j and dropping $(X_B)_r$. But since this implies that $(X_B)_r$ reaches its upper bound, next it must be substituted by using $(X_B)_r = u_r - (X_B)_r'$.
(3) If $\theta = u_j$, then x_j remains nonbasic but at its upper bound. Thus it must be substituted as $x_j = u_j - x_j'$.

The reason for substituting a variable x_j reaching its upper bound is that its complement x_j' will then be nonbasic at zero level; hence the same feasibility condition can be applied to the new tableau.

▶EXAMPLE 2.6-1   Maximize

$$z = 3x_1 + 5x_2 + 2x_3$$

subject to

$$x_1 + 2x_2 + 2x_3 + x_4 \qquad = 10$$
$$2x_1 + 4x_2 + 3x_3 \qquad + x_5 = 15$$
$$0 \leqslant x_1 \leqslant 4, \quad 0 \leqslant x_2 \leqslant 3, \quad 0 \leqslant x_3 \leqslant 3, \quad x_4, \quad x_5 \geqslant 0$$

Starting iteration

Basic vector	x_1	x_2 ↓	x_3	x_4	x_5	Solution
z	-3	-5	-2	0	0	0
x_4	1	2	2	1	0	10
x_5	2	4	3	0	1	15

$$\theta_1 = \min\{10/2, 15/4\} = 15/4, \qquad \theta_2 = \infty, \qquad \theta = \min\{\theta_1, \theta_2, u_2\} = u_2$$

Thus substitute $x_2 = 3 - x_2'$.

First iteration

Basic vector	x_1 ↓	x_2'	x_3	x_4	x_5	Solution
z	-3	5	-2	0	0	15
x_4	1	-2	2	1	0	4
x_5	2	-4	3	0	1	3

$$\theta_1 = \min\{4/1, 3/2\} = 3/2, \qquad \theta_2 = \infty, \qquad \theta = \min\{\theta_1, \theta_2, u_1\} = \theta_1$$

Thus x_5 leaves the solution.

Second iteration

Basic vector	x_1	x_2' ↓	x_3	x_4	x_5	Solution
z	0	-1	$5/2$	0	$3/2$	$39/2$
x_4	0	0	$1/2$	1	$-1/2$	$5/2$
x_1	1	-2	$3/2$	0	$1/2$	$3/2$

$$\theta_1 = \infty, \qquad \theta_2 = \frac{4 - 3/2}{-(-2)} = 5/4, \qquad \theta = \min\{\theta_1, \theta_2, u_2\} = \theta_2$$

Thus introduce x_2' and drop x_1. Then substitute $x_1 = 4 - x_1'$.

Final iteration

Basic vector	x_1'	x_2'	x_3	x_4	x_5	Solution
z	$1/2$	0	$7/4$	0	$5/4$	$83/4$
x_4	0	0	$1/2$	1	$-1/2$	$5/2$
x_2'	$1/2$	1	$-3/4$	0	$-1/4$	$5/4$

Optimal solution is $x_1 = 4 - 0 = 4$, $x_2 = 3 - 5/4 = 7/4$, $x_3 = 0$, $x_4 = 5/2$, and $x_5 = 0$, with $z = 83/4$.◀

2.6.2   Dual Simplex Method for Bounded Variables†

The problem is given in the form of Section 2.6.1; it is assumed, however, that the initial tableau is *dual* feasible. This will be true in the initial tableau if for a maximization (minimization) case the vector

† H. M. Wagner (1957).

$C \leqslant 0 \ (C \geqslant 0)$. If an element c_j of the variable x_j violates this condition, then dual feasibility can be restored by using the substitution $x_j = u_j - x_j'$. (If u_j is not finite, use $M > 0$ and sufficiently large.)

By using the starting (dual feasible) tableau as in Section 2.6.1, the steps of the algorithm at any iteration are given as follows.

Step 1   If any of the current *basic* variables $(X_B)_i$ is above its *upper* bound, substitute it out by using the relationship $(X_B)_i = u_i - (X_B)_i'$. Go to step 2.

Step 2   If all basic variables are feasible, stop; the current solution is feasible and optimal. Otherwise, select the leaving variable x_r as the *basic* variable having the most negative value. (This is the same as the feasibility condition of the "regular" dual method.) Go to step 3.

Step 3   Select the entering variable as the nonbasic variable x_j having

$$\left| \frac{z_j - c_j}{\alpha_{rj}} \right| = \min \left\{ \left| \frac{z_k - c_k}{\alpha_{rk}} \right|, \alpha_{rk} < 0, k \in NB \right\}$$

If all $\alpha_{rk} \geqslant 0$, $k \in NB$, no feasible solution exists. (This is the same as the optimality condition of the "regular" dual method.) Go to step 4.

Step 4   Carry out the Gauss–Jordan elimination technique. Then go to step 1.

The reader should notice that the only difference between the "bounded" and "regular" dual methods occurs in step 1, where basic variables that exceed their upper bounds are substituted out.

▶EXAMPLE 2.6-2   Maximize

$$z = 2x_1 + x_2 - x_3$$

subject to

$$x_1 - 2x_2 \qquad + x_4 \qquad = 3$$
$$-x_1 + x_2 + x_3 \qquad + x_5 = 4$$
$$0 \leqslant x_1 \leqslant 1, \quad 0 \leqslant x_2 \leqslant 2, \quad 0 \leqslant x_3 \leqslant 3, \quad 0 \leqslant x_4 \leqslant 4, \quad 0 \leqslant x_5 \leqslant 5$$

Since c_1 and c_2 are positive, substitute $x_1 = 1 - x_1'$ and $x_2 = 2 - x_2'$. This yields the following dual-feasible starting tableau:

Basic vector	x_1'	x_2'	x_3	x_4	x_5	Solution
z	2	1	1	0	0	4
x_4	−1	2	0	1	0	6
x_5	1	−1	1	0	1	3

(Initial tableau)

$x_4 > 4$, substitute $x_4 = 4 - x_4'$. This yields:

Basic vector	x_1'	x_2'	x_3	x_4'	x_5	Solution
z	2	1	1	0	0	4
x_4'	1	$\boxed{-2}$	0	1	0	-2
x_5	1	-1	1	0	1	3

x_4' leaves and x_2' enters. This yields

Basic vector	x_1'	x_2'	x_3	x_4'	x_5	Solution
z	5/2	0	1	1/2	0	3
x_2'	$-1/2$	1	0	$-1/2$	0	1
x_5	1/2	0	1	$-1/2$	1	4

The solution is $x_1 = 1$, $x_2 = 2 - 1 = 1$, $x_3 = 0$, $x_4 = 4$, and $x_5 = 4$.◀

Problems

2-1 Consider the following problem:

maximize

$$z = 2x_1 + x_2 - 3x_3 + 4x_4$$

subject to

$$8x_1 + x_2 + 5x_3 + 4x_4 \leqslant 10$$
$$4x_1 + x_2 + 6x_3 - 2x_4 \leqslant 5$$
$$x_1, x_2, x_3, x_4 \geqslant 0$$

(a) Find all the *feasible* extreme points of the system.
(b) By using the information above, find the optimum solution.

2-2 In the feasibility condition of the simplex method (Section 2.3.2), why is it necessary to consider positive $(B^{-1}P_j)_i$ only in computing θ? Suppose $(B^{-1}b)_i = 0$ for at least one i. Is this a definite indication that $\theta = 0$?

2-3 Consider the following two inequalities:

$$5x_1 + 3x_2 \qquad + 2x_4 \leqslant 15$$
$$-2x_1 + 4x_2 - 2x_3 + 3x_4 \leqslant 12$$
$$x_1, x_2, x_3, x_4 \geqslant 0$$

Find the largest value of x_1 such that *all* the above conditions are satisfied. Repeat the same for x_2, x_3, and x_4. How do the answers relate to the feasibility condition of the simplex method?

2-4 Suppose in Problem 2-3 the following conditions are added: $0 \leqslant x_1 \leqslant 10, 0 \leqslant x_2 \leqslant 1, 0 \leqslant x_3 \leqslant 15$. How do these affect the answers in Problem 2-3?

2-5 Solve the following problem using the slacks for a starting basic feasible solution:

maximize

$$z = 2x_1 + x_2 + 2x_3 - 3x_4$$

subject to

$$4x_1 + x_2 + x_3 \qquad \leqslant 10$$
$$x_1 + x_2 \qquad + x_4 \leqslant 5$$
$$x_1, x_2, x_3, x_4 \geqslant 0$$

2-6 In Problem 2-5, it is possible to use x_3 and x_4 as starting basic variables since their constraint coefficient form an identity matrix. Solve the problem by using this starting solution.

2-7 Solve the following problem by using artificial starting basic solution:

minimize

$$z = 2x_1 - 3x_2 - 4x_3$$

subject to

$$x_1 + 5x_2 - 3x_3 \geqslant 15$$
$$x_1 + x_2 + x_3 = 5$$
$$5x_1 - 6x_2 + 4x_3 \leqslant 10$$
$$x_1, x_2, x_3 \geqslant 0$$

2-8 Solve Problem 2-7 by the two-phase method.

2-9 *Two-phase method*   If at the end of Phase I, the optimal basis includes artificial variables at zero level, then in Phase II the rules of the simplex method must be changed to ensure that the artificial variables remain at zero level.

Given that $x_k = 0, k = 1, 2, \ldots, r$, are the basic artificial variables and x_j is the entering variable in Phase II, show that:

(a) If $\alpha_{kj} > 0$ for any k, then the artificial variable x_k will become nonbasic, and can be dropped from the tableau in the next iteration.

(b)   If $\alpha_{kj} = 0$ for any k, then the artificial variable x_k will remain basic at zero level.

(c)   If $\alpha_{kj} < 0$ for any k, then, according to the feasibility condition, x_k cannot leave the solution, and if the leaving variable currently has a positive value then x_k will necessarily become positive in the next iteration. Under this condition, the only way to prevent x_k from becoming positive is to select it as the leaving variable. This will make it nonbasic in the next iteration and hence it can be dropped from the tableau.

2-10   Apply the above rules to the following problem:

maximize

$$z = x_1 + 5x_2 + 2x_3$$

subject to

$$3x_1 + x_2 + 4x_3 \leqslant 6$$
$$2x_1 + 5x_2 + x_3 \leqslant 4$$
$$x_1 - 2x_2 - 4x_3 = 0$$
$$x_1, x_2, x_3 \geqslant 0$$

(*Hint*: The starting tableau for Phase I is optimal since the artificial variable of the third constraint is equal to zero.)

2-11   Solve the following set of simultaneous linear equations by the simplex method:

$$3x_1 + 2x_2 = 5$$
$$x_1 + 2x_2 = 4$$

(*Hint*: x_1 and x_2 are unrestricted in sign. Substitute $x_i = x_i' - x_i''$, where $x_i', x_i'' \geqslant 0$. Then use Phase I of the two-phase method.)

2-12   Solve the following problem by the revised simplex method:

minimize

$$z = 2x_1 + 3x_2$$

subject to

$$x_1 + 2x_2 \geqslant 4$$
$$2x_1 + 3x_2 \geqslant 6$$
$$x_1, x_2 \geqslant 0$$

2-13   Write the dual of the following problem:

maximize

$$z = x_1 - 2x_2 + 3x_3$$

subject to

$$x_1 + 3x_2 - 2x_3 \leqslant 10$$
$$-3x_1 + 2x_2 + 3x_3 \geqslant 5$$
$$4x_1 - 2x_2 + x_3 = 4$$
$$x_1, x_2 \geqslant 0$$
$$x_3 \quad \text{unrestricted}$$

Show that if the inequalities are converted to equations by using slack and surplus variables, then the associated dual is exactly equivalent to the one derived from the primal in inequalities form.

2-14 Write the dual of the following problems:

(i) Maximize

$$z = CX$$

subject to

$$AX = b$$
$$X \geqslant 0$$

(ii) Maximize

$$z = CX$$

subject to

$$AX = b$$
$$X \quad \text{unrestricted}$$

2-15 Prove that Theorem 2-2 (Section 2.5.1) applies to the primal problems in Problem 2-14.

2-16 Solve the following inequalities simultaneously by the simplex method:

$$x_1 + 2x_2 - 3x_3 \leqslant 5$$
$$-x_1 + 2x_2 + x_3 \leqslant 4$$
$$3x_1 - 4x_2 + 2x_3 \leqslant 6$$
$$x_1, x_2, x_3 \geqslant 0$$

(*Hint*: Augment the trivial objective function, maximize $z = 0x_1 + 0x_2 + 0x_3$ and solve its dual.)

2-17 If the kth constraint of a primal problem is multiplied by $\lambda \neq 0$, how does this affect the optimal dual solution?

2-18 Consider the linear program:

maximize

$$z = CX$$

subject to

$$AX \leqslant b$$
$$X \geqslant 0$$

If the objective coefficients vector C_B associated with a basic feasible solution is changed to D_B, show that the reduced costs $z_j - c_j$ remain at zero level for all *basic* variables.

2-19 Solve the following problems by the dual simplex method.

(i) Maximize

$$z = x_1 - 2x_2 + 3x_3$$

subject to

$$x_1 + 2x_2 - x_3 \geqslant 4$$
$$-3x_1 + 2x_2 + 4x_3 \geqslant 6$$
$$x_1, x_2, x_3 \geqslant 0$$

(ii) Minimize

$$z = -2x_1 + 3x_2 + x_3$$

subject to

$$x_1 + 3x_2 - x_3 \leqslant 10$$
$$3x_1 + x_2 - 2x_3 \geqslant 5$$
$$x_1, x_2, x_3 \geqslant 0$$

2-20 Solve Problem 2-12 by the column tableau format of the simplex method.

2-21 Solve Problem 2-19 by the *l*-dual simplex method.

2-22 Consider the problem

maximize

$$z = CX$$

subject to

$$AX \leqslant b$$
$$0 \leqslant X \leqslant U$$

Suppose B defines the basis at a current iteration and assume that X_l and X_u are respectively the variables at zero and upper bounds. Prove that the current basic solution is given by

$$X_B = B^{-1}(b - A_u U_u)$$

where A_u is the submatrix of A associated with X_u and U_u is a vector whose elements are those of U corresponding to X_u.

2-23   In Problem 2-22, if C_B is the basic objective vector, prove that $\{z_j - c_j\}$ for X_u, given by $-C_B B^{-1} A_u + C_u$, is the objective vector associated with X_u.

2-24   Solve the following problem by the *primal* simplex method for upper bounded variables.

maximize

$$z = x_1 + 3x_2 + 2x_3$$

subject to

$$\begin{aligned}
x_1 + 4x_2 - 3x_3 &\leqslant 15 \\
-x_1 + 4x_3 &\leqslant 5 \\
x_1 + 2x_2 - x_3 &\leqslant 9
\end{aligned}$$

$$0 \leqslant x_1 \leqslant 5, 0 \leqslant x_2 \leqslant 2, \quad 0 \leqslant x_3 \leqslant 2$$

2-25   Solve the following problem by the *dual* simplex method for upper bounded variables:

maximize

$$z = 3x_1 - 5x_2 + 2x_3$$

subject to

$$\begin{aligned}
x_1 - 2x_2 + 2x_3 &\leqslant 5 \\
2x_1 + x_2 - 4x_3 &\geqslant 12
\end{aligned}$$

$$0 \leqslant x_1 \leqslant 4, \quad 0 \leqslant x_2 \leqslant 3, \quad 0 \leqslant x_3 \leqslant 3$$

CHAPTER 3

Zero–One Implicit Enumeration

3.1 Introduction

Consider the following problem:

minimize

$$z = f(x_1, x_2, \ldots, x_n)$$

subject to

$$g_i(x_1, x_2, \ldots, x_n) \leqslant b_i, \qquad i \in M = \{1, 2, \ldots, m\}$$
$$x_j = (0, 1), \qquad j \in I \subset N = \{1, 2, \ldots, n\}$$
$$x_j \geqslant 0, \qquad j \in N - I$$

The functions f and g_i are assumed to be well defined for all feasible solution points.

When $N = I$, the problem is a *pure* zero–one problem, that is, all the variables x_j are binary. Otherwise, when $I \subset N$, one is dealing with a *mixed* zero–one problem.

The concept of implicit enumeration is first developed for the pure problem in which all the objective and constraint functions are linear. This is later generalized to include the nonlinear (pure) problem.

In the mixed zero–one problem, the presentation is limited to the case where all the objective and constraint functions are linear. Although the

basic algorithms for the mixed problem were developed before the concept of
implicit enumeration was introduced into the literature, it is shown later
in this chapter that these algorithms are actually based on a similar concept.

Before presenting the concept of implicit enumeration, the importance
of the zero–one problem is emphasized by showing that every (general)
integer problem can be converted to an equivalent binary problem. Also,
a theorem relating the solution of the (mixed or pure) zero–one problem
to the extreme points of the solution space of the equivalent linear program
is presented. This theorem will prove useful in some of the developments
in Chapter 7.

3.2 Zero–One Equivalence of the Integer Problem

Suppose the variable x_k is integer and assume $x_k \leqslant u_k$, where u_k is a
known upper bound. Then

$$x_k = \sum_{p=0}^{K} 2^p y_{kp}, \qquad y_{kp} = 0 \text{ or } 1$$

where K is determined such that $2^{K+1} \geqslant u_k + 1$. By substituting for x_k in
terms of y_{kp}, the desired result is at hand. In this case, the converted
problem becomes a mixed or pure zero–one problem.

One obvious disadvantage of the above procedure, however, is that the
number of binary variables may become too large to allow acceptable
computation time.

The (mixed or pure) zero–one linear problem has an important property,
characterized by the following theorem:

▶THEOREM 3-1 In a linear zero–one problem, mixed or pure, assume
that the integer constraint is replaced by the continuous range $0 \leqslant x_j \leqslant 1$;
then (i) the resulting continuous solution space contains no feasible points
(that is, no points satisfying the integer conditions) in its *interior*, and
(ii) the optimum feasible solution of the integer problem occurs at an
extreme point of the (continuous) convex space.

Proof Part (i) follows from the fact that a feasible point must satisfy
the restriction $x_j = 0$ or $x_j = 1$. Since these are boundary planes of the
continuous solution space, it is impossible, by definition, that any such
point can be an interior point.

Part (ii) is obvious for the pure zero–one problem. In the mixed case,
suppose $x_j = x_j{}^*$, where $x_j{}^* = 0$ or 1, $j \in I$, are the optimal feasible integer
values. By fixing these values, the resulting mixed zero–one problem becomes
a regular linear program in the remaining continuous variables. Thus, by
linear programming theory, the optimum solution to the original zero–one

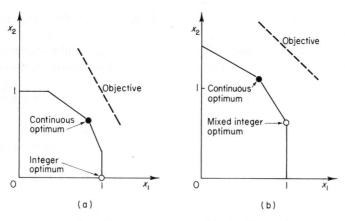

Figure 3-1   (a) Pure 0–1 problem; (b) mixed 0–1 problem.

problem must occur at an extreme point of the (continuous) convex solution space, specifically because $x_j = x_j{}^*$ are *boundary* planes of the convex set.◀

The implication of the theorem is that the zero–one problem can be treated as a continuous problem. The fact that the optimum integer solution occurs at an extreme point suggests that a method similar to the simplex algorithm can be used to solve the problem. The complication here is that the simplex solution may not necessarily satisfy the integrality condition. This point is illustrated graphically in Fig. 3-1 for the mixed and pure cases. Modifications, however, can be introduced that will allow the development of a simplex-like algorithm (see Section 7.2.3-D).

3.3   Concept of Implicit Enumeration

In general, the solution space of an integer program can be assumed to possess a finite number of possible feasible points. A straightforward method for solving integer problems is to *exhaustively* (or *explicitly*) enumerate all such points. In this case, the optimal solution is determined by the point(s) that yields the best (maximum or minimum) value of the objective function.

The obvious drawback of the above technique is that the number of solution points may become impractically too large, with the result that the solution cannot be determined in a reasonable amount of time. The idea of *implicit* (or *partial*) enumeration calls for considering only a portion (hopefully small) of all possible solution points while automatically discarding the remaining ones as nonpromising. To illustrate this point, consider determining all the feasible solutions for the following inequality:

$$3x_1 - 8x_2 + 5x_3 \leqslant -6, \qquad x_j = (0, 1), \quad j = 1, 2, 3$$

Simple inspection shows that in any feasible solution x_2 must be fixed at level one. This means that any binary combination (x_1, x_2, x_3) having $x_2 = 0$ cannot yield a feasible solution. Thus the four combinations $(0, 0, 0)$, $(1, 0, 0)$, $(0, 0, 1)$, and $(1, 0, 1)$ are automatically discarded as nonpromising and are said to be implicitly enumerated.

Given that $x_2 = 1$ is a necessary requirement for feasibility, the inequality is oversatisfied by $-6 - (-8) = 2$ units. Thus x_1 or x_2 (or both) can be allowed to be at level one only if their contributions to the left-hand side of the inequality do not exceed 2. Since the coefficient of x_1 is equal to 3, x_1 must be fixed at level zero, which means that $(1, 1, 0)$ and $(1, 1, 1)$ are implicitly enumerated as nonpromising. Now given $x_1 = 0$ and $x_2 = 1$, the combination $(0, 1, 1)$ is discarded because the coefficient of x_3 is equal to 5. The one remaining combination $(0, 1, 0)$ is the only feasible solution to the inequality.

In order to appreciate the impact of using implicit enumeration, Fig. 3-2 gives a graphical representation of ideas used in the above example. The first conclusion is that x_2 must be fixed at level one. Thus all the "branches" emanating from the branch $x_2 = 0$ (shown by dotted lines) are discarded. In this case, $x_2 = 0$ is said to be *fathomed*. For $x_2 = 1$, the inequality shows that $x_1 = 1$ cannot lead to a feasible solution and hence $(x_2 = 1, x_1 = 1)$ is fathomed. Next, given $x_2 = 1$ and $x_1 = 0$, the branch $x_3 = 1$ leads to an infeasible solution, but the branch $x_3 = 0$ yields a feasible solution. Since all 2^3 combinations have been considered, the process ends.

Two important points for the successful implementation of implicit enumeration can be drawn from the above discussion:

(1) There is a need for an enumeration scheme that ensures that *all* the solution points are enumerated (implicitly or explicitly) in a nonredundant fashion.

(2) A number of "fathoming" tests must be designed to exclude as many nonpromising solutions as possible.

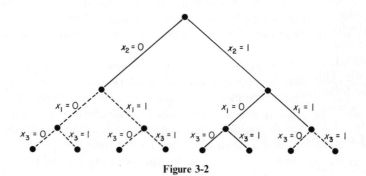

Figure 3-2

In the above example, it is easy to keep track of the enumerated solutions because there are only eight of them. In the general case, an efficient and flexible bookkeeping method would be required for keeping track of all the solutions that have been considered (either implicitly or explicitly) and for generating the remaining ones in a nonredundant fashion. Efficiency here indicates that the scheme should not tax the computer memory, and flexibility means that it should store and retrieve information easily.

The fathoming tests provided for the above example are specially designed to take advantage of its structure. For the general case, these tests must be designed to cover a wider class of problems. One must observe that the fathoming tests can lead to a powerful or a weak algorithm depending mainly on how cleverly these tests are designed to detect the nonpromising solution points.

The first potentially successful work in the area of implicit enumeration was reported by Balas (1965) for solving the zero–one linear problem. Important modifications in Balas' ideas were later given by Glover (1965c), whose work was the basis for other developments by Geoffrion (1967, 1969) and Balas (1967b), among others.

Balas' original enumeration scheme was later refined by Glover (1965c). Geoffrion (1967) then showed how Balas' algorithm can be superimposed on Glover's enumeration scheme. In order to consolidate the ideas of Balas, Glover, and Geoffrion in the form that is currently used, the next section presents Glover's scheme and its modifications by Geoffrion. Section 3.5 will then give the fathoming tests for Balas' 0–1 (linear) algorithm.

3.4 Enumeration Scheme†

Implicit enumeration does not consider "complete" binary combinations. Rather, it starts with one (or more) variable being fixed at a binary value and then gradually builds the "solution" by augmenting new variables at fixed values. As an illustration, in the example of Section 3.3 the enumeration starts by fixing x_2 at level one, then x_1 is augmented at level zero, and so on. In the course of each of these augmentations, it becomes evident that (complete) solution points can be discarded without being considered explicitly. This idea is the basis for Glover's enumeration scheme.

The following definitions are introduced to facilitate the presentation of the scheme. For convenience, the notation $+j$ $(-j)$ is used to indicate that $x_j = 1$ $(x_j = 0)$. For example, the information $x_5 = 0$, $x_2 = 1$, and $x_3 = 1$ can be summarized in one vector as $\{-5, +2, +3\}$.

---

† Glover (1965c), Geoffrion (1967).

(i) *Partial solution* (*J*)   This is an *ordered* set, which assigns binary specifications to a subset *J* of *N*. In the example in Section 3.3, $J = \{+2, -1\}$ is a partial solution in which $x_2 = 1$ and $x_1 = 0$.

(ii) *Free variables* (*N − J*)   A variable that is not assigned a binary value by the partial solution *J* is free to assume a value of zero or one.

(iii) *Completion of J*   This gives a complete assignment to all the variables in *N* and is determined by the assignment in *J* together with a binary specification for each free variable. In the above example, $J = \{+2, -1\}$ has two completions: $\{+2, -1, -3\}$ and $\{+2, -1, +3\}$.

(iv) *Fathomed partial solution*   A partial solution *J* is said to be fathomed if all its completions can be discarded as nonpromising. In the example of Section 3.3, $J = \{-2\}$ and $J = \{+2, +1\}$ are fathomed partial solutions. (See Fig. 3-2.)

The flowchart for Glover's enumeration scheme is given in Fig. 3-3. The mechanics of the scheme is first illustrated by using the same example in Section 3.3. Figure 3-4 will illustrate the steps graphically so the reader can appreciate the effect of the scheme as compared with the complete

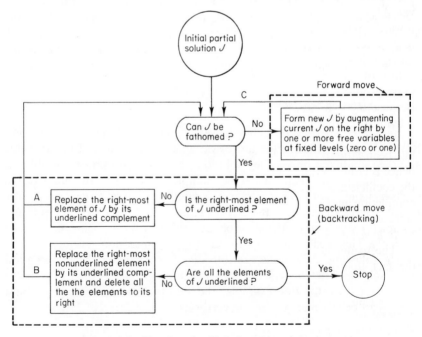

Figure 3-3   Flowchart for Glover's enumeration scheme.

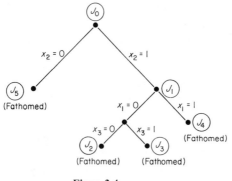

Figure 3-4

enumeration tree in Fig. 3-2. For the purpose of the example, a partial solution is said to be fathomed only if has no feasible completions.

One must realize that the scheme is designed for use with the digital computer. Thus what may appear as unnecessary detail is actually needed for the purpose of writing the computer codes. This point will be specially evident when the enumeration scheme is applied to the numerical example.

Assume the initial partial solution is empty, that is, $J_0 = \varnothing$. Thus the variables x_1, x_2, and x_3 are free. Since $x_2 = 1$ is a promising step toward achieving feasibility, the *forward move* (see the flowchart) is executed to produce $J_1 = \{+2\}$. However, since the coefficient of x_2 is smaller than the right-hand side by $8 - 6 = 2$ only, feasibility can be achieved by fixing the values of the two variables x_1 and x_3 at zero level. Thus, the forward move produces $J_2 = \{+2, -1, -3\}$ and J_2 is labeled as a feasible solution. Because J_2 is a *complete* binary assignment of all three variables, it is (obviously) fathomed. According to Fig. 3-3, the *backward move* produces $J_3 = \{+2, -1, +\underline{3}\}$.† The underline commemorates that J_2 (in which $x_3 = 0$) has been fathomed. J_3 (which is infeasible) is a complete assignment and hence is fathomed. The backward move thus gives $J_4 = \{+2, +\underline{1}\}$. Because the coefficient of the free variable x_3 is positive, J_4 does not have a feasible completion and hence is fathomed. This yields $J_5 = \{-\underline{2}\}$ which is fathomed since $x_2 = 1$ is a necessary condition for feasibility. Because all the elements of *fathomed* J_5 are underlined, the enumeration is complete.

The reader may wonder why the scheme does not take advantage of the fact that x_2 *must* be equal to one in any feasible solution of the

---

† In terms of Fig. 3-4, J_3 is obtained by tracing the branch $x_3 = 0$ upward and then moving downward along the branch $x_3 = 1$. The name "backtracking," which is sometimes given to the backward move, stems from the nature of the back movement along the branch $x_3 = 1$.

inequality. This point can be accounted for by using this sequence of partial solutions:

$$J_1 = \{+\underline{2}\}$$
$$J_2 = \{+\underline{2}, -1, -3\}$$
$$J_3 = \{+\underline{2}, -1, +\underline{3}\}$$
$$J_4 = \{+\underline{2}, +\underline{1}\}$$

Since J_4 is fathomed, and since all its elements are underlined, the enumeration is complete and the process ends. The idea here is that the underline in $J_1 = \{+\underline{2}\}$ is equivalent to asserting that the partial solution $\{-2\}$ has been fathomed; that is, the completions in which $x_2 = 0$ cannot lead to a feasible solution.

Glover's scheme scans the partial solutions on the basis of the LIFO (last-in, first-out) rule. This stems from the way the partial solutions are generated. For example, in Fig. 3-4, when J_1 is considered for fathoming, the scheme (implicitly) stores J_5 for later exploration. Also when J_2 is considered for fathoming, J_4 and J_3 are stored for later exploration. At this point, the stored "list" includes J_5, J_4, and J_3. Although J_3 is the *last* to be stored, according to the scheme, it is the *first* to be taken out of the list for possible fathoming.

Glover's LIFO rule is strikingly similar to a utilization by Beale and Small (1965) for economizing the storage requirements needed to solve the general integer problem by the branch-and-bound methods (see Section 4.3.2-B). The LIFO rule did not originate with Beale and Small, however. It is based on an earlier idea conceived by Little *et al.* (1963) for developing a special algorithm for the traveling salesman problem (see Section 7.4).

It is shown now that all 8 $(=2^3)$ solutions of the above numerical example have been enumerated (implicitly or explicitly). According to Fig. 3-4, the fathomed partial solutions are J_2, J_3, J_4, and J_5. Since in an n-variable problem the number of completions of an s-element partial solution is 2^{n-s}, the number of completions enumerated in the above example is $2^{3-3} + 2^{3-3} + 2^{3-2} + 2^{3-1} = 8$.

The following theorem establishes that so long as the fathoming tests are finite, any enumeration algorithm based on Glover's scheme must also be finite.

▶**THEOREM 3-2**   (i) All future generated completions do not duplicate any of the preceding ones. (ii) The scheme terminates only after all 2^n possible solutions are enumerated.◀

The proof of this theorem is in the work of Glover (1965c). A slightly different proof of essentially the same theorem can be found in the work of Geoffrion (1967). These proofs are based on (almost intuitive) arguments drawn from the logic of the flowchart in Fig. 3-3.

The LIFO rule dictated by the logic of Glover's scheme may not be the most effective way to scan the list of stored partial solutions. In the next section, it will be seen that the order in which this list is scanned may be effective in improving the efficiency of the enumeration. Tuan (1971) observed that it is unnecessary to follow the LIFO rule strictly by always complementing (and underlining) the rightmost (nonunderlined) element of the partial solution. To illustrate his idea, consider the partial solution $J = \{+2, -\underline{5}, +3, +6, +1, +\underline{4}\}$. Let j_1 be the element of J that should be underlined by the LIFO rule. Next, define J'' as the string of elements of J, inclusive of j_1, that is *not* interrupted by an underlined element. For example, in the above J, $j_1 = +1$ and $J'' = \{+3, +6, +1\}$. Tuan's idea stipulates that any of the elements in J'' may be complemented and underlined. This means that the partial solutions generated from J may be one of the following: $\{+2, -\underline{5}, +3, +6, -\underline{1}\}$, $\{+2, -\underline{5}, +3, +1, -\underline{6}\}$, or $\{+2, -\underline{5}, +6, +1, -\underline{3}\}$. Tuan's idea recognizes that the order of the elements 3, 6, and 1 is unimportant. What is important is that the positions of the underlined elements commemorating the fathoming of previous partial solutions should never be disturbed. Naturally, the specific selection of the element to be underlined results in a different structure of the solution tree. This is precisely the main goal of not following the LIFO rule since this flexibility could result in improved opportunities for fathoming.

Tuan's idea as introduced above does not require major changes in Glover's scheme except simply to move the underlined element to the rightmost position of the partial solution. One may think that the idea can be extended to give maximum flexibility by allowing the selection of *any* nonunderlined elements in the partial solution, rather than restricting the selection to one of those in J''. Although this is logically acceptable since it implies the selection of any of the partial solutions in the stored list, this extended flexibility can be accomplished only at the expense of a more complex bookkeeping method. This means that it will no longer be possible to represent the solution tree by a single vector having a maximum of n elements. Indeed, this idea is the basis for the general bookkeeping procedure of the branch-and-bound method, which will be presented in Chapter 4.

One notices that in executing the forward move in Glover's scheme, if every augmented variable is assigned the value one, then an underlined element will *always* coincide with a negative element (zero-valued variable).

In this case, a fathomed partial solution is commemorated by the fact that its rightmost element is negative with the result that it would not be necessary to underline any element. This point is used in Section 3.5.1.

Glover's scheme is general since it does not depend on the specific format of the 0–1 problem. Thus the objective and constraints may be linear or nonlinear, and the problem may be of the maximization or the minimization type. This emphasizes the flexibility of the scheme in the sense that it can be utilized for any well-defined binary problem. In the next section, the use of the scheme is illustrated by constructing algorithms for the linear and the polynomial 0–1 problems.

3.5   Fathoming Tests

The fathoming tests are designed to exclude as many nonpromising completions as possible. These tests are basically heuristic and hence could be as simple as to allow the *explicit* enumeration of all 2^n solutions or strong enough to exclude automatically (or enumerate implicitly) the majority of the solution points. The weakness or strength of the tests depends on how well behaved the functions of the problem are. It is shown below that a linear binary problem usually provides the strongest tests.

3.5.1   Linear Zero–One Problem†

Balas' problem can be defined as:

minimize
$$z = \sum_{j \in N} c_j x_j, \qquad c_j \geq 0, \quad j \in N = \{1, 2, \ldots, n\}$$
subject to
$$\sum_{j \in N} a_{ij} x_j + S_i = b_i, \qquad i \in M = \{1, 2, \ldots, m\}$$
$$x_j = (0, 1), \qquad j \in N$$
$$S_i \geq 0, \qquad i \in M$$

Any 0–1 linear problem can be put in the above form by observing that: (i) a maximization problem can be converted into a minimization form by multiplying the objective function by a negative sign; (ii) any of the negative coefficients c_j in the minimization objective function can be made positive by substituting $x_j = 1 - x_j'$ in the entire problem, where $x_j' = (0, 1)$; and (iii) all the constraints can be put in the form $\leq$.

It is shown below that the above form enhances the effectiveness of the fathoming tests.

† Balas (1965).

The idea of Balas' algorithm is that as the enumeration progresses, the feasible solution having the best objective value (so far) is stored. Since the enumeration terminates only when all 2^n possible solution points have been considered, the last stored feasible solution (if any) is the optimum. If none exists, the problem has no feasible solution.

To formalize the above idea, let J_0 be an initial partial solution and define $z_{\min}$ as the current best value of the objective function associated with a feasible solution. Initially, if no obvious initial partial solution is available, take $J_0 = \varnothing$ and set $z_{\min} = \infty$. Thus, $z_{\min}$ sets an upper bound on the optimum objective value. When the enumeration is complete, the last value of $z_{\min}$ and its associated solution gives the optimum. If $z_{\min} = \infty$ at the end of the enumeration, the problem has no feasible solution.

Because all the objective coefficients are nonnegative, it is optimal to specify that any variable which is not assigned a value by the partial solution should be at zero level. In other words, it is optimal to assume that all the free variables are zero unless the feasibility of the problem requires otherwise. Thus, if $J_0 = \varnothing$, then the associated values of the slack variables S_i are given by $S_i^0 = b_i$, $i \in M$. If $S_i^0 \geqslant 0$ for all $i \in M$, it follows that $J_0 = \varnothing$ yields the optimum; that is, all the variables must be at zero level.

At the tth iteration, let J_t be the current partial solution and define

$$S_i^t = b_i - \sum_{\substack{j \in J_t \\ j > 0}} a_{ij}, \qquad i \in M$$

$$z^t = \sum_{\substack{j \in J_t \\ j > 0}} c_j$$

as the associated values of the slacks and the objective function. Assume for the moment that $S_i^t < 0$ for at least one $i \in M$. The objective of Balas' tests is to check that either (i) J_t has no feasible completion, or (ii) even though J_t may have a feasible completion, it cannot yield a value of z that is superior to $z_{\min}$, the current best objective value. Clearly, in both cases J_t is fathomed and backtracking (or the backward move) must be executed. If, on the other hand, the tests fail to fathom J_t, then one or more free variables are elevated to *level one* as specified by the forward move in search for a *better feasible* completion. These variables must ·be selected intelligently in the sense that there must be evidence that if they are elevated to level one, possible improvements in both the optimality and the feasibility of the problem may result.

The above ideas will now be formalized by expressing the details of the tests mathematically. The subscript t is used to represent the tth iteration.

Test 1   Define

$$A_t = \{j \in N - J_t \,|\, a_{ij} \geq 0 \text{ for all } i \text{ such that } S_i^t < 0\}$$

The set A_t represents the free variables, which when elevated to level one cannot possibly improve on the infeasibility of the current solution. This follows since if x_j, $j \in A_t$, is elevated to level one, then the infeasibility of the problem (as indicated by the values of the slacks) will worsen. This means that all x_j, $j \in A_t$, must be excluded as nonpromising.

Test 2   Let $N_t^1 = N - J_t - A_t$. If $N_t^1 = \varnothing$, this means that none of the free variables can be elevated to level one, which means that J_t is fathomed and the backtracking step is invoked. Otherwise, define

$$B_t = \{j \in N_t^1 \,|\, z^t + c_j \geq z_{\min}\}$$

The set B_t identifies those free variables that, although may improve the infeasibility of the problem, cannot possibly lead to an improved value of the objective function (as compared with $z_{\min}$). Thus x_j, $j \in B_t$, must be excluded as nonpromising.

Test 3   Determine $N_t^2 = N_t^1 - B_t$. If $N_t^2 = \varnothing$, backtrack. Otherwise, define

$$C_t = \left\{ i \in M \,\middle|\, S_i^t < 0, \sum_{j \in N_t^2} a_{ij}^- > S_i^t \right\},$$

where $a_{ij}^- = \min(0, a_{ij})$. If $C_t \neq \varnothing$, this means that by utilizing all x_j, $j \in N_t^2$, advantageously, at least one infeasible slack $(S_i^t < 0)$ will continue to be infeasible. It then follows that J_t cannot have a feasible completion (that is, J_t is fathomed) and the backward move must be executed. If, on the other hand, $C_t = \varnothing$, then J_t cannot be abandoned, and one proceeds to Test 4 to determine the variable to be augmented to J_t in the forward move.

Before presenting Test 4, a remark on the power of Test 3 is in order. Glover and Zionts (1965) noticed that this test can be passed (that is, $C_t = \varnothing$) even though a higher cost feasible solution may result. Therefore, they suggested that the test be modified as follows. Let

$$D_t = \left\{ j \in N_t^2 \,\middle|\, \frac{c_j}{|a_{ij}|} \geq \frac{z_{\min} - z^t}{|S_i^t|}, \; S_i^t < 0, \, a_{ij} < 0 \right\}$$

If $D_t = N_t^2$, then all the completions of J_t should be abandoned since each variable x_j, $j \in N_t^2$, when elevated to level one will contribute to the objective function by more than its permissible ratio for the violated constraint.

Test 4   Select the free variable x_{j^*}, $j^* \in N_t^2$, to be augmented to J_t at level one such that

$$v_{j^*} = \max_{j \in N_t^2} \{v_j\}$$

where

$$v_j = \sum_{i \in M} \min(0, S_i^t - a_{ij}), \qquad j \in N_t^2$$

with v_j an empirical measure of the infeasibility in the solution resulting from setting $x_j = 1$. In case of a tie, j^* is selected so that c_{j^*} is the smallest c_j.

The result of augmentation is $J_{t+1} = J_t \cup \{j^*\}$, which executes the forward move. If $v_{j^*} = 0$, this means J_{t+1} is feasible. Since the previous tests do not allow an inferior feasible solution, the current solution must yield a better value of $z_{\min}$. Thus $z_{\min}$ is set equal to $z^t + c_{j^*}$. Now, because all $c_j \geqslant 0$, it is evident that no other completion of J_{t+1} (in which any x_j, $j \in N - J_{t+1}$, is set equal to one) can yield a better value of $z_{\min}$. This means that J_{t+1} is fathomed and the backward move is invoked.

If, on the other hand, $v_{j^*} < 0$, J_{t+1} must be attempted for fathoming by starting with Test 1 again.

The above four tests when superimposed on Glover's enumeration scheme result in what is usually referred to as *Balas' additive algorithm*. A flowchart summarizing the algorithm is given in Fig. 3-5. The algorithm is additive basically because it requires only addition and subtraction operations for its execution.

Balas' algorithm, as given above, scans the stored partial solutions by using Glover's LIFO rule. A (perhaps) more effective way is to use Tuan's

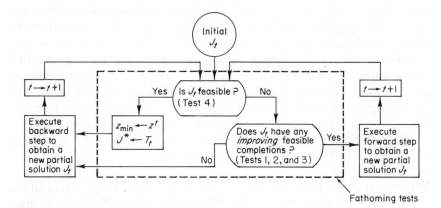

Figure 3-5

rule, which was outlined in Section 3.4. Let J_t'', as defined in the same section, be the set of elements that are eligible for underlining. For example, if $J_t = \{-2, +\underline{3}, +5, +4, +7\}$ is fathomed, then $J_t'' = \{+5, +4, +7\}$. Now, in carrying out the backward move, *any* of the elements $j \in J_t''$ can be complemented and underlined and then moved to the rightmost position of J_t to commemorate fathoming. The basis for selecting a specific element in J_t'' is to maintain conditions that are favorable to producing feasible completions quickly. In Test 4 above, a variable j is set equal to one if it results in the "least amount of infeasibility" as measured by v_j. This idea can also be adopted here so that $j \in J_t''$ is complemented (that is, set equal to zero) and underlined if the resulting partial solution produces the least amount of infeasibility. Thus the selected element p must satisfy

$$w_p = \max_{j \in J_t''} \{w_j\}$$

where

$$w_j = \sum_{i \in M} \min(0, S_i^t + a_{ij}), \qquad j \in J_t''$$

with w_j the measure of infeasibility when x_j is set back to zero level. Again, in case of a tie, p is selected as the variable having the largest objective coefficient among the tied variables.

An interesting observation by Balas is that the additive algorithm may be based on what he calls the *pseudodual simplex* method. He justifies his observation by the fact that the problem starts infeasible $(J_0 = \varnothing)$ but better than optimal $(z^0 = 0$, since all $x_j = 0)$. But the relationship between the additive algorithm and linear programming is stronger than observed by Balas. Suppose the above 0–1 problem is converted to a linear program LP by changing the restrictions $x_j = (0, 1)$ to $0 \leqslant x_j \leqslant 1$, for all $j \in N$. Let Y_i, $i \in M$, be the ith slack variable in LP. An investigation of the type of computations in the additive algorithm reveals that the assignment of binary values in the 0–1 problem (as per J_t and the assumption that all free variables are at zero level) can be achieved exactly in the LP problem by keeping Y_i, $i \in M$, *basic*, with the variables x_j *nonbasic*, $j \in N$, either at lower (zero) bound or at upper (one) bound. (See Section 2.6.)

The net result of the above observation is the following theorem relating the additive algorithm and linear programming.

▶**Theorem 3-3** (TAHA, 1971b) A sufficient condition for the solution of the above LP problem to be feasible for the associated 0–1 problem is that the slacks Y_i, $i \in M$, of LP be basic and feasible.◀

A stronger theorem relating integer and linear programming solutions is due to Cabot and Hurter (1968). Let LP^ε be the same as LP except that the right-hand side b_i, $i \in M$, is perturbed to $b_i + \varepsilon$, where $0 < \varepsilon < 1$; then:

▶**THEOREM 3-4** (CABOT AND HURTER, 1968) A necessary and sufficient condition for the solution of LP^ε problem to be feasible for the associated 0–1 problem is that the slacks Y_i, $i \in M$, of LP^ε be basic and positive.◀

One notices that the stronger conditions of the Cabot–Hurter theorem cannot provide computational advantages to the additive algorithm. This follows primarily from the fact that the implicit enumeration procedure is not sensitive to small variations in the right-hand side of the constraints.

▶**EXAMPLE 3.5-1** Minimize

$$z = (2\ 4\ 6\ 8\ 12)(x_1, x_2, \ldots, x_5)^{\mathrm{T}}$$

subject to

$$\begin{pmatrix} -8 & 4 & 1 & -1 & -5 \\ -6 & 3 & 2 & 0 & -1 \\ 2 & -9 & -3 & 2 & -3 \end{pmatrix} \begin{pmatrix} x_1 \\ x_2 \\ \vdots \\ x_5 \end{pmatrix} \leq \begin{pmatrix} -5 \\ -2 \\ -4 \end{pmatrix}$$

$$x_j = (0, 1), \quad j = 1, 2, \ldots, 5$$

Iteration 0 $J_0 = \varnothing$, $z^0 = 0$, $z_{\min} = \infty$, $S^0 = (-5, -2, -4)^{\mathrm{T}}$; $A_0 = \varnothing$, $B_0 = \varnothing$; hence $N_0^2 = \{1, 2, 3, 4, 5\}$, $C_0 = \varnothing$, $D_0 = \varnothing$; hence N_0^2 cannot be abandoned. Forward move: $v_1 = -6$, $v_2 = -14$, $v_3 = -11$, $v_4 = -12$, $v_5 = \boxed{-2}$. Thus $j^* = 5$ and $J_1 = \{5\}$. This yields $S^1 = (0, -1, -1)^{\mathrm{T}}$ and $z^1 = 12$.

Iteration 1 $A_1 = \{4\}$, $B_1 = \varnothing$, $N_1^2 = \{1, 2, 3\}$; $C_1 = \varnothing$, $D_1 = \varnothing$, N_1^2 cannot be abandoned. Forward move: $v_1 = \boxed{-3}$, $v_2 = -8$, $v_3 = -4$. Thus $J_2 = \{5, 1\}$ with $S^2 = (8, 5, -3)^{\mathrm{T}}$ and $z^2 = 14$.

Iteration 2 $A_2 = \{4\}$, $B_2 = \varnothing$, $N_2^2 = \{2, 3\}$; $C_2 = \varnothing$, $D_2 = \varnothing$, N_2^2 cannot be abandoned. Forward move: $v_2 = \boxed{0}$, $v_3 = \boxed{0}$. Since $c_2 < c_3$, $J_3 = \{5, 1, 2\}$. Because $S^3 = (4, 2, 6)^{\mathrm{T}}$ is feasible, $z_{\min} = 18$ and J_3 is fathomed.

Iteration 3 Backward move: $J_3'' = \{5, 1, 2\}$. $w_1 = -8$, $w_2 = -3$, $w_5 = \boxed{-1}$. Hence x_5 is complemented and $J_4 = \{1, 2, -5\}$ with $z^4 = 6$ and $S^4 = (-1, 1, 3)^{\mathrm{T}}$. (Notice that if Tuan's idea is not used then x_2 is complemented instead of x_5. Notice also that underlining would be unnecessary since the complemented element will always be negative.)

Iteration 4 $A_4 = \{3\}$, $B_4 = \varnothing$, $N_4^2 = \{4\}$; $C_4 = \varnothing$, $D_4 = \varnothing$, and N_4^2 cannot be abandoned. Forward move: $J_5 = \{1, 2, -5, 4\}$ with $S^5 = (0, 1, 1)^{\mathrm{T}}$ and $z^5 = 14$. Since S^5 is feasible, $z_{\min} = 14$ and J_5 is fathomed.

Iteration 5   Backward move: $J_6 = \{1, 2, -5, -4\}$ with $S^6 = (-1, 1, 3)^T$ and $z^6 = 6$.

Iteration 6 $A_6 = \{3\}$, $N_6{}^1 = \varnothing$; and J_6 is fathomed. Let $J_7 \equiv J_6$.

Iteration 7   Backward move: $J_7'' = \{1, 2\}$. $w_1 = -14$, $w_2 = \boxed{-6}$, and x_2 is complemented. Thus $J_8 = \{1, -2\}$ with $S^8 = (3, 4, -6)^T$ and $z^8 = 2$.

Iteration 8 $A_8 = \{4\}$, $B_8 = \{5\}$, $N_8{}^2 = \{3\}$; $C_8 = \{3\} \neq \varnothing$; hence $N_8{}^2$ is abandoned and J_8 is fathomed. Let $J_9 \equiv J_8$.

Iteration 9   Backward move: $J_{10} = \{-1\}$ with $S^{10} = (-5, -2, -4)^T$ and $z^{10} = 0$.

Iteration 10 $A_{10} = \varnothing$, $B_{10} = \varnothing$, $N_{10}{}^2 = \{2, 3, 4, 5\}$; $C_{10} = \{2\} \neq \varnothing$. Hence $N_{10}{}^2$ is abandoned and J_{10} is fathomed. Since all elements of J_{10} are negative, the enumeration scheme is complete and the optimal solution is given by J_5. ◄

In the above example, there was no chance to examine the effectiveness of using the set D. However, computational experience shows that, in general, this test is useful [see Petersen (1967)].

Notice that in the above example the following partial solutions have been fathomed: J_3, J_5, J_6, J_8, and J_{10}. The number of completions explicitly or implicitly enumerated is thus given by $(2^{5-3} + 2^{5-4} + 2^{5-4} + 2^{5-2} + 2^{5-1}) = 2^5$, which are all the possible solutions to the problem.

From the nature of the above fathoming tests, it is evident that the order in which the variables and the constraints are scanned is important in enhancing the efficiency of the algorithm. For example, consider Balas' Test 2. Since in computer codes these variables are scanned from left to right, it is obvious that the effectiveness of the tests is enhanced if these variables are arranged in ascending order of their objective coefficients. A similar reasoning can be given with respect to the arrangement of the constraints in relation to Tests 1 and 3. Normally, the constraints that are suspected to be most binding are placed in the top rows. A typical example is multiple choice constraints of the form $\sum_j x_j = 1$. In Balas' algorithm, the two inequalities substituting this constraint should be placed in the top rows. More information on this point can be found in the work of Glover (1965c) and Petersen (1967).

3.5.2   Surrogate Constraints†

In the exclusion tests of Section 3.5.1, notice that the constraints are utilized one at a time. Glover (1965c) observed that better information can be secured about the feasibility and the optimality of the problem, relative to its free variables, if the constraints are "combined" into a single

---

† Glover (1965c, 1968b), Balas (1967b), Geoffrion (1969).

inequality, which is supposed to provide tighter bounds on the feasible solution space. The single inequality, referred to as the *surrogate* (or substitute) constraint, is formed as a *nonnegative* linear combination of the original constraints of the problem. In other words, given the zero–one problem in matrix form as

$$\min\{CX \,|\, AX \leqslant b, x_j = (0, 1), j \in N\}$$

a surrogate constraint is defined as

$$\mu(AX - b) \leqslant 0$$

where μ is a nonnegative vector. In general, the surrogate constraint is defined relative to the current partial solution J. This means that before the surrogate constraint is constructed, the variables designated by the partial solution are substituted in the original constraints at their specified binary values, and hence the surrogate constraint is defined in terms of the free variables only.

To illustrate the usefulness of surrogate constraints, consider the two inequalities, $-x_1 + 2x_2 \leqslant -1$ and $2x_1 - x_2 \leqslant -1$. Each separate constraint has a feasible solution. However, when considered simultaneously, the system is infeasible. This result may be secured by adding the two constraints, that is, $\mu = (1, 1)$, thus yielding the surrogate constraint $x_1 + x_2 \leqslant -2$. The new constraint immediately ascertains infeasibility.

Glover proves two important properties of the above surrogate constraint:

(i)   If $\overline{X}$ is a feasible solution to the original problem, then $\overline{X}$ is also feasible for the surrogate constraint.

(ii) If the surrogate constraint has no feasible solution, then neither does the original problem.

The proof of these two properties follows directly from the fact that $A\overline{X} \leqslant b$ implies $\mu A\overline{X} \leqslant \mu b$.

These properties are useful since they show that the surrogate constraint can be used to make inferences about the infeasibility of the original problem. This is especially enhanced by the following theorem:

▶**THEOREM 3-5**   (GEOFFRION, 1969)   (i)   The constraint $\beta - \sum_j \alpha_j x_j \geqslant 0$
(> 0), $x_j = (0, 1)$, is said to be binary infeasible (i.e., has no feasible binary solution) if and only if $\max\{\beta - \sum_j \alpha_j x_j \,|\, x_j = 0, 1\}$ $= \beta - \sum_j \min(0, \alpha_j) < 0 \ (\leqslant 0)$.

(ii) In any binary solution of $\beta - \sum_j \alpha_j x_j \geqslant 0 \ (> 0)$, $\beta - \sum_j \min(0, \alpha_j)$ $- |\alpha_{j0}| < 0 \ (\leqslant 0)$ implies $x_{j0} = 0$ or 1 according as $\alpha_{j0} > 0$ or $\alpha_{j0} < 0$, respectively.

The proof of this theorem is trivial. Indeed, part (i) is essentially the same as Balas' Test 3 (Section 3.5.1).◀

To illustrate the first part of the theorem, the constraint $-5 - (-x_1 - 2x_2 + 3x_3) \geq 0$ does not have a binary solution, since $-5 - (-1 - 2) = -2 < 0$. An example of the second part is that in *any* binary solution, the constraint $-1 - (-2x_1 + x_2 + x_3 + x_4) \geq 0$ must have $x_1 = 1$ since $\beta - \sum_j \min(0, \alpha_j) - |\alpha_1| = -1 - (-2) - |-2| < 0$.

Theorem 3-5 indicates that if (i) is satisfied for the surrogate constraint, the associated partial solution is fathomed, since none of its completions is feasible. On the other hand, (ii) provides a simple way for fixing the values of some of the free variables.

Clearly, with μ_i being restricted only to nonnegative values, an infinite number of surrogate constraints can be developed for the same problem. Clearly also, some of these constraints may be superior to the others in the sense that they may yield better information about the optimality and the feasibility of the problem relative to its current free variables.

The above remark led Glover to think of a definition of the strength of a surrogate constraint. Given such a definition, one can then develop a strongest surrogate for the problem. Glover's definition is as follows.

A.   *Glover's Surrogate Constraint*

Assume that the problem is given in matrix form by $\min\{CX \,|\, Ax \leq b, x_j = (0, 1), j \in N\}$. The surrogate constraint $\mu^1(AX - b) \leq 0$ is stronger than the surrogate constraint $\mu^2(AX - b) \leq 0$ relative to the current partial solution J_t (or the free variables $N - J_t$) if

$$\min_X\{CX \,|\, \mu^1(AX - b) \leq 0\} \geq \min_X\{CX \,|\, \mu^2(AX - b) \leq 0\}$$

where the minimization is taken over the *binary* values of the *free* variables with the remaining variables being specified by the partial solution.

An intuitive justification for Glover's definition of strength in surrogate constraints is as follows: A desired property of the surrogate constraint is that it provides the tightest bounds on the variables. In some sense, this is equivalent to providing a more restrictive solution space (although not necessarily in the sense that the more restrictive space is a strict subset of the less restrictive one). Clearly, this result is accomplished if the minimum (maximum) of a given objective function subject to the stronger surrogate constraint is larger (smaller) than its optimum value subject to the weaker one.

In its general case, Glover's surrogate constraint entails the solution of separate binary problems. Except for the case where there are two parent constraints only (for which Glover provides a simple solution method), the computational effort associated with the general problem may be tedious [see Glover (1965c) for the details]. This limits the practical use of Glover's definition.

Subsequent developments by Balas (1967b) and Geoffrion (1969) show how Glover's definition can be modified to allow a simpler derivation, and hence a more useful application, of the surrogate constraint. This section presents Geoffrion's method only. (Balas' method is discussed in the problems for this chapter.) A comparison of the three methods and their relationships to one another is presented by Glover (1968b).

B. *Geoffrion's Surrogate Constraint*

Geoffrion proposes that in addition to combining the constraints $AX \leqslant b$ in the form $\mu AX \leqslant \mu b$ as suggested by Glover, the constraint $CX \leqslant z_{\min}$ or $z_{\min} - CX \geqslant 0$ must be augmented with *unit weight*. The justification for augmenting this additional constraint is that the resulting surrogate constraint may yield an objective value CX that is better than the best available value $z_{\min}$.

Geoffrion defines the strength of his constraint as follows. Given $\mu^1 \geqslant 0$ and $\mu^2 \geqslant 0$, the surrogate constraint

$$\mu^1(b - AX) + (z_{\min} - CX) \geqslant 0$$

is stronger than

$$\mu^2(b - AX) + (z_{\min} - CX) \geqslant 0$$

relative to the partial solution J_t if

$$\max_{X=\{0,1\}} \{\mu^1(b - AX) + z_{\min} - CX \,|\, J_t\} \leqslant \max_{X=\{0,1\}} \{\mu^2(b - AX) + z_{\min} - CX \,|\, J_t\}$$

(The intuitive justification of strength is the same as in Glover's constraint.) Thus the strongest surrogate constraint according to Geoffrion's definition is given by the following *minimax* problem:

$$\min_{\mu \geqslant 0} \max_{X=\{0,1\}} \{\mu(b - AX) + (z_{\min} - CX) \,|\, J_t\} \geqslant 0$$

The interesting result about Geoffrion's definition (the minimax problem) is that the determination of μ is equivalent to the solution of an ordinary linear programming problem. The equivalence is established as follows.

The maximand part associated with the constraint can be written as follows: Let x_j^t be the binary value of x_j according to J_t; then

$$\max\left\{\sum_{i=1}^{m}\mu_i\left(b_i - \sum_{j\in J_t}a_{ij}x_j^t - \sum_{j\notin J_t}a_{ij}x_j\right) + \left(z_{\min} - \sum_{j\in J_t}c_jx_j^t - \sum_{j\notin J_t}c_jx_j\right)\,\middle|\,\mu_i \geqslant 0,\right.$$

$$\left.\text{for all}\quad i\in M,\quad x_j = (0,1), j\notin J_t,\text{ and } x_j = x_j^t, j\in J_t\right\}$$

$$=\sum_{i=1}^{m}S_i^t\mu_i + z_{\min} - z^t$$

$$+\max\left\{\sum_{j\notin J_t}(-1)\left(\sum_{i=1}^{m}a_{ij}\mu_i + c_j\right)x_j\,\middle|\,x_j = (0,1), j\notin J_t\right\}$$

$$=\left(\sum_{i=1}^{m}S_i^t\mu_i + z_{\min} - z^t\right)$$

$$+\max\left\{\sum_{j\notin J_t}(-1)\left(\sum_{i=1}^{m}a_{ij}\mu_i + c_j\right)x_j\,\middle|\,0 \leqslant x_j \leqslant 1, j\notin J_t\right\}$$

$$=\left(\sum_{i=1}^{m}S_i^t\mu_i + z_{\min} - z^t\right)$$

$$+\min\left\{\sum_{j\notin J_t}y_j\,\middle|\,y_j \geqslant 0, y_j \geqslant -\left(\sum_{i=1}^{m}a_{ij}\mu_i + c_j\right), j\notin J_t\right\}$$

In the above derivation, the second equality follows from the fact that the unit cube $0 \leqslant x_j \leqslant 1$, $j\notin J_t$, contains no feasible *interior* points (see Theorem 3-1) and that linear programming theory guarantees that any resulting solution occurs at an extreme point, which is integer. The third equality follows from duality theory (Section 2.5). As a result of this, the nonlinear expression in μ_i and x_j is now converted into a linear expression in μ_i and y_j.

The minimax problem is now written as:

minimize

$$w^t = \sum_{i=1}^{m}S_i^t\mu_i + \sum_{j\notin J_t}y_j + (z_{\min} - z^t)$$

subject to

$$-\sum_{i=1}^{m}a_{ij}\mu_i - y_j \leqslant c_j$$

$$\mu_i, y_j \geqslant 0,\qquad i\in M,\quad j\notin J_t$$

Let $w_{min}{}^t$ be the optimum objective value of the above problem. It follows that:

(i)   If $w_{min}{}^t < 0$, then from part (i) of Theorem 3-5 J_t has no better feasible completion and hence is fathomed. In general, it will not be necessary to compute $w_{min}{}^t$, since in carrying out the simplex iterations the same result is at hand as soon as the objective value w_t becomes < 0.

(ii) If $w_{min}{}^t \geq 0$, then J_t is not fathomed, but the associated optimal values of μ_i are used to yield the surrogate constraint. This constraint may be augmented to the problem, and part (ii) of Theorem 3-5 is then used to fix the values of some of the free variables. In addition, one or more of Balas' tests (Section 3.5.1) may be applied to the augmented problem.

Geoffrion's surrogate constraint may be strengthened further by replacing the inequality $z_{min} - CX \geq 0$ by $z_{min} - c_l - CX \geq 0$, where $c_l = \text{lcm}\{c_1, c_2, \ldots, c_n\} \geq 1$, provided all c_j are integers. This means that the next best objective value must be equal to at most $z_{min} - c_l$, provided there exist feasible solutions having this value.

An interesting observation is that the above dual problem (in μ_i and y_j) is essentially the dual of the continuous version of the original 0–1 problem, given J_t. (See Problem 3-7.) Thus, if $w_{min}{}^t \geq 0$ and if its dual solution is all integer (0 or 1), this is taken as an improved solution to the original problem, and the value of z_{min} is updated accordingly. In any case, even if such values are not integers, a rounded dual solution should be attempted since it may yield an improved solution, provided it is feasible. The indicated dual relationship can also be used to show that a lower bound on the optimum objective value is equal to $z_{min} - w_{min}{}^t$. (See Problem 3-8.) This lower bound may be used with Balas' Test 2 (Section 3.5.1) to exclude more nonpromising variables.

An extension of the above development to the mixed zero–one case is discussed in Problem 3-9.

▶**EXAMPLE 3.5-2**   Minimize

$$z = (3, 2, 5, 2, 3)(x_1, x_2, \ldots, x_5)^T$$

subject to

$$\begin{pmatrix} -1 & -1 & 1 & 2 & -1 \\ -7 & 0 & 3 & -4 & -3 \\ 11 & -6 & 0 & -3 & -3 \end{pmatrix} \begin{pmatrix} x_1 \\ x_2 \\ \vdots \\ x_5 \end{pmatrix} \leq \begin{pmatrix} 1 \\ -2 \\ -1 \end{pmatrix}$$

$$x_j = 0 \text{ or } 1, \quad \text{for all } j$$

Before the surrogate constraint can be applied, it is necessary first to secure a feasible solution, and hence a value of z_{min}. Thus, applying Balas' tests in Section 3.5.1, one gets the feasible solution

$$J_1 = \{5\}, \qquad S^1 = (2, 1, 2), \qquad z_{min} = 3$$

Now J_1 is fathomed, which yields

$$J_2 = \{-5\}, \qquad S^2 = (1, -2, -1), \qquad z^2 = 0$$

The problem defining the μ's for the surrogate constraint is given by:

minimize

$$w^2 = \mu_1 - 2\mu_2 - \mu_3 + y_1 + y_2 + y_3 + y_4 + (z_{min} - 1 - z^2)$$

subject to

$$
\begin{aligned}
\mu_1 + 7\mu_2 - 11\mu_3 - y_1 & \leqslant 3 \\
\mu_1 \qquad + 6\mu_3 \qquad - y_2 & \leqslant 2 \\
-\mu_1 - 3\mu_2 \qquad\qquad - y_3 & \leqslant 5 \\
-2\mu_1 + 4\mu_2 + 3\mu_3 \qquad\qquad - y_4 & \leqslant 2 \\
\mu_i, y_j & \geqslant 0
\end{aligned}
$$

The optimal solution to this problem yields

$$w_{min}{}^2 = -0.99 + (z_{min} - 1 - z^2) = 1.01$$

while $\mu_1 = 0$, $\mu_2 = 0.48$, and $\mu_3 = 0.03$.

Since $w_{min}{}^2 > 0$, J_2 is not fathomed and a surrogate constraint is constructed. By letting $x_5 = 0$ according to J_2, the surrogate constraint is given by

$$0.48\{-2 - (-7x_1 + 3x_3 - 4x_4)\} + 0.03\{-1 - (11x_1 - 6x_2 - 3x_4)\}$$
$$+ \{(3 - 1) - (3x_1 + 2x_2 + 5x_3 + 2x_4)\} \geqslant 0$$

or

$$-0.03x_1 + 1.82x_2 + 6.44x_3 - 0.01x_4 \leqslant 1.01$$

Applying Theorem 3-5, part (ii), to the resulting surrogate constraint, it follows that x_2 and x_3 must be fixed at zero level in order for a feasible completion to exist. It is important to notice that this information would not have been obtained from any one of the three main constraints. This shows that the surrogate constraint is better in the sense that it traps useful information.

Now, to commemorate that $x_2 = x_3 = 0$ for any feasible completion of

$J_2 = \{-5\}$, J_3 is written as $\{-5, -2, -3\}$.† The resulting problem now consists of two variables only, namely, x_1 and x_4. Balas' Test 4 augments J_3 by $x_4 = 1$ to yield $J_4 = \{-5, -2, -3, 4\}$, with $z^4 = 2$ and $S^4 = (-1, 2, 2)^{\mathrm{T}}$.

A new surrogate constraint is now defined by solving the problem:

minimize
$$w^4 = -\mu_1 + 2\mu_2 + 2\mu_3 + y_1 + (3 - 1 - 2)$$
subject to
$$\mu_1 + 7\mu_2 - 11\mu_3 - y_1 \leqslant 3$$
$$\mu_i, y_1 \geqslant 0$$

The solution yields $w_{\min}{}^4 = -\infty < 0$, and thus by Theorem 3-5, part (i), J_4 is fathomed yielding $J_5 = \{-5, -2, -3, -4\}$.

Continuing in the same fashion, the algorithm shows that no better solution than $z_{\min} = 3$ exists.◀

Geoffrion (1969) reports his computational experience with a variety of problems having up to 90 variables. A comparison between his method and other methods shows the clear superiority of utilizing the surrogate constraint. The savings in computation time seems to be more pronounced as the number of variables increases. This is important since it indicates that the method could be effective for large problems. Geoffrion's main conclusion is that the use of surrogate constraints in solving set covering, optimal routing, and knapsack problems:

> mitigates what appears to be exponential dependence of computing time on the number of variables (since the number of possible solutions equals 2^n) to what appears to be low-order polynomial dependence . . . If these results are at all indicative, then the algorithm can be expected to cope routinely with quite large structured integer linear programs.

3.5.3 Zero–One Polynomial Programs‡

The zero–one polynomial problem is defined as:

minimize
$$z' = \sum_{j=1}^{N} c_j f_{0j}(y)$$

---

† This situation is equivalent to saying that the free variables x_2 and x_3 were previously augmented at level *one*; but having fathomed the resulting partial solutions necessitates replacing these elements by their complements. Notice that, in general, it may be necessary to act as if a free variable were augmented at *zero* level. In this case, it will no longer be true that the negative elements of a partial solution will coincide with the underlined elements, and it becomes essential to follow Glover's original enumeration scheme (Section 3.4).

‡ Taha (1972a, b).

subject to

$$\sum_{j=1}^{n} a_{ij}' f_{ij}(y) + S_i = b_i', \qquad i \in M \equiv \{1, 2, \ldots, m\}$$

where

$$f_{ij}(y) = \prod_{k_j} y_{k_j}^{\alpha_{jk_j}}, \qquad i \in \{0\} \cup M$$

$$y_{k_j} = (0, 1), \qquad j \in N \equiv \{1, 2, \ldots, n\}$$
$$k_j \in K_j \equiv \{k_{j1}, \ldots, k_{jp_j}\}$$

$$S_i \geqslant 0, \qquad i \in M$$

The exponent α_{jk_j} is assumed to be positive for all j and k_j.

Since $y_{k_j} = (0, 1)$ and $\alpha_{jk_j} > 0$, then $y_{k_j}^{\alpha_{jk_j}} = y_{k_j}$. Letting $x_j = \prod_{k_j} y_{k_j}$, then x_j also is a binary variable. Define

$$J^+ = \{j \mid c_j' > 0\}$$
$$J_1^- = \{j \mid c_j' < 0, K_j \cap K_r = \varnothing, r \in J^+\}$$
$$J_2^- = \{j \mid c_j' < 0, K_j \cap K_r \neq \varnothing, r \in J^+\}$$
$$= N - J^+ - J_1^-$$

The 0–1 polynomial problem may then be written as:

minimize

$$z = \sum_{j \in N} c_j x_j$$

subject to

$$\left. \begin{array}{ll} \sum_{j \in N} a_{ij} x_j + S_i = b_i, & i \in M \\[2mm] x_j = (0, 1), & j \in N \\[2mm] S_i \geqslant 0, & i \in M \end{array} \right\} \begin{array}{l} \text{main} \\ \text{constraints} \end{array}$$

$$x_j = \left\{ \begin{array}{ll} \prod_{k_j} y_{k_j}, & j \in J^+ \cup J_2^- \\[3mm] 1 - \prod_{k_j} y_{k_j}, & j \in J_1^- \end{array} \right\} \begin{array}{l} \text{secondary} \\ \text{constraints} \end{array}$$

The relationship between c_j and c_j' is realized by observing the substitution effected by the secondary constraints. The idea here is to make as many coefficients as possible positive in the objective function. This will allow taking advantage of the condition that once a better objective value is secured, then the associated partial solution is fathomed provided all the free variables have nonnegative c_j coefficients. The conversion to positive c_j, however, may not be correct for $j \in J_2^-$, since the interaction between the polynomial terms associated with $j \in J^+$ and $j \in J_2^-$ may not, in general, result in a monotone nondecreasing objective function. For example, the function $z = y_1 y_2 - 2 y_1 y_2 y_3$ should *not* be converted to $z = x_1 + 2x_2 - 2$, where $x_1 = y_1 y_2$ and $x_2 = 1 - y_1 y_2 y_3$, since as x_2 decreases (to zero) x_1 *automatically* increases (to one). Thus z is *not* monotone nondecreasing in x_1 and x_2.

It is noted that the main constraint and the objective function constitute an ordinary linear zero–one problem. The immediate thought then is to convert the secondary constraints into linear equivalences. This idea was first proposed by Fortet (1959) and extended to the above problem by Balas (1964). A subsequent, but independent, development of Balas' extension was given by Watters (1967). The equivalent linear constraints are determined by replacing each secondary constraint by

$$\sum_{k_j} y_{k_j} - (|K_j| - 1) \leqslant \left\{ \begin{matrix} x_j \\ 1 - x_j \end{matrix} \right\} \leqslant \frac{1}{|K_j|} \sum_{k_j} y_{k_j} \quad \begin{matrix} j \in J^+ \cup J_2^- \\ j \in J_1^- \end{matrix}$$

where $|K_j|$ is the cardinality (number of elements) of the set K_j. The transformation, in conformance with the secondary constraints, guarantees that for $j \in J^+ \cup J_2^-$ $(j \in J_1^-)$, $x_j = 1$ $(x_j = 0)$ when every $y_{k_j} = 1$ and $x_j = 0$ $(x_j = 1)$ when at least one $y_{k_j} = 0$. The resulting problem then becomes a regular linear zero–one program for which Balas' additive algorithm is directly applicable (with the possible modification of allowing for negative c_j).

The drawback of this method is that it may lead to an increase in the number of variables (both y_{k_j} and x_j are considered explicitly). Also, the increase in the number of constraints may make the size of the problem intractable. The first difficulty is especially important since, as clear from the nature of implicit enumeration, the number of variables is the prime factor in determining the computation time.

Developments by Glover and Woolsey (1973) show how the linear transformation can be effected with a fewer number of variables and/or constraints. A more direct method was proposed by Taha (1972a), which attacks the problem by considering its main constraints explicitly while accounting for the secondary constraints implicitly.

The new procedure is based on the following obvious result: Define

$$G_1 = \{\text{binary solutions} \,|\, \text{main constraints}\}$$
$$G_2 = \{\text{binary solutions} \,|\, \text{main } and \text{ secondary constraints}\}$$

It follows that $G_2 \subseteq G_1$.

The above observation indicates that knowing the set of all feasible solutions G_1, it is possible to determine *all* the feasible solutions to the polynomial problem by excluding those of G_1 that do not satisfy the secondary constraints. The applicability of this idea is enhanced further by noticing that the special structure of the secondary constraints should allow a simple checking of infeasibility. This point is explained later. Also, it may appear at first thought that it will be necessary to determine G_1 completely before the problem is solved. But it will be shown that the search for a feasible solution in G_1 can be conditioned so that only those that are potentially promising (both from the optimality and feasibility viewpoints) with respect to the polynomial problem are considered. Before explaining how this can be accomplished, the procedure for determining G_1 is introduced first.

A.   Determination of the Set G_1

The determination of all the feasible solutions described by G_1 is accomplished through simple changes of the fathoming techniques of Balas' algorithm (Section 3.5.1). Notice first that all the main constraints must be originally of the type $\leqslant$.

For the purpose of determining all the feasible solutions, a partial solution J_t (feasible or infeasible) is said to be fathomed if it is proved to have no feasible completions. If J_t is infeasible (as usual, it is assumed that J_t behaves as its completion, in which all the free variables are at zero level), then it is fathomed in exactly the same way proposed by Balas in Section 3.5.1. Notice, however, that since the objective is to determine *all* the feasible solutions, then Balas' Test 2 associated with the objective function can be nullified or bypassed.

Suppose now that J_t is feasible. Then, unlike Balas' case, another feasible completion may be determined by elevating at least one free variable to level one. Define

$$Q_t = \{j \mid x_j \text{ is free and } a_{ij} \leqslant S_i^t \text{ for all } i\}$$

as the set representing those free variables that when elevated, *one at a time*, to level one will yield a feasible partial solution. Let

$$u_k = \max_{j \in Q_t} \left\{ \sum_{i \in M} (S_i^t - a_{ij}) \right\}$$

Then a "good" choice of the new (feasible) partial solution is $J_{t+1} = J_t \cup \{k\}$. The selection of $k \in Q_t$ is made such that there is a better chance that another $j \in Q_t$ will yield a feasible partial solution when augmented to J_{t+1}.

Suppose now that for $t^* > t$, $Q_{t^*} = \varnothing$, then either J_{t^*} has no feasible completion or there exists one (or more) in which *at least two* free variables must be at level one. In either case, the fathoming of J_{t^*} is achieved by augmenting it by a free variable to be selected according to Balas' Test 4

(Section 3.5.1). Let the resulting partial solution be $J_{t*} \cup \{r\}$. Since $Q_{t*} = \varnothing$, then $J_{t*} \cup \{r\}$ must be infeasible. Thus Balas' tests are used to fathom $J_{t*} \cup \{r\}$, which is essentially equivalent to fathoming J_{t*}.

▶**EXAMPLE 3.5-3** Consider the determination of all feasible solutions in Example 3.5-1.

The steps for determining J_3 are essentially the same as in Example 3.5-1. For convenience, let $J_0 \equiv J_3 = \{5, 1, 2\}$ and $S^0 \equiv S^3 = (4, 2, 6).^T$.

Iteration 1 J_0 is feasible. Thus $Q_1 = \{3, 4\}$. But according to the measure u_k given above, $k = 3$ is augmented to J_0 since

$$u_k = u_3 = \max\{(4 - 1) + (2 - 2) + (6 + 3), (4 + 1) + (2 - 0) + (6 - 2)\}$$
$$= 12$$

Thus let $J_1 = J_0 \cup \{3\} = \{5, 1, 2, 3\}$. This gives $S^1 = (3, 0, 9)^T$.

Iteration 2 J_1 is feasible. $Q_2 = \{4\}$, $J_2 = \{5, 1, 2, 3, 4\}$, and $S^2 = (4, 0, 7)^T$. Since $J_2 = N$, backtracking is applied to J_2. Notice that Tuan's idea (Section 3.5.1) can be used to determine the complemented variable, but such details are not included here.

Iterations 3 and 4 $J_3 = \{5, 1, 2, 3, -4\}$, which is the same solution as J_1. $J_4 = \{5, 1, 2, -3\}$ with $S^4 = (4, 2, 6)^T$, which is the same as J_0.

Iteration 5 $Q_5 = \{4\}$ and $J_5 = \{5, 1, 2, -3, 4\}$ with $S^5 = (5, 2, 4)^T$.

Iterations 6 and 7 $J_6 = \{5, 1, 2, -3, -4\}$ and $J_7 = \{5, 1, -2\}$ with $S^7 = (8, 5, -3)^T$.

Iteration 8 By using Balas' tests, $\{3\}$ is augmented to J_7 to yield $J_8 = \{5, 1, -2, 3\}$. This gives $S^8 = (7, 5, 0)^T$.

The process of completing the problem is left as an exercise for the reader. In addition to J_0, J_1, J_2, J_5, and J_8 given above, there is one remaining feasible partial solution, which is given by $\{-5, 1, 2, 4\}$.◀

B. *The Polynomial Algorithm*

It is now shown how the determination of G_1 can be utilized to solve the polynomial problem. The algorithm is again superimposed on Glover's enumeration scheme (Section 3.4).

The general idea of the algorithm is as follows. Balas' additive algorithm is used to generate improved feasible solutions for the *linear* portion of the problem (henceforth referred to as the *master problem*), that is, any new *feasible* partial solution must yield an objective value z which is better than z_{min}. If the secondary constraints are satisfied, then the solution is also feasible for the polynomial problem. Otherwise, a new feasible solution to the master problem is generated and the process is repeated until the enumeration is complete.

Perhaps the only modification in Balas' additive algorithm as applied to the master problem is that a feasible partial solution is fathomed according to the objective value (Balas' Test 2) only if none of the free variables has a negative objective coefficient; since otherwise a better completion may exist. Again as in Balas' algorithm, it is assumed that all the free variables in the master problem are at zero level but can be assigned a specific value by the partial solution if necessary.

Before showing how the secondary constraints can be used to fathom partial solutions generated by the master problem, the following definition is introduced.

A (*feasible* or *infeasible*) partial solution J_t of the *master* problem is said to be *inconsistent* if it leads to inconsistency in the secondary constraints. For example, consider $J_t = \{3, 2\}$ where, from the secondary constraints, $x_2 = y_1 y_2 y_3$ and $x_3 = 1 - y_1 y_2$. By J_t, $x_2 = 1$ implies $y_1 = y_2 = y_3 = 1$, which implies that x_3 must be zero, a result that is inconsistent with J_t.

Clearly, if J_t is inconsistent, then none of its completions can yield a feasible solution to the polynomial problem. In this case, J_t is fathomed and the backward move is executed.

Suppose J_t is consistent and feasible. (If J_t is consistent but infeasible, an attempt to secure a new feasible and consistent partial solution should be made by using Balas' algorithm.) Let $r \in J_t$ and define J^+ and J_1^- as given above by the secondary constraints of the polynomial problem. The values of y_{k_r} associated with J_t are thus given by

$$y_{k_r}{}^* = \begin{cases} 1, & \text{if } x_r = 1, \quad r \in J_t \cap J^+ \quad \text{or} \quad x_r = 0, \quad r \in J_t \cap J_1^- \\ 0 \text{ or } 1, & \text{otherwise} \end{cases}$$

Let

$$E_t = \{k_t \mid y_{k_r}{}^* = 1 \text{ as specified by } J_t\}$$

The set E_t defines an assignment of value one to a subset of the y_{k_j} variables. Now consider the associated values of the free variables x_q, $q \in N - J_t$:

$$x_q{}^* = \begin{cases} \displaystyle\prod_{k_q \in K_q} y_{k_q}, & \text{if } K_q \subseteq E_t \quad \text{and} \quad q \in (J^+ \cup J_2^-) \cap (N - J_t) \\[2ex] 1 - \displaystyle\prod_{k_q \in K_q} y_{k_q}, & \text{if } K_q \subseteq E_t \quad \text{and} \quad q \in J_1^- \cap (N - J_t) \\[2ex] 0 \text{ or } 1, & \text{if } K_q \nsubseteq E_t \end{cases}$$

Complications will arise if any $x_q{}^*$, $q \in N - J_t$, *must* assume the value one in order to satisfy the secondary constraints. For then it may no longer be true that J_t, together with the free variables at level one, will *potentially* produce an improved feasible solution.

To formalize the above idea, let

$$F_t = \{q \,|\, x_q{}^* \text{ must equal } 1, q \in N - J_t\}$$

(1) If $F_t = \varnothing$, an improved solution is at hand and $z_{\min}$ is updated accordingly. In this case, if $(N - J_t) \cap J_2{}^- = \varnothing$, that is, none of the free variables has $c_j < 0$, then J_t is fathomed. Otherwise, an improved feasible solution may result from elevating one (or more) free x_j having $c_j < 0$ to level one.

(2) If $F_t \neq \varnothing$, then a logical way to obtain a solution satisfying the *secondary* constraints is to consider augmenting J_t on the right with F_t (forward move). Two conditions must be considered: (i) $J_t \cup F_t$ satisfies the main constraints, and (ii) $z^t + \sum_{j \in F_t} c_j < z_{\min}$. The following cases can result.

(a) If conditions (i) and (ii) are satisfied, then $J_{t+1} = J_t \cup F_t$ is feasible with respect to the entire problem. In this case, J_{t+1} is treated as if $F_{t+1} = \varnothing$ as given in (1).

(b) If condition (ii) is *not* satisfied and if $c_j > 0$, for all $j \in N - J_t - F_t$, or $z^t + \sum_{j \in F_t} c_j + \sum_{j \in F_t{}^*} c_j > z_{\min}$, where $F_t{}^* = \{j \,|\, c_j < 0, j \in N - J_t - F_t\}$, then regardless of condition (i), J_t cannot yield an improved objective value, and hence it is fathomed. The reason is that a realization of J_t *always* implies F_t, so that z^t must actually be inflated to $z^t + \sum_{j \in F_t} c_j$.

(c) If condition (ii) is satisfied but not (i), then F_t is not augmented to J_t, and instead a new element of G_1 must be determined, that is, a new feasible partial solution is determined. Two important points must be observed here. First, none of the elements of F_t is to be selected for augmentation to J_t. This means that only the free variables specified by $N - J_t - F_t$ can be considered. This follows since J_t always implies F_t and an augmentation by elements from F_t adds no new information. Second, in applying the optimality test on the "selected" free variable, the value of z associated with J_t must be taken as $z^t + \sum_{j \in F_t} c_j$. Thus the free variable eligible for augmentation to J_t can be selected from the set

$$H_t = \left\{ k \in N - J_t - F_t \,\middle|\, z^t + \sum_{j \in F_t} c_j + c_k < z_{\min} \right\}$$

Clearly, if $H_t = \varnothing$, then J_t is fathomed. Otherwise, if there exists $k^* \in H_t$ such that $J_t \cup \{k^*\}$ is feasible, then J_{t+1} is taken equal to $J_t \cup \{k^*\}$ and the regular tests for a feasible partial solution are repeated on J_{t+1}. If no such k^* exists, then J_{t+1} is formed by augmenting J_t by a free variable in H_t yielding the least measure of infeasibility (Balas' Test 4).

(d) Under all other outcomes of conditions (i) and (ii), a new feasible partial solution for the master problem should be generated.

The reader may notice that part (d) can be improved by developing more detailed tests depending on the outcomes of the conditions together with the types of free variables associated with the partial solution. This is true; but such details are not presented here only to simplify the presentation.

▶**EXAMPLE 3.5-4**   Consider the problem:

minimize

$$z' = 2x_1' + 4x_2' - 6x_3' + 8x_4' - 12x_5'$$

subject to

$$\begin{pmatrix} -8 & 4 & -1 & -1 & -5 \\ -6 & 3 & -2 & 0 & -1 \\ 2 & -9 & 3 & 2 & -3 \end{pmatrix} \begin{pmatrix} x_1' \\ x_2' \\ x_3' \\ x_4' \\ x_5' \end{pmatrix} \leqslant \begin{pmatrix} -6 \\ -4 \\ -1 \end{pmatrix}$$

where $x_1' = y_4 y_5$, $x_2' = y_5$, $x_3' = y_1 y_2$, $x_4' = y_3 y_4$, and $x_5' = y_3 y_5$, and all the variables are binary.

Since x_3' is the only variable with negative c_j that has no common y-variables with any other variable with positive c_j, the above problem may be written as:

minimize

$$z = 2x_1 + 4x_2 + 6x_3 + 8x_4 - 12x_5$$

subject to

$$\begin{pmatrix} -8 & 4 & 1 & -1 & -5 \\ -6 & 3 & 2 & 0 & -1 \\ 2 & -9 & -3 & 2 & -3 \end{pmatrix} \begin{pmatrix} x_1 \\ x_2 \\ x_3 \\ x_4 \\ x_5 \end{pmatrix} \leqslant \begin{pmatrix} -5 \\ -2 \\ -4 \end{pmatrix}$$

where $x_1 = y_4 y_5$, $x_2 = y_5$, $x_3 = 1 - y_1 y_2$, $x_4 = y_3 y_4$, and $x_5 = y_3 y_5$. This means that $J^+ = \{1, 2, 4\}$, $J_1^- = \{3\}$, and $J_2^- = \{5\}$. The steps of the solution are given below:

1. $J_1 = \{5\}$, which is infeasible.
2. $J_2 = \{5, 1\}$, which is consistent but still infeasible.
3. $J_3 = \{5, 1, 2\}$, which is feasible for the main constraints. To check the secondary constraints, J_3 yields $y_3 = y_4 = y_5 = 1$, and $E_3 = \{3, 4, 5\}$. This gives $x_3^* = 0$ or 1 and $x_4^* = 1$; hence $F_3 = \{4\}$.
4. Since $J_3 \cup F_3$ is feasible for the main constraints and since z_{min} is currently set at ∞, it follows that $J_4 = J_3 \cup F_3 = \{5, 1, 2, 4\}$ is feasible for the polynomial problem and $z_{min} = 2$. Because $(N - J_4) \cap J_2^- = \varnothing$, J_4 is fathomed.
5. $J_5 = \{5, 1, 2, -4\}$, which is feasible for the main constraints but inconsistent. Thus J_5 is fathomed.
6. $J_6 = \{5, 1, -2\}$, which is inconsistent.
7. $J_7 = \{5, -1\}$, which is consistent but infeasible. Applying Balas' algorithm shows that J_7 has no feasible completions.
8. $J_8 = \{-5\}$. Because $z_{min} = 2$, none of the free variables can improve solution and the enumeration is complete.

The optimal solution is $x_1 = x_2 = x_4 = x_5 = 1$, and $x_3 = 0$; or $y_1 = y_2 = y_3 = y_4 = y_5 = 1$. ◀

It is important to notice that the surrogate constraints described in Section 3.5.2 can be used directly to determine the feasible solutions to the linear problem. This should enhance the effectiveness of the algorithm and possibly equip it with the same computation capability of the linear model.

A remark about the computational efficiency of the above algorithm is in order. Taha (1972a) shows that for problems having large polynomial terms the ratio of the computation time using the above algorithm to its equivalence using the Fortet–Balas–Watters linearized version varies between $\frac{1}{2}$ and $\frac{1}{20}$. In addition, the computer storage requirement is considerably smaller with Taha's method. However, for problems in which the Fortet–Balas–Watters method does not necessitate the addition of *any* new y-variables explicitly, their method gives superior results. This follows since it requires the addition of new constraints only, which generally enhances the efficiency of Balas' additive algorithm.

C. *Extensions of the Polynomial Algorithm*

The procedure used with the polynomial problem can be extended to encompass a wider class of problems. Basically, the idea is to partition the constraints so that the *main* constraints are linear. All the remaining constraints are then regarded as secondary. Because it is possible to determine all the feasible solutions satisfying the main constraints, suitable tests for excluding those violating the secondary constraints can usually be developed. Illustrative examples of these situations are given as exercises at the end of the chapter.

The polynomial algorithm can be applied also to problems that initially may not be in polynomial form. Typical is the case where the functions of the problem are separable in the binary variables. Separability here implies that the function can be decomposed additively and/or multiplicatively in terms of single-variable functions only. For example, $f(x_1, x_2, x_3) = f_1(x_1)f_2(x_2) + f_2(x_2)f_3(x_3)$ serves as a typical illustration.

The importance of separability becomes obvious by noticing that, given x_i a binary variable, then

$$f_i(x_i) = a_i + (b_i - a_i)x_i$$

where

$$a_i = f_i(0), \qquad b_i = f_i(1)$$

By using this substitution, the separable (binary) problem is converted to a proper 0–1 polynomial problem.

3.5.4 Binary Quadratic Programming

Models in which the objective function is quadratic and the constraints are linear exist in capital budgeting problems (see Example 1.3-1), among others. Although the zero–one polynomial algorithm presented above can be used to solve such a problem, it is expected to be inefficient since the variables substituting the cross products do not appear in the main constraints. Another method is to determine all the feasible solutions of the linear constraints and then determine the optimum by evaluating the associated objective values; but the objective value plays a passive role here. A specialized algorithm that develops fathoming tests based on the objective function (and the constraints) is introduced by Laughhunn (1970) and outlined in Problem 3-16. The main limitation here, however, is that the objective function must be a positive (semi-) definite quadratic form. Such a restriction is unnecessary if the zero–one polynomial algorithm is used.

3.6 Nonlinear Zero–One Problem†

In this section we show how a nonlinear 0–1 problem can be solved by implicit enumeration provided its functions satisfy certain monotony conditions. The enumeration scheme used with the problem is entirely

† Lawler and Bell (1966).

different from Glover's (Section 3.4) since each trial solution generates a complete binary assignment. Also the binary solution are generated in a fixed order.

The nonlinear problem must appear in the form:

minimize

$$z(x) = g_{01}(x) - g_{02}(x)$$

subject to

$$g_{i1}(x) - g_{i2}(x) \leqslant b_i, \qquad i = 1, 2, \ldots, m$$

where

$$x = (x_1, x_2, \ldots, x_n), \quad x_j = (0, 1), \quad j = 1, 2, \ldots, n$$

An important restriction on the above problem is that each of the functions $g_{01}, g_{02}, g_{11}, g_{12}, \ldots, g_{m1}, g_{m2}$ is monotone nondecreasing in each of the variables $x_1, x_2, \ldots, x_n$.

Consider the binary vector x. It is said that $x \leqslant y$ if and only if $x_j \leqslant y_j$ for all j. Thus $(1, 0, 0) \leqslant (1, 0, 1)$, while $(1, 0, 1)$ and $(0, 1, 0)$ are not comparable.

In this algorithm, the 2^n solutions are examined according to a fixed *numerical ordering* (cf. Glover's enumeration scheme). The vector

$$x = (x_1, x_2, \ldots, x_n)$$

is said to have the order

$$r_x = 2^{n-1}x_1 + 2^{n-2}x_2 + \cdots + 2^0 x_n$$

whose numerical value is determined by substituting a binary value for each x_j.

Treating x as a binary number, it follows that $r_{x+1} = r_x + 1$. To illustrate, let $x = (1, 0, 0, 1, 1)$; then $x + 1 = (1, 0, 0, 1, 1) + (0, 0, 0, 0, 1) = (1, 0, 1, 0, 0)$. It is clear that $r_x = 19$ and $r_{x+1} = 20$.

The enumeration starts with $x = (0, 0, \ldots, 0)$ having lowest numerical ordering $(r = 0)$ and terminates when $x = (1, 1, \ldots, 1)$ having the highest ordering $(r = 2^n)$ is reached. In the process of enumerating these solutions, proper fathoming tests are used to exclude as many nonpromising points as possible.

Let x^* be the *first* vector succeeding an arbitrary vector x in the numerical ordering such that $x \nleqslant x^*$. This means that $x \leqslant x + 1 \leqslant \cdots \leqslant x^* - 1$. The steps for generating x^* are as follows:

  (i)   Subtract 1 from x to obtain $x - 1$;
  (ii) Logically "or" x and $x - 1$ to obtain $x^* - 1$;
  (iii) Add 1 to $x^* - 1$ to obtain x^*.

For example, let $x = (0, 1, 1, 0, 1)$: (i) $x - 1 = (0, 1, 1, 0, 0)$; (ii) $x^* - 1 = (0, 1, 1, 0, 1)$; (iii) $x^* = (0, 1, 1, 1, 0)$.

The relationship between $x, x + 1, \ldots, x^* - 1$, and x^* together with the monotone property of the functions g_{i1} and g_{i2}, $i = 0, 1, \ldots, n$, form the basis for skipping nonpromising "solution" vectors. Specifically, consider $g_{i1}(x) - g_{i2}(x^* - 1)$. For any x' in the interval $[x, x^* - 1]$, $g_{i1}(x)$ and $g_{i2}(x^* - 1)$ minimize $g_{i1}(x') - g_{i2}(x')$. In other words, $g_{i1}(x) - g_{i2}(x^* - 1)$ is an underestimator in the interval $[x, x^* - 1]$. However, because $x \leqslant x + 1 \leqslant \cdots \leqslant x^* - 1$, if the indicated underestimator does not satisfy the constraint, then neither would any of the vectors in the interval $[x, x^* - 1]$. Thus one may skip directly to x^*. Notice that the objective function can be treated as a regular constraint of the type $z(x) \leqslant z_{\min}$, where $z_{\min}$ is the (updatable) best value of $z(x)$ associated with a feasible solution. (Initially $z_{\min}$ is set equal to ∞.)

The above remark is the basis for the following algorithm. Assume $\hat{x}$ is the best solution vector so far attained. Set $x = (0, 0, \ldots, 0)$ and $x^* = x + 1$.

Step 1   If $x^* - 1 = (1, 1, \ldots, 1)$, stop. Otherwise, go to step 2.

Step 2   If $g_{01}(x) - g_{02}(x^* - 1) > z_{\min}$ or if $g_{i1}(x) - g_{i2}(x^* - 1) > b_i$ for any i, set $x = x^*$, compute new x^* and go to step 1. Otherwise, go to step 3.

Step 3   If $g_{i1}(x) - g_{i2}(x) \leqslant b_i$ for all i, that is, x is feasible, and if $z(x) < z_{\min}$, then set $\hat{x} = x$ and $z_{\min} = z(x)$. If $x = (1, 1, \ldots, 1)$, stop. Otherwise, set $x = x + 1$, compute new x^* and go to step 2.

▶**EXAMPLE 3.6-1**   Minimize

$$z(y) = (4y_1 y_3 y_4 + 6y_3 y_4 y_5 + 12y_1 y_5) - (2y_1 y_2 + 8y_1 y_3)$$

subject to

$$(8y_1 y_2 + 4y_1 y_3 y_4 + y_3 y_4 y_5 + y_1 y_3) - (5y_1 y_5) \leqslant 4$$
$$(6y_1 y_2 + 3y_1 y_3 y_4 + 2y_3 y_4 y_5) - (y_1 y_5) \leqslant 4$$
$$(0) - (2y_1 y_2 + 9y_1 y_3 y_4 + 3y_3 y_4 y_5 + 2y_1 y_3 + 3y_1 y_5) \leqslant -8$$
$$y_j = (0, 1), \quad j = 1, 2, \ldots, 5$$

The solution is given by the successive steps shown in Table 3-1.

TABLE 3-1

Number	y	$(y^* - 1)$	Remarks
1	(0, 0, 0, 0, 0)	(0, 0, 0, 0, 0)	Constraint 3 is not satisfied; skip to y^*
2	(0, 0, 0, 0, 1)	(0, 0, 0, 0, 1)	Constraint 3 is not satisfied; skip to y^*
3	(0, 0, 0, 1, 0)	(0, 0, 0, 1, 1)	Constraint 3 is not satisfied; skip to y^*
4	(0, 0, 1, 0, 0)	(0, 0, 1, 1, 1)	Constraint 3 is not satisfied; skip to y^*
5	(0, 1, 0, 0, 0)	(0, 1, 1, 1, 1)	Constraint 3 is not satisfied; skip to y^*
6	(1, 0, 0, 0, 0)	(1, 1, 1, 1, 1)	y is infeasible; set $y = y + 1$
7	(1, 0, 0, 0, 1)	(1, 0, 0, 0, 1)	Constraint 3 is not satisfied; skip to y^*
8	(1, 0, 0, 1, 0)	(1, 0, 0, 1, 1)	Constraint 3 is not satisfied; skip to y^*
9	(1, 0, 1, 0, 0)	(1, 0, 1, 1, 1)	y is infeasible; set $y = y + 1$
10	(1, 0, 1, 0, 1)	(1, 0, 1, 0, 1)	Constraint 3 is not satisfied; skip to y^*
11	(1, 0, 1, 1, 0)	(1, 0, 1, 1, 1)	y is feasible; $\hat{y} = (1, 0, 1, 1, 0)$, $z_{\min} = -4$; set $y = y + 1$
12	(1, 0, 1, 1, 1)	(1, 0, 1, 1, 1)	y is infeasible; set $y = y + 1$
13	(1, 1, 1, 1, 0)	(1, 1, 1, 1, 1)	Constraint 1 is not satisfied; skip to y^*; since $y^* - 1 = (1, 1, 1, 1, 1)$, stop ◀

A remark about the efficiency of the above algorithm can be made here. Computational experience reported by Taha (1972a) shows that the Lawler–Bell algorithm may be extremely inefficient. The comparison was made with the polynomial algorithm in Section 3.5.4. Some test problems that were solved in a few seconds by the polynomial algorithm could not be completed in one hour by the Lawler–Bell algorithm.

3.7 Mixed Zero–One Problem

The mixed zero–one linear problem is defined as:

maximize

$$z = \sum_{j \in P} c_j x_j + \sum_{j \in Q} d_j y_j$$

subject to

$$\sum_{j \in P} a_{ij} x_j + \sum_{j \in Q} e_{ij} y_j + S_i = b_i, \qquad i \in M \equiv \{1, \dots, m\}$$

$$x_j \geqslant 0, \qquad j \in P \equiv \{1, \dots, p\}$$

$$y_j = (0, 1), \qquad j \in Q \equiv \{1, \dots, q\}$$

$$S_i \geqslant 0, \qquad i \in M$$

There are three special enumeration methods for solving the above problem due to Driebeek (1966), Benders (1962), and Lemke and Spielberg (1967). The Lemke–Spielberg method is closely related to Benders'. A fourth method was developed by R. E. Davis *et al.* (1971). This method, however, may be regarded as a special case of the Beale–Small algorithm (1965) for the general mixed integer problem (see Section 4.3.2-B).

3.7.1   The Penalty Algorithm†

Driebeek was the first to conceive the use of penalties in solving mixed integer linear programs. His idea is to first solve the problem as a continuous linear program, that is, with the conditions $y_j = (0, 1)$ replaced by $0 \leqslant y_j \leqslant 1$, for all $j \in Q$. In general, some of the y_j variables may be fractional in the optimum continuous solution. If these variables are forced to assume integer values then the objective value of the linear program will decrease. The same result holds if a binary variable y_j at zero (one) level in the continuous solution is forced to assume the value one (zero).

Let $z_{\max}$ be the optimum objective value of the linear program and let δ be the decrease from $z_{\max}$ resulting from imposing restrictions on y_j such that a specific binary assignment for all y_j, $j \in Q$, is realized. Then the associated optimum objective value is $z_{\max} - \delta$. Now if z^* is a known *lower* bound on the optimum objective value of the mixed 0–1 problem, then the specific binary assignment resulting in δ may be discarded if $z_{\max} - \delta \leqslant z^*$. By enumerating all 2^q binary combinations and updating z^* each time an improved solution is attained, then at the termination z^* and its associated solution give the optimum solution. However, in order for this procedure to reach the sophisticated level of a potentially efficient algorithm, some of the 2^q combinations must be discarded automatically, that is, enumerated implicitly.

In the above outline, one notices that δ can be determined exactly only by solving a linear program for each binary assignment. This may be extremely costly from the computational standpoint. What Driebeek proposes is to estimate a *lower* bound $\underline{\delta}$ on the value of δ, by using the information in the continuous optimum linear program only. Although $z_{\max} - \underline{\delta}$ is not as strong as $z_{\max} - \delta$, it is evident that a binary assignment yielding $z_{\max} - \underline{\delta} \leqslant z^*$ must be discarded. The weakening of the condition can be tolerated since, as will be shown, $\underline{\delta}$ can be computed easily and quickly.

The lower bound $\underline{\delta}$ defines the so-called *penalty*, and the next section shows how penalties are determined. However, if Driebeek's method is followed exactly, one would deal with an *enlarged* continuous problem having $(m + 2q)$ *explicit* constraints and $(p + 3q + m)$ variables in order to compute the penalties. This is actually unnecessary as the simplified version (due to Taha) will show. It must be stressed, however, that the basic steps of the algorithm (after the penalties are computed) are essentially those of Driebeek.

---

† Driebeek (1966).

A.   *Computation of the Penalties*

The linear program associated with the mixed 0–1 problem [that is, with $y_j = (0, 1)$ replaced by $0 \leqslant y_j \leqslant 1, j \in Q$] is best solved by the bounded variables (primal or dual) simplex method (Sections 2.6.1 and 2.6.2). At the optimum tableau, some of the binary variables y_j are basic at fractional or integer values. The remaining y_j variables are nonbasic either at zero (lower bound) or at one (upper bound) level. This follows from the fact that the constraint $0 \leqslant y_j \leqslant 1$ is implicit. Thus, in estimating the penalty $\underline{\delta}$, two types of integer restrictions are taken into account: (i) A nonbasic y_j at zero (one) level is forced to assume the value one (zero). (ii) A fractional $y_j = y_j{}^*, 0 < y_j{}^* < 1$, is forced to take a binary value. A special case of type (ii) occurs when a basic y_j is integer but is forced to assume the other binary value. These two situations will first be considered separately and then combined to give the penalty $\underline{\delta}$.

Suppose the *optimum* objective equation of the continuous linear program is written as

$$z + \sum_{k \in NB} (z_k - c_k)w_k = z_{\max}$$

where $z_k - c_k$, as defined in Section 2.3.3, is the *reduced cost* associated with the nonbasic variable $w_k, k \in NB$. (The variables w represent any of the variables x, y, or S in the original problem.)

From linear programming theory, the optimal value of z varies with the continuous values of a basic binary variable y_j according to a concave function with linear segments. A typical illustration is shown in Fig. 3-6.

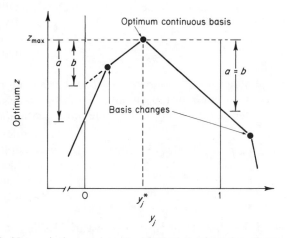

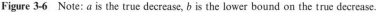

Figure 3-6   Note: a is the true decrease, b is the lower bound on the true decrease.

Because of the cancavity of the function around $y_j{}^*$, the true decrease in the optimum value of z as a result of assigning a binary value to y_j is *at least* equal to the decrease computed when $z_k - c_k$, $k \in NB$, remains unchanged, that is, when no change in basis takes place. This is the idea for computing the penalty as will now be detailed.

(i) *Penalties for nonbasic binary variables*   From the definition of $z_k - c_k$ and the assumption of no change in basis, the penalty for increasing (decreasing) a binary variable w_k to level one (zero) is directly equal to $z_k - c_k$. Thus, if $K \subseteq NB$ is the set of nonbasic *binary* variables *whose current integer values are changed to complementary values*, then the total penalty associated with this change is

$$\delta_1 = \sum_{k \in K} (z_k - c_k)$$

(ii) *Penalties for basic binary variables*   Let the constraint equation associated with basic y_j in the optimal continuous tableau be given by

$$y_j + \sum_{k \in NB} \alpha_{jk} w_k = y_j{}^*, \qquad 0 < y_j{}^* < 1$$

If y_j is decreased to zero level, then this is equivalent to imposing the restriction $y_j \leq 0$ on the continuous optimum solution. From the above constraint, $y_j \leq 0$ is equivalent to

$$- \sum_{k \in NB} \alpha_{jk} w_k \leq -y_j{}^*$$

Similarly, forcing y_j to be at level one is equivalent to the constraint $y_j \geq 1$, or

$$\sum_{k \in NB} \alpha_{jk} w_k \leq y_j{}^* - 1$$

By augmenting the above two constraints *one at a time* to the continuous optimum tableau and applying the dual simplex method, the optimum objective value must decrease by *at least*

$$\Delta_j = \begin{cases} \Delta_j{}^0 = y_j{}^* \min\limits_{k \in NB} \left\{ \left| \dfrac{z_k - c_k}{\alpha_{jk}} \right| \; \alpha_{jk} > 0 \right\}, & \text{if } y_j = 0 \\[4ex] \Delta_j{}^1 = (y_j{}^* - 1) \max\limits_{k \in NB} \left\{ \left| \dfrac{z_k - c_k}{\alpha_{jk}} \right| \; \alpha_{jk} < 0 \right\}, & \text{if } y_j = 1 \end{cases}$$

Now, if R is the set of *basic binary* variables in the continuous optimum, then a lower bound on the decrease in the objective value as a result of forcing all y_j, $j \in R$, to take specific binary values is given by

$$\delta_2 = \max_{j \in R}\{\Delta_j\}$$

Because the above penalties are measured under the assumption of no change in basis, it follows that the overall penalty associated with a complete binary assignment of all y_j is given by

$$\underline{\delta} = \max\{\delta_1, \delta_2\}$$

The penalty $\underline{\delta}$ may now be used to compute the upper bound $z_{max} - \underline{\delta}$ on the optimum objective value associated with any binary assignment to the variables y_j.

B.   *The Algorithm*

Let z^* be the current best objective value associated with a feasible solution. If no initial feasible solution is available, take $z^* = -\infty$. Solve the continuous linear program and record its optimum objective value z_{max}. Then compute the penalties associated with the binary variables.

Step 0   Select a desired level for each y_j variable. A good initial choice is to select the binary value yielding the lower penalty. Go to step 1.

Step 1   Compute the penalty $\underline{\delta}$ associated with the current binary assignment.

(a)   If $z_{max} - \underline{\delta} \leqslant z^*$, discard the current integer combination as non-promising and go to step 2. Otherwise,

(b)   Substitute the binary values of y_j in the mixed integer problem and solve the resulting linear program. If no feasible solution exists or if the objective value drops below z^*, discard the current integer combination and go to step 2. Otherwise, the resulting solution yields an improved value of z^*. Record this value and go to step 2.

Step 2   If all 2^q binary combinations have been enumerated, stop; z^*, when finite, gives the optimum. Otherwise, generate a new binary combination and go to step 1.

The efficiency of Driebeek's algorithm is dependent in the first place on how effective the penalties are in producing the smallest upper bound, and also on how good initial z^* is. One notices, however, that because the penalties are computed once and for all from the continuous tableau, they have the disadvantage that better values cannot be generated as more binary combinations are enumerated.

►EXAMPLE 3.7-1   The following example is taken from Driebeek's paper to allow a comparison between his original algorithm and the modified version presented in this book.

Maximize

$$z = -3x_1 - 3x_2 - y_2 + 10$$

subject to

$$-x_1 + x_2 + y_1 - 0.4y_2 = 0.7$$
$$x_1, x_2 \geqslant 0$$
$$y_1, y_2 = (0, 1)$$

The continuous optimum obtained after replacing $y_1, y_2 = (0, 1)$ by $0 \leqslant y_1, y_2 \leqslant 1$ is obviously given as:

		x_1	x_2	y_2
z	10	3.0	3.0	1.0
y_1	0.7	−1.0	1.0	−0.4

The penalties are computed as follows. Notice that y_1 is basic and y_2 is nonbasic:

$y_1 = 0$: $(0.7)(3/1.0) = 2.1$

$y_1 = 1$: $(0.7 - 1)\left(\max\left\{\dfrac{3}{-1.0}, \dfrac{1}{-0.4}\right\}\right) = 0.75$

$y_2 = 0$:   penalty equal to zero since y_2 is already nonbasic at zero level

$y_2 = 1$:   penalty $= 1$

This is summarized in tabular form as follows:

y_i \ i	1	2
0	2.1	0
1	0.75	1.0

*

The asterisk (∗) implies that y_1 is basic in the continuous optimum tableau.

Let $z^* = -\infty$, and select the initial binary assignment $(1, 0)$ since it yields the least sum of penalties. Because $z^* = -\infty$, the combination $(1, 0)$ cannot be examined based on the penalties. Thus substituting $y_1 = 1$ and $y_2 = 0$ in the mixed integer problem gives the linear program,

maximize

$$z = -3x_1 - 3x_2 + 10$$

subject to

$$-x_1 + x_2 = -0.3$$

$$x_1, x_2 \geq 0$$

The optimal solution is $x_1 = 0.3$ and $x_2 = 0$ with $z = 9.1$. Since it is feasible, $z^* = 9.1$.

Next consider the combination $(1, 1)$, which gives $\delta_1 = 1$ and $\delta_2 = 0.75$. Hence, $\underline{\delta} = \max\{\delta_1, \delta_2\} = 1$. Since $z_{max} - \underline{\delta} = 10 - 1 = 9 < z^*$, the combination $(1, 1)$ is discarded. Similarly, the combination $(0, 1)$ has

$$\underline{\delta} = \max\{1, 2.1\} = 2.1$$

and the combination $(0, 0)$ has $\underline{\delta} = \max\{0, 2.1\} = 2.1$. In both cases, $z_{max} - \underline{\delta} < z^*$ and hence $(0, 1)$ and $(0, 0)$ are also discarded. This exhausts all the binary combinations, and the optimal solution is given by $z^* = 9.1$, $y_1^* = 1$, $y_2^* = 0$, $x_1^* = 0.3$, and $x_2^* = 0$.

Figure 3-7 gives the solution tree with its $2^2 = 4$ branches. Unlike Glover's enumeration scheme, all the branches originate from the same node, with each branch representing a complete binary assignment.◄

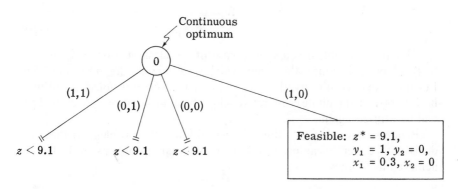

Figure 3-7

Driebeek reports good computational results with problems having a small number of binary variables. These results are based on his original algorithm. It should be expected that the modified version presented here would yield improved results, since it will no longer be necessary to deal with an enlarged linear program as Driebeek proposes.

The reader may already realize that Driebeek's algorithm can be modified further by using the concept of partial solutions together with Glover's enumeration scheme. This is true; but the computation of penalties should be somewhat different since the objective now is to provide an upper bound on the objective value associated with all the completions of a given partial solution. Indeed, the above idea is the basis for the algorithm by R. E. Davis *et al.* (1971). But as mentioned earlier, their algorithm can be viewed as a special case of a general algorithm by Beale and Small (1965), which will be presented in Section 4.3.2-B. The reader will appreciate this point by recalling that in Section 3.4 it is stated that Glover's scheme for binary problems (which is used by R. E. Davis *et al.*) and Beale–Small's scheme for the general problem are based on very similar ideas. Also, the Beale–Small algorithm utilizes the concept of penalties to "fathom" solutions.

3.7.2   The Partitioning Algorithm†

The mixed zero–one problem can be put in matrix form as follows:

maximize

$$z = CX + DY$$

subject to

$$AX + EY \leqslant b$$
$$X \geqslant 0$$
$$Y = (0, 1) \tag{P}$$

The above problem is a special case of the class of problems solvable by Benders' partitioning algorithm, which is explained fully in Section 2.5.6. For the sake of continuity, the algorithm for the above problem is outlined here. In order to facilitate cross-referencing, the same notation in Section 2.5.6 will be utilized again.

Benders' algorithm partitions the variables of the problem so that one deals with either linear programs in X or binary problems in Y. This is achieved as follows.

---

† Benders (1962).

If the vector Y is fixed, then (P) may be written as:

maximize

$$v = CX + DY$$

subject to

$$AX \leqslant b - EY$$
$$X \geqslant 0 \qquad\qquad (PX)$$

Then the dual of (PX) is:

minimize

$$w = u(b - EY)$$

subject to

$$uA \geqslant C$$
$$u \geqslant 0 \qquad\qquad (D1)$$

The solution space of (D1) is independent of Y. If (PX) has a feasible solution, then so does (D1); and by augmenting a regularity constraint $\sum_{i \in M} u_i \leqslant M$, $M > 0$ and very large, if necessary, the optimum solution to (D1), when it exists, must occur at an extreme point u^k of its feasible solution space. This means that if u^k, $k = 1, 2, \ldots, K$, is the set of extreme points of (D1), then (P) can be written as:

maximize

$$z = DY + \min_{u^k}\{u^k(b - EY) \,|\, u^k \geqslant 0, k = 1, 2, \ldots, K\}$$

subject to

$$Y = (0, 1) \qquad\qquad (P^*)$$

Now consider the problem:

maximize z

subject to

$$z \leqslant DY + u^k(b - EY), \qquad k = 1, 2, \ldots, r$$
$$u^k \geqslant 0 \qquad\qquad (P^r)$$

where $1 \leqslant r \leqslant K$. The constraints associated with (P^r) are usually known as *Benders' cuts*.

When $r = K$, then (P^r) becomes the same as (P^*) and hence solves (P). But observe that if u^k, $k = 1, 2, \ldots, r$, are known, then (P^r) is a pure zero–one problem (except for the continuous variable z whose effect can be absorbed by slightly modifying the implicit enumeration algorithm). Thus the idea is that when (P^K) is solved as a binary program, it yields optimal Y in (P). By utilizing these optimal binary values in (PX) and solving it as a linear program, this should give optimal X in (P).

It is not really necessary to determine all the extreme points u^k prior to solving (P). Rather, these extreme points are generated as needed by using the following algorithm. Let $\underline{z}$ $(\bar{z})$ be a lower (upper) bound on the optimum objective value z^* of (P) and define z^r as the optimum objective value of (P^r).

Step 0 Let u^0 be any feasible solution of (D1), not necessarily an extreme point. Set $r = 1$ and go to step 1.

Step 1 Solve (P^r) in z and Y, that is,

maximize z

subject to

$$z \leqslant DY + u^k(b - EY), \qquad k = 0, 1, \ldots, r - 1$$
$$Y = (0, 1)$$

(Notice that the index on k starts from zero.) Let (z^r, Y^r) be the optimum. Then $\bar{z} = z^r$. Go to step 2.

Step 2 Solve (D1) in X given $Y = Y^r$ and let u^r be the optimal solution. Then

$$\underline{z} = DY^r + u^r(b - EY^r)$$

Go to step 3.

Step 3 If $\underline{z} = \bar{z}$, Y^r is optimum for (P); go to step 4. Otherwise, set $r = r + 1$ and go to step 1.

Step 4 By using $Y = Y^r$, solve (PX) and let x^r be the optimum solution. Then $X = X^r$, $Y = Y^r$, and $z^* = \underline{z} = \bar{z}$ gives the optimum solution to (P). Stop.

▶**EXAMPLE 3.7-2** Consider Example 3.7-1. Problems (PX) and (D1) are given as follows:

maximize

$$v = (-3, -3)(x_1, x_2)^T + (10 - y_2)$$

subject to

$$(-1, 1)(x_1, x_2)^T = 0.7 - (1, -0.4)(y_1, y_2)^T$$
$$x_1, x_2 \geqslant 0$$
$$y_1, y_2 = (0, 1) \qquad\qquad (PX)$$

minimize

$$w = u_1(0.7 - y_1 + 0.4y_2)$$

subject to

$$-3 \leqslant u_1 \leqslant 3$$
$$u_1 \text{ unrestricted in sign} \qquad\qquad \text{(D1)}$$

Let $u_1{}^0 = 1$. Then (P^1) becomes:

maximize z

subject to

$$z \leqslant (10 - y_2) + (0.7 - y_1 + 0.4y_2) = 10.7 - y_1 - 0.6y_2$$

This yields $y_1{}^1 = y_2{}^1 = 0$ and $z^1 = 10.7 = \bar{z}$.
Using $Y = (0, 0)$ in (D1), this gives

minimize

$$w = 0.7u_1$$

subject to

$$-3 \leqslant u_1 \leqslant 3$$
$$u_1 \text{ unrestricted in sign}$$

This gives the solution $u_1^1 = -3$ and $\underline{z} = 10 + (-2.1) = 7.9$. Since $\underline{z} < \bar{z}$, the search continues.
Now (P^2) is given by:

maximize z

subject to

$$z \leqslant 10.7 - y_1 - 0.6y_2$$
$$z \leqslant (10 - y_2) + (-3)(0.7 - y_1 + 0.4y_2) = 7.9 + 3y_1 - 2.2y_2$$
$$y_1, y_2 = (0, 1)$$

This yields $y_1{}^2 = 1$, $y_2{}^2 = 0$, and $z^2 = 9.7 = \bar{z}$.
Again, substituting in (D1), one gets

minimize

$$w = -0.3u_1$$

subject to

$$-3 \leqslant u_1 \leqslant 3$$
$$u_1 \text{ is unrestricted in sign}$$

This gives the optimal solution $u_1 = 3$ and $\underline{z} = 10 + (-0.9) = 9.1$. Since $\underline{z} < \bar{z}$, (P^3) is solved. This is given by

maximize z

subject to

$$z \leqslant 10.7 - y_1 - 0.6y_2$$
$$z \leqslant 7.9 + 3y_1 - 2.2y_2$$
$$z \leqslant (10 - y_2) + 3(0.7 - y_1 + 0.4y_2) = 12.1 - 3y_1 + 0.2y_2$$
$$(y_1, y_2) = 0$$

This yields $y_1{}^3 = 1$, $y_2{}^3 = 0$, and $z^3 = 9.1 = \bar{z}$. Since $\underline{z} = \bar{z}$, $Y = (1, 0)$ is optimal.

Solving (PX) in (x_1, x_2) after substituting $Y = (1, 0)$, the optimal solution yields $x_1 = 0.3$ and $x_2 = 0$. This is the same optimal solution obtained in Example 2.7-1. ◀

The reader may wonder why Benders' algorithm is classified as an implicit enumeration method. This may not be apparently obvious, but in the next section it will be shown that the same ideas can be used directly with implicit enumeration.

Perhaps it will help also to compare Driebeek's and Benders' general ideas to show how Benders' method indirectly enumerates all binary solution points as in Driebeek's.

In both methods, the vector Y is fixed and the resulting linear program in X is manipulated in search for the optimum. The basic difference stems from the fact that Driebeek's search is guided by a direct enumeration of all binary assignments, while Benders' produces an equivalent result indirectly by enumerating the extreme point of (D1). When all such extreme points are enumerated (implicitly or explicitly), then equivalently all the binary assignments also have been enumerated. Moreover, the concept of penalties (or estimated upper bounds) in Driebeek's algorithm is replaced in Benders' method by the use of Benders' cuts in (P^r).

Indeed, the above relationship suggests that it may be possible to improve Driebeek's algorithm by utilizing Benders' cuts. For example, once a feasible Y is found, this can be used to generate a Benders' cut, which is then augmented to the original problem. By resolving the augmented problem as a linear program, one can then revise the penalty table. This is the type of feedback information that Driebeek's algorithm lacks. Another suggestion is to use Benders' algorithm to generate a first feasible Y. This is specially appealing since Driebeek's procedure may not, in general, produce an initial feasible solution quickly.

3.7.3   The Modified Partitioning Algorithm†

As in Driebeek's algorithm, the idea of Lemke and Spielberg is to control the search for the optimum by enumerating (implicitly or explicitly) all the binary combinations of the Y vector. The enumeration is done by using the concept of partial solutions. As usual, the objective coefficients of the binary variables can be conditioned so that all the free variables associated with a partial solution are at zero level. Thus, each partial solution implicitly specifies a complete binary assignment to all binary variables.

The interesting feature of the Lemke–Spielberg development is the use of the information in the partial solution to develop an associated Benders' cut whose variables are all binary. These cuts may then be used to fathom partial solutions by employing the appropriate tests in Section 3.5. These ideas are developed in detail below.

Let z be a lower bound on the optimum objective value of the mixed zero–one problem. Then from Benders' cut, one gets

$$z \leqslant DY + u^k(b - EY), \qquad \text{for all}\quad k$$

If a numerical value of z is available, then for a given u^k the above constraint is completely binary.

A numerical value of z is computed as follows. Let $\overline{Y}$ be a binary assignment determined by the current partial solution and the tentative assignment that all the free variables are at zero level. From problem (P*) in Section 3.7.2, it is evident that a corresponding z is determined by

$$z = D\overline{Y} + \min\{u(b - E\overline{Y}) \,|\, uA \geqslant C, u \geqslant 0\}$$

Under the assumption that $\{u \,|\, uA \geqslant C, u \geqslant 0\}$ is (or can be made) a bounded set, the optimum value of u determined from this problem together with z can now be substituted in Benders' cut to produce a binary constraint. Thus, every time a new partial solution is generated, a corresponding binary constraint can be constructed. Naturally, the value of z is updated each time an improved lower bound is determined. In each updating, all previously generated constraints are revised in terms of the current best z. The collection of these binary constraints may then be used to fathom partial solutions. This process is continued until the enumeration is complete.

The Lemke–Spielberg constraint presented here may be regarded as a surrogate constraint for the mixed zero–one problem. The interesting feature about this surrogate constraint is that it is purely binary so that the fathoming tests developed previously for the zero–one problem (Section 3.5) can be applied to it directly.

† Lemke and Spielberg (1967).

▶EXAMPLE 3.7-3   Consider Example 3.7-1. The binary variables are y_1 and y_2. Let $J_0 = \varnothing$ be the initial partial solution. Thus y_1 and y_2 are free at zero level. The value of $\underline{z}$ associated with J_0 is determined as

$$\begin{aligned} \underline{z} &= 10 + \min\{0.7u_1 \mid -3 \leqslant u \leqslant 3\} \\ &= 10 + 0.7(-3) \\ &= 7.9 \end{aligned}$$

with optimum $u_1 = -3$.

The Benders' cut associated with $\underline{z} = 7.9$ and $u_1 = -3$ is

$$10 - y_2 + (-3)(0.7 - y_1 + 0.4y_2) \geqslant 7.9$$

or

$$3y_1 - 2.2y_2 \geqslant 0$$

This constraint indicates that y_1 must be at level one in any feasible solution. Thus take $J_1 = \{+\underline{1}\}$, which commemorates that the partial solution $\{-1\}$ has been fathomed.

Computing the new $\underline{z}$ associated with J_1, this gives

$$\begin{aligned} \underline{z} &= 10 + \min\{(0.7 - 1)u_1 \mid -3 \leqslant u_1 \leqslant 3\} \\ &= 10 + (-0.3)(3) \\ &= 9.1 \end{aligned}$$

with optimum $u_1 = 3$.

The resulting Benders' cut is

$$10 - y_2 + (3)(0.7 - y_1 + 0.4y_2) \geqslant 9.1$$

or

$$-3y_1 + 0.2y_2 \geqslant -3$$

Since $\underline{z}$ given J_1 is better than $\underline{z}$ given J_0, the Benders' cut associated with J_0 is revised to make use of this information, giving

$$3y_1 - 2.2y_2 \geqslant 1.2$$

The available binary constraints at this point are

$$3y_1 - 2.2y_2 \geqslant 1.2, \qquad -3y_1 + 0.2y_2 \geqslant -3$$

The first constraint shows that y_2 must be fixed at zero in any feasible solution. This gives $J_2 = \{+\underline{1}, -\underline{2}\}$. Since J_2 is a complete binary assignment, it is fathomed. Because all the elements of fathomed J_2 are underlined, the enumeration is complete. Thus, optimal $y_1 = 1$ and optimal $y_2 = 0$. As in Benders' partitioning algorithm, the optimal binary solution is used in (PX) to find the optimal values of x_1 and x_2. ◀

3.8 Concluding Remarks

The main conclusion in this chapter is that the concept of implicit enumeration is general in the sense that it is applicable to *any* well-defined binary program. The conclusion follows from the fact that implicit enumeration is, in effect, equivalent to considering *all* possible solution points of the problem. Although the concept of implicit enumeration has such general applicability, the practicality of using it to solve problems lies primarily in the computational efficiency. As indicated in the chapter, the efficiency of the algorithms is a function of the fathoming tests, which are designed to exclude automatically as many nonpromising solution points as possible.

One of the important characteristics of implicit enumeration algorithms is that the problem of machine round-off errors is practically nonexistent, in contrast with the other methods in which this problem represents a severe difficulty.

Problems

3-1 *Zero–one problem* (Taha, 1971b). Show that a minimization zero–one problem, mixed or pure, can be converted to a continuous problem in which the objective function is concave and the solution space is a convex polyhedron. (*Hint*: If x_j is fractional, that is, $0 < x_j < 1$, a very high penalty $p_j = M(1 - x_j)$, $M > 0$ and very large, is incurred.)

3-2 Solve the following problem by Balas' algorithm:

minimize

$$z = -5y_1 + 7y_2 + 10y_3 - 3y_4 + y_5$$

subject to

$$-y_1 - 3y_2 + 5y_3 - y_4 - 4y_5 \geqslant 0$$
$$-2y_1 - 6y_2 + 3y_3 - 2y_4 - 2y_5 \leqslant -4$$
$$-y_2 + 2y_3 + y_4 - y_5 \geqslant 2$$
$$y_1, \ldots, y_5 = (0, 1)$$

3-3 The multiple choice constraint

$$\sum_{j=1}^{n} x_j = 1, \qquad x_j = (0, 1) \quad \text{for all} \quad j$$

is used in many zero–one problems. Show how such constraints can be included in the Balas algorithm *implicitly*; that is, develop a fathoming test so that these constraints may not be included in the main constraints as inequalities.

3-4   Generalize the development in Problem 3-3 to constraints of the form:

$$\sum_{j=1}^{n} a_j x_j = b, \qquad x_j = (0, 1) \quad \text{for all} \quad j$$

where $a_1, \ldots, a_n$, and b are nonnegative. What difficulties may arise if some of the coefficients are negative?

3-5   Using the idea of Problem 3-3, develop an implicit enumeration scheme for the assignment problem defined by

minimize

$$z = \sum_{i=1}^{n} \sum_{j=1}^{n} c_{ij} x_{ij}$$

subject to

$$\sum_{i=1}^{n} x_{ij} = 1, \quad j = 1, 2, \ldots, n$$

$$\sum_{j=1}^{n} x_{ij} = 1, \qquad i = 1, 2, \ldots, n$$

$$x_{ij} = (0, 1)$$

Comment on the computational efficiency of the method.

3-6   Apply Geoffrion's surrogate constraint method to Problem 3-2 and compare the computations.

3-7   *Surrogate constraint*   (Geoffrion, 1969).   Prove that the linear program defining Geoffrion's surrogate constraint (Section 3.5.3-B) is actually the dual of the continuous version of the original zero–one problem in which some of the variables are fixed according to the current partial solution J_t.

3-8   Use the results of Problem 3-7 to show that a *lower* bound on the optimum objective value of the zero–one problem is equal to $z_{\min} - w_{\min}^t$, where $z_{\min}$ and $w_{\min}^t$ are as defined in Section 3.5.3-B. How does this result sharpen Balas' Test 2 (Section 3.5.1)?

3-9   *Surrogate constraints for mixed zero–one case*   (Geoffrion, 1969). Given the mixed 0–1 problem

minimize

$$z = cx + dy$$

subject to

$$Ax + Dy = b > 0$$
$$0 \leqslant x_j \leqslant 1 \qquad \text{and integer for all} \quad j$$
$$y \geqslant 0$$

where d and y are n_1-vectors and D is an $m \times n_1$ matrix. Define a partial solution in terms of a subset of x_j. However, a "completion" should now involve a specification of y in addition to the x_j variables. Define the surrogate constraint relative to the partial solution and show how a strongest surrogate constraint may be developed. Show how the results can be implemented to solve a mixed zero–one problem. (*Hint*: cf. the Lemke–Spielberg constraint, Section 3.7.3, and its relationship to Benders' cut, Section 3.7.2.)

3-10 *Surrogate constraint* (Balas, 1967b). Consider the problem: For $\mu_i \geqslant 0$, $i = 1, 2, \ldots, m$,

$$z(\mu) = \max \sum_{j \notin J_t} c_j x_j$$

subject to

$$x_j \in A_t = \left\{ x_k \middle| \sum_{i=1}^{m} \mu_i \left(\sum_{k \notin J_t} a_{ik} x_k - S_i^t \right) \leqslant 0, 0 \leqslant x_k \leqslant 1, \quad k \notin J_t \right\}$$

All the other symbols are as defined in Section 3.5.3-B. The surrogate constraint

$$\sum_{i=1}^{m} \mu_i \left(\sum_{j \notin J_t} a_{ij} x_j - S_i^t \right) \leqslant 0, \qquad \mu_i \geqslant 0 \quad \text{for all} \quad i$$

is said to be stronger when $\mu = \mu^1$ than when $\mu = \mu^2$ if $\mu = \mu^1$ leads to a more restrictive solution space; that is, $z(\mu^1) < z(\mu^2)$. In view of this definition, a strongest surrogate constraint is associated with $\mu = \mu^0$ such that

$$z(\mu^0) = \min_{\mu \geqslant 0} z(\mu) = \min_{\mu \geqslant 0} \left\{ \max \sum_{j \notin J_t} c_j x_j \middle| x_j \in A_t \right\}$$

Prove that μ^0 is equal to the optimal *dual* values associated with the first m constraints of the following linear program:

maximize

$$z_t' = \left\{ \sum_{j \notin J_t} c_j x_j \middle| \sum_{j \notin J_t} a_{ij} x_j \leqslant S_i^t, i = 1, \ldots, m, 0 \leqslant x_j \leqslant 1, j \notin J_t \right\}$$

Notice that this is the continuous version of the 0–1 problem given the partial solution J_t.

3-11   Solve Problem 3-2 by using Balas' surrogate constraint. Compare the efficiencies of Balas' and Geoffrion's methods. (Notice that Geoffrion's constraint is closely related to Balas' with the exception that Geoffrion appends the objective row to the surrogate constraint.)

3-12   In the polynomial zero–one algorithm (Section 3.5.3), show how the following information may be used to sharpen the fathoming tests: Given $x_p = \prod_{j \in K_p} y_j$ and $x_r = \prod_{j \in K_r} y_j$, if $K_p \subseteq K_r$, then $x_r \leqslant x_p$, where all the y_j are binary variables.

3-13   *Bivalent hyperbolic programming* (Taha, 1971d)   Consider the problem,

minimize

$$z = \frac{c_0 + c_1 x_1 + \cdots + c_n x_n}{d_0 + d_1 x_1 + \cdots + d_n x_n}$$

subject to

$$\sum_{j=1}^{n} a_{ij} x_j \leqslant b_i, \qquad i = 1, 2, \ldots, m$$

$$x_j = (0, 1), \qquad j = 1, 2, \ldots, n$$

The coefficients a, b, c, and d are unrestricted in sign.

(i) Develop an implicit enumeration algorithm for the problem stating all the assumptions.

(ii) Derive a surrogate constraint similar to Geoffrion's and state the conditions under which the constraint is valid. [*Note*: Related developments may be found in the work of Hammer and Rudeanu (1968), Robillard (1971), and Florian and Robillard (1971).]

3-14   Find the minimum of the following unconstrained quadratic function:

$$z = 75x_1 + 100x_2 + 3x_4 + 4x_5 - 100x_1 x_2 - 200x_1 x_3$$
$$- 400x_2 x_3 - 4x_4 x_5 - 8x_4 x_6 - 16x_5 x_6$$

where $x_j = (0, 1)$, $j = 1, 2, \ldots, 6$.

3-15   Suppose the following linear constraints are added to the objective function in Problem 3-14:

$$x_1 + 2x_2 + 4x_3 + x_4 + 2x_5 + 4x_6 \geqslant 2$$
$$x_1 + 2x_2 + 4x_3 - x_4 - 2x_5 - 4x_6 \geqslant -2$$
$$-x_1 - 2x_2 - 4x_3 - x_4 - 6x_5 + 12x_6 \geqslant -6$$

Solve the problem by finding all the feasible solutions satisfying the constraints. (*Note*: This problem is called a *quadratic binary problem* and has several applications especially in capital budgeting. An algorithm specially tailored for the problem, which also includes the objective function in the fathoming tests, is outlined in the next problem.)

3-16 *Quadratic binary algorithm* (Laughhunn, 1970). Consider the problem:

minimize
$$z = \sum_{i,j} x_i \sigma_{ij} x_j$$
subject to
$$\sum_j a_{ij} x_j - S_i = b_i, \qquad i = 1, 2, \ldots, m$$
$$x_j = (0, 1), \qquad j = 1, 2, \ldots, n$$
$$S_i \geqslant 0, \qquad i = 1, 2, \ldots, m$$

where $[\sigma_{ij}]$ is an $m \times n$ positive semidefinite matrix; that is, $\sum_{i,j} x_i \sigma_{ij} x_j \geqslant 0$. Let J_t, z^t, and $z_{\min}$ be defined in the same manner given in the text. Prove that the following exclusion tests are valid:

(i) If for all $j \in N - J_t$,

$$g_j = \sigma_{jj} + d_j + 2 \sum_{\substack{i \in N - J_t \\ i \neq j}} \min(\sigma_{ij}, 0)$$

where $d_j = 2 \sum_{i \in J_t} \sigma_{ij}$, then g_j is a lower bound on the possible increment added to z^t as a result of transforming x_j from a free variable to a fixed variable.

(ii) If $d_j \geqslant 0$ for all $j \in N - J_t$, then the best completion of J_t (in terms of the objective value) is given by $x_j = 0$ for all $j \in N - J_t$.

(iii) None of the completions of J_t can yield a better solution if

$$\sum_{i,j \in J_t} x_i \sigma_{ij} x_j + \sum_{j \in N - J_t} \min(d_j, 0) \geqslant z_{\min}$$

(iv) If J_t is feasible with $z^t < z_{\min}$, and if

$$T_t = \{j \in N - J_t | z^t + g_j < z_{\min}, g_j < 0 \text{ or } a_{ij} > 0 \text{ for some } S_i^t < 0\} = \varnothing$$

then no feasible completion of J_t exists with $z < z_{\min}$:

(v) If $T_t \neq \varnothing$ and $\sum_{j \in H_t} \max(0, a_{ij}) + S_t^i < 0$ for some $S_t^i < 0$, where

$$H_t = T_t \{j \in N - J_t | z^t + g_j \geqslant z_{\min}\}$$

then no feasible completion exists with $z < z_{\min}$.

3-17 Solve the following problem by the Lawler–Bell algorithm:

minimize
$$z = x_1 x_7 + 3x_2 x_6 + x_3 x_5 + 7x_4$$

subject to

$$x_4 + x_5 + 6x_6 \geqslant 8$$
$$3x_1x_3 + 6x_4 + 4x_5 \leqslant 20$$
$$4x_1 + 2x_3 + x_6x_7 \leqslant 15$$
$$x_1x_6 + x_2 + 3x_5 \geqslant 7$$
$$x_j = (0, 1) \quad \text{for all} \quad j$$

3-18   Using the notation of the Lawler–Bell algorithm in Section 3.6, show that any linear constraint satisfies

$$g_{i1}(x^* - 1) - g_{i2}(x) = g_{i1}(x') - g_{i2}(x')$$

for some vector x' in the interval $[x, x^* - 1]$.

3-19   *Generalized origin in implicit enumeration* (Salkin, 1970). In Balas' zero–one algorithm, the enumeration always starts at the origin, that is, all variables are assumed to be at zero level. This seems natural since all c_j are nonnegative. But such a starting point may not be "close enough" to the optimum solution, with the result that the enumeration may be inefficient. An alternative idea is to locate the starting point as close as possible to the continuous optimum. In this case, the new origin is implemented by substituting $x_j = 1 - x_j'$ if $x_j = 1$ in the starting point and $x_j = 0$ otherwise. The same idea can be implemented every time a feasible solution is encountered so that the origin of the problem is constantly changed. Give the details of how such an "improved" starting point may be determined and assess its merits, if any.

3-20   Solve the following problem as a mixed zero–one algorithm by using Driebeek's algorithm (Section 3.7.1):

minimize
$$z = 3x_1 + 2x_2 + 8y_1 + 2y_2$$
subject to
$$2x_1 - x_2 + 3y_1 - y_2 \geqslant 2$$
$$x_1 + x_2 + 3y_1 + y_2 \geqslant 3$$
$$0 \leqslant y_1 \leqslant 3$$
$$0 \leqslant y_2 \leqslant 3$$
$$x_1, x_2 \geqslant 0$$
$$y_1, y_2 \quad \text{integers}$$

3-21   Solve Problem 3-20 by using Benders' partitioning algorithm (Section 3.7.2).

3-22   Solve Problem 3-20 by using the modified partitioning algorithm (Section 3.7.3).

CHAPTER 4

Branch-and-Bound Methods

4.1 The Concept of Branch-and-Bound

The solution space of a general integer programming problem can be assumed bounded. The number of integer points that should be investigated is thus finite. The simplest (and, perhaps, naive) way to solve the integer problem is to enumerate all such points, discarding infeasible ones and always keeping track of the feasible solution with the best objective value. When the enumeration is complete, the optimum solution, when one exists, is associated with the best objective value.

If the above idea is to be attractive computationally, it must be refined so that some (hopefully most) of the nonpromising integer points are discarded without being tested. This is the prime purpose of the branch-and-bound methods.

From the presentation in Chapter 3, it is evident that the branch-and-bound method is based on the same broad principle of the zero–one implicit enumeration. In fact, as will be explained further in Section 4.2, implicit enumeration is essentially a branch-and-bound method. The main difference is that in the general integer problem, each variable may assume more than two values, and hence, the tests for excluding nonpromising points become more sophisticated, more complex, and perhaps less effective.

The branch-and-bound procedure does not deal directly with the integer problem. Rather, it considers a continuous problem defined by relaxing the integer restrictions on the variables. Thus the solution space of the integer problem is only a *subset* of the continuous space. The prime reason for dealing with the continuous problem is that it is simpler to manipulate especially when it is a linear program.

If the optimal continuous solution is all integer, then it is also optimum for the integer problem. Otherwise, the branch-and-bound technique is applied by implementing two basic operations:

(i) *Branching* This partitions the continuous solution space into sub-spaces (subproblems), which are also continuous. The purpose of partitioning is to eliminate parts of the continuous space that are not feasible for the integer problem. This is achieved by imposing (mutually exclusive) constraints that are *necessary* conditions for producing integer solutions, but in a way that no feasible integer point is eliminated. In other words, the resulting collection of subproblems completely defines every feasible integer point of the original problem. Because of the nature of the partitioning operation, it has been given the name *branching*.

(ii) *Bounding* Assuming the original problem is of the maximization type, the optimal objective value for each subproblem created by branching obviously sets an upper bound on the objective value associated with any of its *integer feasible* values. This bound is essential for "ranking" the optimum solutions of the subsets, and hence in locating the optimum integer solution. This operation obviates the reason for the name *bounding*.

Each of the created subproblems may now be solved as a continuous problem. When the solution of a subproblem is integer, the subproblem is not branched (partitioned); otherwise, further branching is necessary. The optimum integer solution is available when the subproblem having the *largest* upper bound among *all* subproblems yields an integer solution. (Notice that a maximization problem is assumed.)

An important question immediately poses itself: How does this procedure help in enumerating feasible points *implicitly*? If the branching and bounding operations are implemented "intelligently," then "many" subproblems containing one (or more) feasible integer points can be discarded automatically. This is equivalent to implicit enumeration. For example, after a branching is effected, if the next subproblem to be examined (for possible branching) is always selected as the one having the *largest* upper bound among all unexamined subproblems, then the *first* subproblem to yield a feasible integer solution is optimum and all the remaining subproblems are discarded. A typical tree representation of this porcedure is shown in Fig. 4-1. The

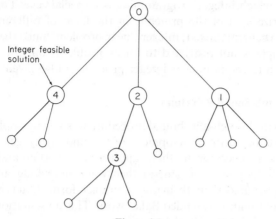

Figure 4-1

nodes represent the subproblems and their numbers represent the order of examination (according to the upper bound conditon). Node 4 yields the first feasible solution; hence it is the optimum. All the remaining (unnumbered) nodes are thus discarded.

Unfortunately, there is no simple way of representing the tree in Fig. 4-1 for the purpose of computer coding (cf. Glover's enumeration scheme, Section 3.4, where the entire tree is represented by a single vector). The reasons are: (i) a very large number of branches and nodes may be created; and (ii) each node will have an associated continuous subproblem that must be stored for later scanning. This means that the complexity of the computer code is dependent not only on the very large number of branches that may be created, but also on the need to economize the use of the computer storage requirements as much as possible.

The reader may wonder about the logic of the branch-and-bound procedure. First, the integer problem is replaced by a continuous space by relaxing the integer conditions. Then branching is used to eliminate parts of the continuous space that are not feasible with respect to the integer problem by reactivating (some of) the integer restrictions. This appears as a "built-in" redundancy. In principle, this may be true; but the fact of the matter is that the present state of the art in discrete mathematics does not provide sufficient information to solve an integer problem directly. By using the above "trick," one is assured that every subproblem is continuous and can thus draw upon the tremendous advances in the theory of continuous mathematics to solve these problems.

In the next section, a mathematical formulation of the branch-and-bound principle is presented. This principle is applicable to a wide class of lems, of which integer programming is a special case. This is followed by particularizations of the principle in the form of different algorithms applied to the general (mixed) integer linear problem. Since the branch-and-bound principle is not restricted to linear problems only, a discussion of its application to the nonlinear integer problem is also given.

4.2   Branch-and-Bound Principle†

Although the branch-and-bound procedure was validly applied to several particular situations, the development of the underlying general principle came only as a successor to these applications. Bertier and Roy (1965) were the first to present a general theory for branching and bounding. Balas (1968) restated their theory in a simpler form. Later Mitten (1970) generalized and slightly extended Balas' work. This presentation draws upon the ideas of both Balas and Mitten.

A typical branch-and-bound problem is described as follows. Let S be an arbitrary set. Associated with each element in S is a specified real-valued (return or objective) function

$$z: \quad S \to R$$

which orders all the elements $s_i \in S$. The problem is to determine the element $s^* \in S$ that is best in terms of the return function z. Assuming a maximization problem, then s^* corresponds to

$$z(s^*) = \max\{z(s_i) \,|\, s_i \in S\}$$

The number of elements may be too large to allow (trivial) exhaustive enumeration. This is where the branch-and-bound principle proves useful.

The solution of the above problem involves fulfilling three basic conditions:

    (i)   There exists an arbitrary superset T of S, that is, $T \supset S$ with a real-valued (return) function

$$w: \quad T \to R$$

        which extends z such that

$$t_j \in S \Rightarrow w(t_j) = z(t_j)$$

        where t_j is an element of T. This means that when t_j is a common element in S and T, the values of the two return functions must be the same. The selection of the set T and its function w must render (relatively) easier computations than those associated with the direct use of S and Z. Clearly, a good bit of art is needed here to select "the" most efficient T and w.

---

† Balas (1968), Mitten (1970).

(ii) *Branching*   A branching rule B can be defined to generate a family of subsets $\{T_i^k\}$ from the subset $T^k \subseteq T$, that is,

$$B(T^k) = \{T_1^k, \ldots, T_q^k\}$$

where it is assumed that $|T^k| \geqslant 2$, and provided that

$$\bigcup_{i=1}^{q} T_i^k = T^k - \{t_{j_k}\}$$

The element t_{j_k} is defined such that

$$w^k = w(t_{j_k}) = \max\{w(t_j)\,|\,t_j \in T^k\}$$

In general, $\bigcap_{i=1}^{q} T_i^k \neq \varnothing$ is a permissible situation, that is, the T_i^k's need not be mutually exclusive. However, if they are, then branching is equivalent to *partitioning* T^k. This sharpens the procedure.

The basic design of the branching rule B depends mainly on the selection of T and w; but in general different rules can be developed for given T and w. "Clever" devices can also be developed that will enhance the effectiveness of branching. This point is illustrated in Section 4.3.

(iii) *Bounding* T, w, and B are selected such that an *upper* bound on the objective value of every $t_j \in T^k$ can be (easily) determined for each $T^k \subset T$. This upper bound is given by w^k as defined in (ii).

These three conditions provide the basic ingredients for developing a branch-and-bound algorithm. The efficiency of computation can be enhanced further, however, by adding the concept of a *lower* bound on the objective value of every $s_j \in S$. Define a subset T^k (generated by B) as *active* if it has not yet been examined, and let $A \subset T$ be the family of all active T^k. A lower bound $\underline{z}$ on every $s_j \in S$ is determined as the value of w associated with any $t_j \in S$. This may be obtained by inspection or by some other sophisticated methods. The lower bound can thus be used to discard all active subsets T^k whose optimal objective value (upper bound) w^k does not exceed $\underline{z}$. In other words, given

$$\overline{w} = \max\{w^k\,|\,T^k \text{ active}\}$$

then a subset T^k is stored only if

$$\underline{z} < w^k \leqslant \overline{w}$$

One of the advantages of the lower bound is that any T^k having $w^k \leqslant \underline{z}$ need not be stored for later consideration. This should result in reducing computer storage requirements, which is an important factor in almost all branch-and-bound methods.

The use of the lower bound (together with the upper bound) allows flexibility in implementing the branching operation. For example, it will not be necessary always to examine the subset associated with the largest w^k among all active subsets. Rather, any such subset can be examined. The advantage here is that one may be able to devise rules to select an active subset that provides a good lower bound quickly. Naturally, the value of $\underline{z}$ is updated every time a better (higher) value is encountered. One must note, however, that the branch-and-bound procedure will terminate in this case only if every active subset T^k having w^k such that $\underline{z} < w^k \leqslant \overline{w}$ is examined or discarded. At this point, the optimum solution given S is associated with $\underline{z}$.

The idea of the *lower bound* is illustrated below by the additive algorithm of Section 3.5.1. This will also serve to show that 0–1 implicit enumeration algorithm can be classified as a branch-and-bound method.

In the additive algorithm, the set S defines the *feasible* binary points, while the set T defines all 2^n (feasible and infeasible) binary points (hence $T \supset S$). The objective function w associated with T is taken the same as z associated with S. Each unfathomed partial solution represents an active subset, and the branching is exercised when a new variable is augmented. This creates two new subsets associated with the zero and one values of the augmented variable.

Because the 0–1 problem of the additive algorithm is of the *minimization* type, the upper (lower) bound as defined above for the maximization problem should now be replaced by an equivalent lower (upper) bound. The upper bound (given by $z_{\min}$ in the additive algorithm) is determined as soon as a feasible solution is encountered. This bound is used to fathom partial solutions with larger objective values. At the termination, the optimum, when it exists, is associated with $z_{\min}$. Under the assumption that all $c_j \geqslant 0$, the lower bound of the additive algorithm is defined by z^t. This is used in Balas' Test 2 for possible fathoming of the partial solution J_t.

The preceding illustration shows that the selection of w, T, and B is very important in developing the details of the branch-and-bound algorithm. This point is further illustrated in the next section by other examples.

The algorithms in the next section are designed for the *mixed* (linear) integer problem. However, they extend trivially to the *pure* integer case.

4.3   General (Mixed) Integer Linear Problem

The first known branch-and-bound algorithm was developed in 1960 by Land and Doig as an application to the mixed and pure integer problem. This algorithm will be presented in Section 4.3.1. Several improvements were introduced to enhance the computational efficiency of the algorithm.

They mainly concentrate on the design of less costly rules for branching and bounding. This generally leads to savings in computer time and storage requirements. Some of these improvements are presented in Section 4.3.2. A comparison between the efficiencies of these methods is also given.

4.3.1   The Land–Doig Algorithm†

The Land–Doig algorithm follows the steps of the general branch-and-bound algorithm of Section 4.2. The mixed integer linear problem is written as:

maximize

$$z = \sum_{j \in N} c_j x_j$$

subject to

$$\sum_{j \in N} a_{ij} x_j \leqslant b_i, \qquad i \in M$$

$$x_j \geqslant 0, \qquad j \in N$$

$$x_j \quad \text{integer}, \quad j \in I \subseteq N$$

Using the notation of Section 4.2, the set S can be defined for the above problem to include all its feasible solutions, while the set T is defined to include the continuous feasible space that is obtained by relaxing the integrality conditions. The return function w, extending z, will take the same form of z except that it is now defined on the continuous range of all x_j, $j \in N$. In other words, the resulting problem forms a regular linear program. It is shown below that this selection of T and w will prove convenient for performing the branching and bounding steps.

The Land–Doig algorithm starts by solving the associated linear program. Clearly, the resulting optimum solution provides an upper bound on all the elements in S. Assuming that the optimum solution of the linear problem is infeasible with respect to S, the continuous problem must now be branched. It is here now that some explanation is needed to show how proper branching can be effected given T and w.

The idea is to select any variable x_k, $k \in I$, that is noninteger in the continuous optimum solution. Assume that its associated value is given by $x_k = x_k^*$. Since $k \in I$, the *least* decrease in the value of w resulting from forcing x_k to be integer must be associated with $[x_k^*]$ or $[x_k^*] + 1$, where $[x_k^*]$ is the largest integer value not exceeding x_k^*. In other words, by branching to the subsets of T associated with $[x_k^*]$ and $[x_k^*] + 1$, one is

---

† Land and Doig (1960).

certain that the resulting bounds will remain proper with respect to the *optimum* element in S.

The above result can be justified as follows. It can be proved that the relationship between x_k and w is concave with its peak point occurring at x_k^*. One such relationship is shown in Fig. 4-2, where x_k^{max} and x_k^{min} represent the limits of the feasible values of x_k as specified by the continuous

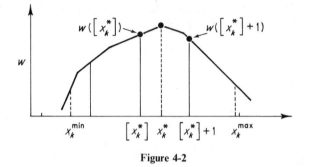

Figure 4-2

solution space (cf. Fig. 3-6, Section 3.7). Notice that, in general, $w([x_k^*])$ need not be larger than $w([x^*] + 1)$. Thus, in order for x_k to be integer, the value of the objective w must be reduced to at least

$$\max\{w([x_k^*]), w([x_k^*] + 1)\}.$$

The important observation, however, is that $w([x_k^*])$ acts as an upper bound for all integer $x_k < [x_k^*]$, while $w([x_k^*] + 1)$ acts as an upper bound for all integer $x_k > [x_k^*] + 1$. Consequently, it is sufficient at this point to branch to the two subsets of T that are defined by imposing the integer restriction $x_k = [x_k^*]$ and $x_k = [x_k^*] + 1$ on the continuous problem, since one is certain that no superior feasible points in S are bypassed. This result is true because T is a convex polyhedron.

The determination of $w([x_k^*])$ and $w([x_k^*] + 1)$ is achieved by solving the respective linear programming problems $\max\{w | T$ and $x_k = [x_k^*]\}$ and $\max\{w | T$ and $x_k = [x_k^*] + 1\}$. These provide the upper bounds on the two subsets.

In order to keep track of the generated branches, a tree as shown in Fig. 4-3 may be used. A node represents a given linear program, while the branches represent the constraints leading to the realization of the linear program associated with the node. Another way of interpreting the nodes is to consider them as the family of subsets generated by the branching rule. Thus, in Fig. 4-3, w^0 is the top node associated with the solution of the original linear program. The two nodes w_m^k and w_M^k gives the values of w corresponding to $x_k = [x_k^*]$ and $x_k = [x_k^*] + 1$, respectively.

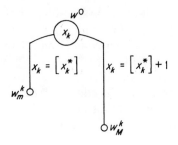

Figure 4-3

The active set now includes the two nodes $w_m{}^k$ and $w_M{}^k$. In the Land–Doig algorithm, the next node to be examined is always the one having the largest upper bound among all active nodes. Suppose $w_m{}^k > w_M{}^k$, then according to this rule the node $w_m{}^k$ is examined. If its optimum solution is in S, it is optimum. Otherwise, let x_r be an integer variable having a fractional value at this node; then x_r is used to effect branching. (As a rule of thumb, if more than one integer variable is fractional, the variable with the largest fraction is selected for branching.)

Applying the same branching rule used at w^0 to x_r, this yields the tree in Fig. 4-4, where

$$w_m{}^r (w_M{}^r) = \max\{w \mid T, x_k = [x_k{}^*] \text{ and } x_r = [x_r{}^*] \ (x_r = [x_r{}^*] + 1)\}$$

Although the branching at $w_m{}^k$ provides the proper upper bounding for all the subsets emanating from this node, it may not provide the proper upper bound on all the branches (subsets) emanating from the node x_k with $x_k = [x_k{}^*] - 1$. Remember that $w_m{}^k$ bounds all the branches emanating from

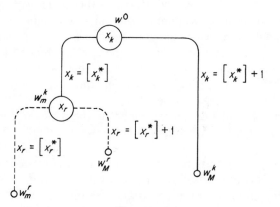

Figure 4-4

w^0, where $x_k < [x_k{}^*]$. But this says nothing about the relationship between $w(x_k)$ for $x_k \leqslant [x_k{}^*] - 1$, and $w_m{}^r$ and $w_M{}^r$ associated with $w_m{}^k$. Consequently, in order to secure the proper bounds on all the subsets, a single branch $x_k = [x_k{}^*] - 1$ must be added at w^0. The result of this is shown in Fig. 4-5, where $w_a{}^k$ is the optimum value of w given T and $x_k = [x_k{}^*] - 1$. The general rule here for adding the additional branch is as follows: Given that $x_j = v$ is the branch leading to the node chosen for examination ($v = [x_k{}^*]$ in the above case), then the two branches $x_j = v - 1$ and $x_j = v + 1$ must

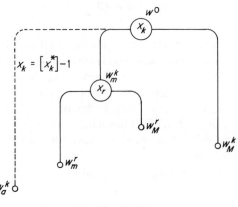

Figure 4-5

emanate from the node at the top end of the branch $x_j = v$ in order to ensure proper bounding on all active subsets. Since *one* of the two branches $x_j = v - 1$ and $x_j = v + 1$ is always in the tree, only the other one is added.

The active set A is now defined by the nodes $w_a{}^k$, $w_m{}^r$, $w_M{}^r$, and $w_M{}^k$. Assume $w_M{}^r = \max\{w|\text{active set } A\}$; then the node $w_M{}^r$ is examined. Suppose that the linear programming solution at $w_M{}^r$ is *not* in S (if it is in S, the process ends). Let x_t be a violated integer variable. Then using the above arguments, *three* branches are added to the tree, as shown by dotted lines in Fig. 4-6. The branch $x_r = [x_r{}^*] + 1$ is taken here as representing $x_r = v$. Since $x_r = [x_r{}^*]$ ($=v - 1$) is already in the tree, only the branch $x_r = [x_r{}^*] + 2$ ($=v + 1$) should be added.

The active set A in Fig. 4-6 is defined by the nodes $w_a{}^k$, $w_m{}^r$, $w_m{}^t$, $w_M{}^t$, $w_a{}^r$, and $w_M{}^k$. Again, let $w_M{}^k = \max\{w|\text{active set } A\}$; then $w_M{}^k$ is examined next.

The above procedure is repeated at $w_M{}^k$ and the process of branching and bounding is continued until either a feasible solution in S is obtained (which is the optimum) or every branch terminates with an infeasible solution. In the latter case no feasible solution exists for the problem.

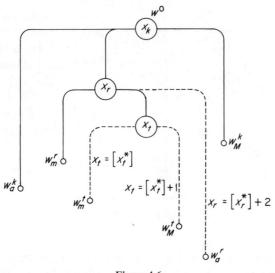

Figure 4-6

An important remark about the lower bound can be made here. If upon branching, the solution associated with a node yields a feasible point in S, then the corresponding value of w (w_b, say) can be used as a lower bound $\underline{z} = w_b$ on all future nodes. Thus, in future branching, any node yielding a value of w less than or equal to $\underline{z}$ can be discarded. This should prove useful in truncating the calculations advantageously.

▶**EXAMPLE 4.3-1**   Maximize

$$z = 3x_1 + x_2$$

subject to

$$17x_1 + 11x_2 \leqslant 86.5$$
$$x_1 + 2x_2 \leqslant 10.2$$
$$x_1 \qquad \leqslant 3.87$$
$$x_1, x_2 \qquad \text{nonnegative integers}$$

The tree in Fig. 4-7 gives the steps of the calculations, where each node is numbered according to its order of introduction in the tree. The solutions associated with the nodes are shown in the continuous solution space of Fig. 4-8.

The computations start with the continuous optimum node ⓪. Selecting $x_2 = 1.87$ to apply the integrality condition, the branch $x_2 = 2$ yields node ①, while $x_2 = 1$ yields node ②. According to the bounding rule, node ①

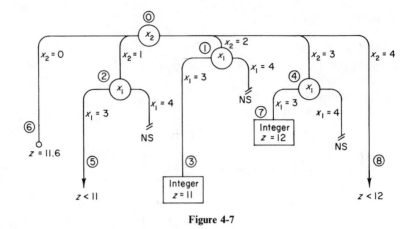

Figure 4-7

is branched with $x_1 = 3$ and $x_1 = 4$. But $x_1 = 4$ (together with $x_2 = 2$) is infeasible, while $x_1 = 3$ yields the integer solution (3, 2) with $z = 11$. This now acts as a *lower bound* on all the integer solutions. Before bounding is applied to the active nodes, a new branch $x_2 = 3$ must be added at node ⓪, which yields node ④. Node ② is now selected for branching, which adds branch $x_1 = 4$ with no feasible solution and branch $x_1 = 3$, which is discarded since it yields a value of z less than 11, the lower bound. Again $x_2 = 0$ is added at node ⓪ to yield node ⑥. Node ④ is now branched. This gives the infeasible branch $x_1 = 4$ and the integer solution (3, 3) with $z = 12$, which is the new lower bound. To ensure proper bounding, again $x_2 = 4$ is added at node ⓪, but this yields a value of $z < 12$. Node ⑦ is now eligible for branching. But since it is a feasible point, it is optimal.◀

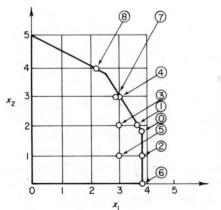

Figure 4-8 Continuous solution space. ⓪: (3.87, 1.87), $z = 13.48$; ①: (3.8, 2), $z = 13.4$; ②: (3.87, 1), $z = 12.61$; ③: (3, 2), $z = 11$ (lower bound); ④: (3.15, 3), $z = 12.54$; ⑤: (3, 1), $z = 10$; ⑥: (3.87, 0), $z = 11.61$; ⑦: (3, 3), $z = 12$ (optimal); ⑧: (2.2, 4), $z = 10.2$.

It is interesting to notice in the above example that if x_1 is selected at node ⓪ to force the integer condition, the solution tree becomes much more simplified, as shown in Fig. 4-9. The immediate advantages are: (1) the convergence of the algorithm is fast, and (2) the computer storage requirements are smaller. Both points are of immense importance in solving integer programs.

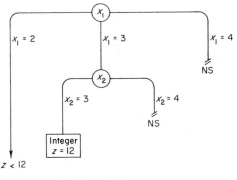

Figure 4-9

The above observation emphasizes the importance of selecting the branching variable at a given node. In the above developments, the rule is to choose the basic variable with the largest fraction. This is of no help in this example since both x_1 and x_2 have the same fractions. Later, in Section 4.3.3-B, a number of heuristics are presented that will usually select a variable uniquely.

Since the linear programming problems at the different nodes are generated by imposing a constraint of the form $x_j = v$, where v is an integer value, on a previously solved problem, some type of sensitivity (or parametric) analysis can be used to secure the new solutions with the least additional computations. Also, if it can be shown that an upper bound on the objective value associated with newly generated nodes is smaller than a lower bound on the integer solutions, then it would not be necessary to solve the associated linear programs completely. Opportunities like these should prove useful in enhancing the computational efficiency of the algorithm.

4.3.2   Improved Land–Doig Algorithm

The above Land–Doig algorithm has three basic drawbacks:

1.   It is possible that a large number of branches could originate from the same node. The problem here is that this number normally cannot be predicted in advance, which complicates the tree search and may also lead to a severe taxation of the computer memory.

2.   If the algorithm is taken literally, it will be necessary to solve a linear program for each branch in order to determine the proper *upper* bound. This appears to be a costly method, especially since some of the resulting nodes may never be branched.

3.   The determination of a *lower* bound comes only as a *byproduct* of solving the linear program at the different nodes. This means that in spite of its importance in truncating computations, there is no systematic method that is designed to secure a "tight" lower bound at an early stage of the procedure.

The above three points show that, from the computational standpoint, the Land–Doig rules for branching and bounding are not adequate. Section 4.3.1-A gives an improved branching rule, which creates exactly two branches from each eligible node. Section 4.3.1-B then shows how an *upper* bound can be estimated for each node without solving a complete linear program. Finally, Section 4.3.1-C shows how a *lower* bound can be sought at the early stages of the calculations. Interestingly, these improvements are developed based on the same definitions of T and w used with the Land–Doig algorithm. This shows that it is possible to design different branching and bounding rules for the same T and w.

A.   *Branching Rule*†

Suppose that x_k is an integer variable whose fractional value at a given node is x_k^*. Then a necessary condition for x_k to be integer is that

$$x_k \leq [x_k^*] \qquad \text{or} \qquad x_k \geq [x_k^*] + 1$$

This means that the range $[x_k^*] < x_k < [x_k^*] + 1$ is not feasible with respect to the integer problem. The idea then is that at each node only two branches are needed, namely, one corresponding to $x_k \leq [x_k^*]$ and the other corresponding to $x_k \geq [x_k^*] + 1$. Because these additional constraints are inequalities (cf. the Land–Doig algorithm), the entire range of possible integer points is covered by these two branches. As a result, no further branching from the same node would be necessary.

In solving the linear program associated with a node, the additional constraints defining the node need not be considered explicitly. Rather, the bounded variables techniques are used (see Section 2.6).

▶ EXAMPLE **4.3-2**   Consider Example 4.3-1. Although there is a distinct advantage in branching on x_1 at the first node, it is more instructive to branch on x_2 since it creates more nodes.

A summary of the computations is given in Table 4-1. The associated tree is shown in Fig. 4-10.

† Dakin (1965).

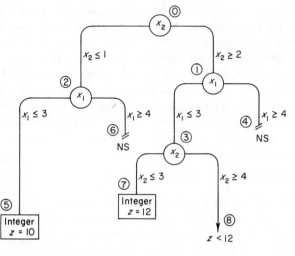

Figure 4-10

Table 4-1

Node	Additional constraints	Solution			Property of solution
		x_1	x_2	z	
0	—	3.87	1.87	13.48	Noninteger
1	$x_2 \geqslant 2$	3.8	2	13.4	Noninteger
2	$x_2 \leqslant 1$	3.87	1	12.6	Noninteger
3	$x_2 \geqslant 2$ $x_1 \leqslant 3$	3	3.23	12.23	Noninteger
4	$x_2 \geqslant 2$ $x_1 \geqslant 4$	—	—	—	No solution
5	$x_2 \leqslant 1$ $x_1 \leqslant 3$	3	1	10	Integer (*lower bound*)
6	$x_2 \leqslant 1$ $x_1 \geqslant 4$	—	—	—	No solution
7	$x_1 \leqslant 3$ $x_2 \leqslant 3$ $x_2 \geqslant 2$	3	3	12	Integer (*optimal*)
8	$x_1 \leqslant 3$ $x_2 \geqslant 4$	—	—	<12	Discard

The first lower bound is encountered at node ⑤, but it is too loose to be effective for this problem. Node ⑦ yields an improved lower bound which is used to discard node ⑧.

Although bounds were imposed on x_2 for the first time at ⓪, it is possible that x_2 can be restricted by new bounds again at a succeeding node as exemplified by node ③ in Fig. 4-10. A comparison between Fig. 4-7 and 4-10 shows that Dakin's modification can indeed lead to a much simpler solution tree and hence a less complex (and more efficient) computer coding. However, for the case where all the integer variables are binary, the two trees will be the same.◀

B. *Computation of Upper Bounds*†

In the Land–Doig algorithm and its modification by Dakin, most (if not all) of the linear programs at the different nodes are solved completely. This is costly from the computational standpoint, especially when the only information needed at a node may be only the optimum objective value.

Beale and Small (1965) overcame this problem by developing a procedure that estimates an upper bound on the exact optimal objective value at each node. The associated calculations are only a byproduct of the simplex method computations at the branched node and hence are very simple. The estimated upper bounds then replace the exact upper bounds in the Land–Doig algorithm in selecting the next node to be examined.

The Beale–Small method utilizes the concept of penalties conceived by Driebeek for the mixed zero–one problem (Section 3.7.1). Perhaps the main difference is that it is now applicable to the general integer problem, and, unlike Driebeek's method where the penalties are computed once and for all, the penalties are recomputed at each node.

Suppose the optimum linear program at a node is defined as follows:

maximize

$$z = \bar{c}_0 - \sum_{j \in NB} \bar{c}_j x_j$$

subject to

$$x_i = x_i^* - \sum_{j \in NB} \alpha_{ij} x_j, \qquad i \in M$$

$$x_i \geq 0, \qquad\qquad\qquad i \in M$$

$$x_j \geq 0, \qquad\qquad\qquad j \in NB$$

where $\bar{c}_j = z_j - c_j$ is the reduced cost associated with the nonbasic variable $x_j, j \in NB$ (see Section 2.3.3) and $x_i, i \in M$, is the set of basic variables. The

† Beale and Small (1965), Tomlin (1971).

current optimum solution is thus given by

$$z = \bar{c}_0$$
$$x_i = x_i^*, \qquad i \in M$$
$$x_j = 0, \qquad j \in NB$$

Suppose x_k, $k \in M$, is an *integer* (basic) variable whose value x_k^* is fractional. Hence

$$x_k^* = [x_k^*] + f_k, \qquad 0 < f_k < 1$$

The branching procedure in the Land–Doig algorithm as modified by Dakin is to impose the restrictions

$$x_k \leqslant [x_k^*] \qquad \text{or} \qquad x_k \geqslant [x_k^*] + 1$$

These constraints cannot improve the optimum objective value, and the objective now is to determine the penalties that estimate a *lower* bound on the exact decrease in the optimum objective values.

Figure 4-11, which is similar to Fig. 3-6 (Section 3.7.1), gives the variation of *optimum* z as a function of x_k, where $[x_k^*] \leqslant x_k \leqslant [x_k^*] + 1$.

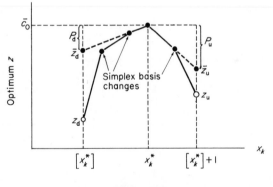

Figure 4-11

The solid line segments represent the exact variation with the breaking points signifying changes in the basis (of the simplex method). The points z_d and z_u are the exact optimum values of z given $x_k \leqslant [x_k^*]$ and $x_k \geqslant [x_k^*] + 1$, respectively.

Let P_u and P_d be the up and the down penalties that will be used to estimate the upper bound when x_k is "upped" to at least $[x_k^*] + 1$ or "downed" to at least $[x_k^*]$. The proposal of Beale and Small is to compute P_u and P_d under the assumption that the current basis (associated with x_k^*) is unchanged. This is indicated in Fig. 4-11 by the extended dotted lines. In this case, the required upper bounds on the true values z_u and z_d are given by

$$\bar{z}_u = \bar{c}_0 - P_u \geq z_u, \qquad \bar{z}_d = \bar{c}_0 - P_d \geq z_d$$

The values of P_u and P_d can be obtained directly from the simplex tableau as follows. The equation associated with integer x_k can be written as

$$x_k = x_k^* - \sum_{j \in NB} \alpha_{kj} x_j$$

As shown previously, in order for x_k to be integer, one of the two conditions $x_k \leq [x_k^*]$ and $x_k \geq [x_k^*] + 1$ must be satisfied. Let $J^+ (J^-) = \{j \in NB \mid \alpha_{kj} > 0 \ (< 0)\}$. Then $x_k \leq [x_k^*]$ yields

$$x_k - [x_k^*] = f_k - \sum_{j \in J^+} \alpha_{kj} x_j - \sum_{j \in J^-} \alpha_{kj} x_j \leq 0$$

or

$$- \sum_{j \in J^+} \alpha_{kj} x_j - \sum_{j \in J^-} \alpha_{kj} x_j \leq -f_k \qquad (1)$$

Now augmenting (1) to the current simplex tableau and applying the dual simplex method, then, under the assumption of no change in basis (see Fig. 4-11), P_d is given by

$$P_d = \min_{j \in J^+} \{\bar{c}_j f_k / \alpha_{kj}\}$$

To determine P_u, consider $x_k \geq [x_k^*] + 1$. This yields

$$x_k - [x_k^*] = f_k - \sum_{j \in J^+} \alpha_{kj} x_j - \sum_{j \in J^-} \alpha_{kj} x_j \geq 1$$

or

$$\sum_{j \in J^+} \alpha_{kj} x_j + \sum_{j \in J^-} \alpha_{kj} x_j \leq f_k - 1 \qquad (2)$$

Again, using the dual simplex calculations, it follows that

$$P_u = \min_{j \in J^-} \{\bar{c}_j (f_k - 1)/\alpha_{kj}\}.$$

In the expression for P_d and P_u, the quantities f_k/α_{kj} and $(f_k - 1)/\alpha_{kj}$ represent the increases in the value of a nonbasic variable that are required to accomodate the changes in x_k. Taken differently, the quantities

$$S_u = \min_{j \in J^-} \{\bar{c}_j / -\alpha_{kj}\} = P_u/(1 - f_k)$$

$$S_d = \min_{j \in J^+} \{\bar{c}_j / \alpha_{kj}\} = P_d/f_k$$

represent, respectively, the *marginal* cost of increasing and decreasing x_k above and below x_k^*.

Once the penalties are computed, the estimated upper bounds $\bar{z}_u$ and $\bar{z}_d$ can replace the exact z_u and z_d in executing the Land–Doig algorithm; that is, select the next node to be examined as the one having largest estimated upper bound. Also, a node can be discarded if

$$\max\{\bar{z}_u, \bar{z}_d\} = \bar{c}_0 - \min\{P_u, P_d\} \leqslant \underline{z}$$

Beale and Small, however, suggest the use of the LIFO rule for scanning the unexamined nodes (cf. Glover's enumeration scheme, Section 3.4). This is achieved as follows:

1.   Store the branch associated with $\max\{P_u, P_d\}$ in a "list" and solve the linear program associated with the other branch.

2.   Compute the penalties associated with the resulting node and repeat (1).

3.   When a situation is reached where no better solution can be obtained from the current branch, scan the stored list in the opposite order, that is, using the last-in, first-out rule. [The LIFO rule is particularly effective with tape storage units where subproblems are stored and retrieved linearly. Moreover, the last branch stored in the list is likely to produce feasible integer solutions quickly because of the (relatively) large number of integrality conditions leading to this branch.]

As in the implicit enumeration algorithms of Chapter 3, which are based on Glover's enumeration scheme, the algorithm terminates only when each unexamined node yields an optimum objective value which is not better than $\underline{z}$.

The above information can be utilized to tighten the bound on the nonbasic and basic solutions at the different nodes. The idea was suggested by Driebeek. For a nonbasic variable x_j, it is evident that each unit of increase above zero would cost $\bar{c}_j$. Thus, given $\underline{z}$ as defined above, it follows that x_j must be bounded by

$$0 \leqslant x_j \leqslant (\bar{c}_0 - \underline{z})/\bar{c}_j \equiv \mu_j, \qquad j \in NB$$

if it were to yield a value better than $\underline{z}$. Thus μ_j can be used as the upper bound on x_j. This is especially useful if $\mu_j < 1$, in which case x_j must be fixed at zero level *for this branch*.

The basic variable x_k can be treated similarly. Recall from the above development of P_d and P_u that $S_d = P_d/f_k$ and $S_u = P_u/(1 - f_k)$ are the marginal costs for decreasing and increasing x_k around x_k^*. Thus, given $\underline{z}$, the acceptable bounds for x_k are given by

$$x_k^* - (\bar{c}_0 - \underline{z})/S_d \leqslant x_k \leqslant x_k^* + (\bar{c}_0 - \underline{z})/S_u$$

which may make it possible to tighten the bounds on x_k for the current branch.

▶**EXAMPLE 4.3-3**   Maximize

$$z = 18x_1 + 14x_2 + 8x_3 + 4x_4$$

subject to

$$15x_1 + 12x_2 + 7x_3 + 4x_4 + x_5 \leqslant 43$$
$$x_j \geqslant 0 \text{ and integer for all } j$$

The complete tree associated with the solution is shown in Fig. 4-12. The number in the top part of each node represents the order in which a node is branched, while the number in the bottom part represents the *exact* value of the associated objective function. The variable defining the branches emanating from the node is shown in the middle. Finally, the penalties are given in parentheses on each branch.

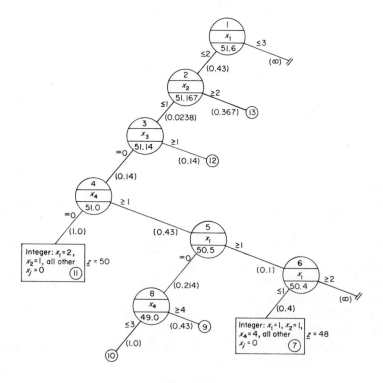

Figure 4-12

For the sake of illustration, the details of computing the penalties for the branches of nodes 1 and 2 will be given here. The optimum simplex tableau at node 1 is given by

	x_2	x_3	x_4	x_5	x_6
z	51.6	0.4	0.4	0.8	1.2	1.2
x_1	2.867	0.8	0.467	0.267	0.067	0.067

Since $x_1 = 2.867$, the two branches $x_1 \le 2$ and $x_1 \ge 3$ must emanate from node 1. Because all α_{1j}, $j = 2, \ldots, 6$, are positive, then $P_u = \infty$, which means that $x_1 \ge 3$ is abandoned. Obviously, one must now consider $x_1 \le 2$ leading to node 2.

The tableau at node 2 (which is obtained from the tableau at node 1 by using the dual simplex method for upper bounded variables, Section 2.6.2) is given by

	x_1'	x_3	x_4	x_5	x_6	
z	51.167	0.5	0.167	0.667	1.167	1.167
x_2	1.083	-1.25	0.583	0.333	0.083	0.083

$x_1 + x_1' = 2$

Thus $f_2 = 0.083$, and

$$P_u = 0.5(0.083 - 1)/(-1.25) = 0.367$$

$$P_d = 0.083\left(\min\left\{\frac{0.167}{0.583}, \frac{0.667}{0.333}, \frac{1.167}{0.083}\right\}\right) = 0.0238$$

Since $P_d < P_u$, node (3) is associated with $x_2 \le 1$. The process is continued in the order shown until the optimum solution $x_1 = 2$, $x_2 = 1$, all other $x_j = 0$, and $z = 50$ is attained.

The first lower bound is encountered at node (7) with $z = 48$. This is used to discard nodes (9) and (10) since if they were to yield a better solution, they must have an optimum $z \ge 49$. This is impossible according to their associated penalties. The second (improved) lower bound is obtained at node (11) with $z = 50$. Since all the coefficients in the original objective function are positive integers with the least common multiple equal to 2, any better solution must have $z \ge 50 + 2$. This automatically eliminates nodes (12) and (13) as nonpromising.◀

It is possible to take advantage of the fact that some nonbasic variables must be integers. Suppose x_q is such a nonbasic variable. Then, if x_q is to be greater than zero, it must be at least equal to one. The penalty associated with such an increase is given by $\bar{c}_q$ in the simplex tableau. Thus, if $\bar{c}_0 - \bar{c}_q \leqslant \underline{z}$, then x_q must be fixed at zero level in the current branch.

The above idea was used more effectively by Tomlin (1971) to develop stronger penalties. Consider the equation associated with an integer x_k in the optimum tableau of the current node. If $x_k \leqslant [x_k{}^*]$ is to be satisfied, some nonbasic variable x_q having $\alpha_{kq} > 0$ must be increased (above zero) in order for this equation to remain satisfied. Similarly, if $x_k \geqslant [x_k{}^*] + 1$ is to be satisfied, another nonbasic variable x_p having $\alpha_{kp} < 0$ must be increased. This means that whether x_k is increased or decreased, an associated nonbasic variable must always be increased. But if such a nonbasic variable is also integer, then its value must be increased to at least one. This means that the associated penalty must be at least equal to $\bar{c}_q$ (or $\bar{c}_p$), and the revised *stronger* penalties are then given as

$$
P_u' = \min_{j \in J^-} \begin{cases} \dfrac{\bar{c}_j(f_k - 1)}{\alpha_{kj}}, & j \notin I \\[3mm] \max\left\{\bar{c}_j, \dfrac{\bar{c}_j(f_k - 1)}{\alpha_{kj}}\right\}, & j \in I \end{cases}
$$

$$
P_d' = \min_{j \in J^+} \begin{cases} \bar{c}_j\left(\dfrac{f_k}{\alpha_{kj}}\right), & j \notin I \\[3mm] \max\left\{\bar{c}_j, \bar{c}_j\left(\dfrac{f_k}{\alpha_{kj}}\right)\right\}, & j \in I \end{cases}
$$

The new expressions now combine the fact that, in addition to basic x_k being integer, some nonbasic variables may also be integer.

Note that the "strength" of the penalties P_u and P_d depends primarily on how small the coefficients α_{kj} are. Thus Tomlin observed that, if a constraint can be found that is tightest (this is equivalent to saying that the absolute values of α_{kj} are smallest) *but does not violate the integer condition*, then the effectiveness of P_u and P_d can be further enhanced. Such a constraint was developed by Gomory (1960b) in connection with the cutting methods and, indeed, is derived by combining constraints (1) and (2) above, and then modifying the resulting constraint so it will trap better information about the integer variables (see Section 5.2.3 for the exact development). The Gomory constraint is given by

$$
S = -f_k + \sum_{j \in NB} \alpha_{kj}{}^* x_j \geqslant 0
$$

where S is an auxiliary nonnegative variable, and

$$
\alpha_{kj}{}^* =
\begin{cases}
\alpha_{kj}, & \text{if} \quad j \in J^+, \quad j \notin I \\[2ex]
\dfrac{f_k \alpha_{kj}}{(f_k - 1)}, & \text{if} \quad j \in J^-, \quad j \notin I \\[2ex]
f_{kj}, & \text{if} \quad f_{kj} \leqslant f_k, \quad j \in I \\[2ex]
\dfrac{f_k(f_{kj} - 1)}{(f_k - 1)}, & \text{if} \quad f_{kj} > f_k, \quad j \in I
\end{cases}
$$

where f_{kj} is defined such that $\alpha_{kj} = [\alpha_{kj}] + f_{kj}$, $0 \leqslant f_{kj} \leqslant 1$. The above constraint represents a necessary condition for satisfying the integrality condition. Thus, by the dual simplex method, a minimum penalty of

$$
P^* = \min_{j \in NB} \left\{ \frac{\bar{c}_j f_k}{\alpha_{kj}{}^*} \right\}
$$

must be incurred if the new integrality constraint is to be satisfied. By substituting for $\alpha_{kj}{}^*$ as given above and then combining $P_u{}'$ and $P_d{}'$ into one minimizing expression, it follows immediately that

$$
P^* \geqslant \min\{P_u{}', P_d{}'\}
$$

This shows that the resulting penalty is stronger. If $\bar{c}_0 - P^* \leqslant \underline{z}$, the current node is discarded. Otherwise, it must be branched, and the decision as to which branch should be selected is made on the basis of the relative values of $P_u{}'$ and $P_d{}'$.

▶EXAMPLE 4.3-4   Example 4.3-3 will be solved again by using Tomlin's penalties to illustrate the superiority of Tomlin's method over that of Beale and Small.

The solution tree is shown in Fig. 4-13. Tomlin's method requires the solution of seven linear programs as compared to Beale and Small's, where ten linear programs are solved. Notice that the penalties at some branches in Fig. 4-12 are generally smaller than their corresponding ones in Fig. 4-13 as contemplated by Tomlin's theory. The comparison shows that, in general, Tomlin's methods should yield superior results.◀

In the above example, P^* is used primarily to check whether the associated node can be discarded as nonpromising relative to the current lower bound, $\underline{z}$. But even if a lower bound is not yet available, P^* can still be used to decide which basic variable is to be used for branching at the

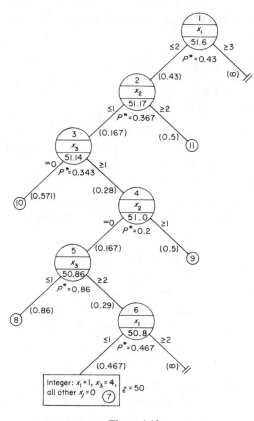

Figure 4-13

current node, in the sense that the basic variable yielding the largest P^* may result in the *smallest* (estimated) upper bound.

The general concept of penalties is interesting, but their computational effectiveness in solving large (practical) problems has not been established. This point is detailed further in Section 4.3.3.

C. *Computation of Lower Bound*†

The lower bound $\underline{z}$ plays an important role in truncating the calculations. Yet the above rules for branching and bounding do not provide a systematic procedure for determining a good lower bound at the early stages of the calculations. Instead, these rules are concerned primarily with the economic utilization of computer time and storage requirements. Although these rules may produce an early feasible solution, this objective

† Taha (1971c).

is only secondary. Even when a feasible solution is encountered, it does not necessarily have to be a good solution (relative to the optimum integer solution).

The procedure proposed here calls for inspecting the vertices of the unit hypercube containing the optimum solution to the continuous linear problem. Because these vertices represent the immediate (integer) neighborhood of the continuous solution, there is a good chance that a *feasible* point among these vertices would be very close to the optimum solution, if not the optimum solution.

The problem of determining the lower bound then reduces to investigating all 2^n vertices of the unit hypercube. However, exhaustive enumeration would *not* be necessary to effect the desired result. The basic idea is to define a (imbedded) zero–one linear problem with its possible solution points represented by the 2^n vertices of the unit hypercube. This problem is easily defined as follows: Let the original continuous problem be given by

$$\max\left\{ \sum_{j \in N} c_j x_j \middle| \sum_{j \in N} a_{ij} x_j \leqslant b_i, i \in M; x_j \geqslant 0, j \in N \right\} \tag{P}$$

Assume the optimum continuous solution to (P) to be given by $\{x_1{}^*, x_2{}^*, \ldots, x_n{}^*\}$ and let

$$x_j = [x_j{}^*] + y_j, \qquad y_j = (0, 1), \qquad j \in N$$

Using this substitution in (P), the zero–one problem is given by

$$\max_{\substack{y_j = (0, 1) \\ j \in N}} \left\{ \sum_{j \in N} c_j \, y_j + \sum_{j \in N} c_j [x_j{}^*] \middle| \sum_{j \in N} a_{ij} y_j \leqslant b_i - \sum_{j \in N} a_{ij}[x_j{}^*], i \in M \right\} \tag{Q}$$

This is essentially equivalent to translating the origin of (P) to ($[x_1{}^*]$, ..., $[x_n{}^*]$).

Any feasible solution to (Q) must be feasible also with respect to (P), provided it is expressed in terms of x_j. However, a strongest lower bound for (P) is given by the optimum solution to (Q).

The optimum solution to (Q) is obtained by using the (relatively efficient) implicit enumeration methods of Chapter 3. Notice that the number of variables in (Q) is the same as in the original problem (P). This is important since the efficiency of the implicit enumeration methods depends primarily on the number of variables.

The above procedure can also be utilized at the different nodes of the Land–Doig algorithm to secure improved lower bounds. However, if the zero–one problem is applied too frequently, this may be too costly .in comparison with the possible improvements in the lower bound. Consequently, one must rely on experience to determine how often the zero–one

problem should be invoked. The experience of this author indicates that it
is beneficial to apply the zero–one procedure at the successive nodes until
a lower bound is encountered for the first time, after which the zero–one
problem is not utilized.

The above procedure can be extended directly to the mixed integer
problem. This is achieved by replacing (Q) by a *mixed* zero–one problem.
The only difference is that the y_j variables corresponding to continuous
x_j in the original problem are not restricted to binary values. Observe,
however, that the *pure* zero–one problem can still be used to produce a
legitimate lower bound for the original mixed problem. This obviously will
not be as strong as the one obtained from the mixed zero–one problem.

▶EXAMPLE **4.3-5** Consider Example 4.3-3. The optimal continuous
solution is $z = 51.6$ and $x_1 = 2.867$, with all the remaining x_j being equal
to zero. Thus the translated origin of (P) is given by (2, 0, 0, 0, 0), and the
resulting (Q) is

maximize

$$z = 18y_1 + 14y_2 + 8y_3 + 4y_4 + 36$$

subject to

$$15y_1 + 12y_2 + 7y_3 + 4y_4 + y_5 \leqslant 13$$
$$y_j = (0, 1), \quad j = 1, 2, \ldots, 5$$

The optimum solution of the zero–one problem is $\{y_j\} = (0, 1, 0, 0, 0)$.
Translated in terms of x_j, this gives $\{x_j\} = \{2, 1, 0, 0, 0\}$ with $z = 50$, which
is now used as a lower bound.

From the objective function of Example 4.3-3, any improved solution
must yield $z \geqslant 50 + 2$. However, this is impossible since optimum z of the
linear program associated with the original integer problem is 51.6. Thus
the zero–one problem gives the optimum solution directly, and no further
branching is necessary.◀

The above example illustrates the possible effectiveness of the proposed
method. Generally, however, the zero–one solution may not be optimum
with respect to (P). Indeed, (Q) may not have a feasible solution at all.
In this case, the procedure calls for attempting to solve zero–one problems
at some of the succeeding nodes until a lower bound is determined.

J. E. Moore (1974) has tested the effectiveness of using the imbedded
zero–one problem to locate a good lower bound on pure integer problems
with the number of variables equal to 30 or 100. He reports that the use

of the imbedded 0–1 problem reduces the (total) average computational time considerably. However, when the pure 0–1 problem is used to find a lower bound for mixed integer problems, the results are discouraging.

The favorable experience reported for the pure integer problem may be too limited to allow a general conclusion. Indeed the effectiveness of the procedure is obviously controlled by the efficiency of the zero–one codes, which at the time being cannot handle more than two hundred variables in a reasonable amount of computational time. This is probably inadequate for large problems. As a result it may be necessary to resort to heuristics in order to find a good initial lower bound quickly. This point is discussed further in the next section.

4.3.3   Heuristics for Solving Large Branch-and-Bound Problems†

The presentations in Sections 4.3.1 and 4.3.2 suggest that the complexity of the tree representation of a branch-and-bound problem is primarily dependent on (1) the choice of the next node to be examined, and (2) the choice of "branching" variable at a given node.

In the original Land–Doig algorithm the next node is the one having the largest objective value among all unexplored nodes, while its modification by Beale and Small utilizes the LIFO rule, which always stores the node (branch) with the larger penalty and acts on the other node immediately. The branching variable is selected either as the one having the largest fraction or as the one yielding the largest combined penalty P^*.

The above rules are primarily developed for the purpose of simplifying "local" computations at each node, without a serious attempt to check whether this would lead to a feasible (hopefully optimal) solution quickly. In other words, these rules do not have the capability to "look ahead" in search for a good feasible solution.

The above discussion shows that the uses of upper bounds, penalties, and the LIFO rule as given in Section 4.3.2 do not relate directly to the ultimate goal of producing the optimum solution quickly. This goal can be achieved not only by reducing the computation time at each node, but also by enforcing conditions that reduce (hopefully minimize) the *number* of nodes that must be examined before optimality is verified.

This section presents computational experience that shows that the effectiveness of the upper bounds, penalties, and the LIFO rule becomes inadequate as the size of the problem becomes larger. A number of heuristics designed to find a good (if not the optimum) solution quickly are then presented. The presentation is based on the excellent paper by Forrest *et al.* (1974).

† Forrest *et al.* (1974).

In order to appreciate the significance of the results below, the classes of problems used in the study include:

(i) a relatively small number (about 100) of zero–one variables imbedded in very large and difficult linear programs with a few thousand constraints and complex matrix structure; and

(ii) a relatively large number (several hundred) of integer variables with complex logical relationships but imbedded in much simpler linear programs with 500–1000 constraints.

These problems are actual practical problems. Forrest *et al.* indicate they have not encountered (nor do they expect to encounter) practical problems that are difficult in terms of both linear and integer variables. The computational experience with the tested problems is based on the commercial code UMPIRE developed by Scientific Control Systems, Ltd., London (1970).

A. *Failure of the Penalty Approach*

Shaw (1970) and Tomlin (1970) report considerable success in solving small problems by using penalties and the LIFO rule. However, as the size of the problem becomes very large, these simple rules become ineffective. One of the main reasons is that penalties do not usually provide a reasonable guide in terms of the true decrease in the objective value as a result of effecting branching. For example, in Fig. 4-11, $P_u > P_d$ implies that it is cheaper to branch on $[x^*]$, rather than on $[x^*] + 1$. But the exact objective value shows that the opposite is true. Although in relatively small problems P_u and P_d usually give a proper indication as to which branch is cheaper, computational experience with large problems shows that this result is not true in general. Indeed, in large problems the changes in basis (as the optimum solution moves away from $x_k = x_k^*$) are so frequent that the variation of optimum z with x_k appears more as a smooth (rather than piecewise continuous) concave function. The result is that the penalties computed from the basis associated with x_k^* are very misleading as to which branch is actually cheaper.

A principle difficulty with the use of penalties relates to the selection of the branching variable at a node since the amount of computations depends very much on these selections. At the early stages of the calculations where a fewer number of integer variables have been branched on, there is greater flexibility in the model, which in turn leads to greater underestimation of the true decrease in the objective value, that is, smaller, and hence misleading, penalties. Because the choices of the branching variables at these early nodes are crucial in simplifying or complicating the search tree, it is unfortunate that such choices are based on the least reliable values of the penalties. Under these conditions, the use of misleading penalties may well lead to an almost arbitrary tree search.

Another "less serious" difficulty results from the use of penalties with the LIFO rule to decide on which branch to store. A simple procedure that partially alleviates this is to use the so-called *node swapping* procedure. This calls for tentatively selecting the branch with the cheaper penalty while storing the other. If the *exact decrease* in the objective value of the selected branch exceeds the penalty of the stored branch, then the subproblem of the stored branch is solved. If its true objective value is higher, then the two branches (nodes) are swapped with the originally selected branch now being stored. Otherwise, the choice according to penalties persists. Although computational experience with one problem shows that node swapping halved the number of branches needed to find and verify optimality, there are indications that this result will not hold in general.

The preceding discussion indicates the unrealiability of the simple use of penalties as a principal guide in constructing the search tree. Moreover, the main advantage of the LIFO rule is that with tape storage units subproblems are added and subtracted linearly. However, the availability of random access (disc) units makes it worthwhile to investigate the use of more flexible rules for selecting the next node to be examined. This also calls for devising new criteria (perhaps including penalties) for evaluating nodes and potential branching variables at a selected node. The heuristics presented below are developed through computational experience to account for these two points. Part B is concerned with node selection and Part C deals with the selection of a branching variable.

B.   *Heuristics for Node Selection*

Three methods are presented in this section: (1) best projection criterion, (2) pseudocosts criterion, and (3) percentage-error criterion. As will be shown below, the development of these criteria depends on the availability of a (good) feasible solution. Each criterion is now considered separately.

(1)   *Best projection criterion*   Let z^0 be the optimum objective value associated with the linear program defining the first node of the tree and assume $\underline{z}$ is a known (or estimated) lower bound on the optimum integer solution. For node k, define

$$s_k = \sum_{i \in I} \min\{f_i, 1 - f_i\}$$

where I is the set of integer variables x_i and f_i is defined such that, given $x_i = x_i^*$ at node k, then $x_i^* = [x_i^*] + f_i$. s_k is called the "sum of integer infeasibilities," where $\min\{f_i, 1 - f_i\}$ gives the smaller deviation of x_i^* from $[x_i^*]$ and $[x_i^*] + 1$.

The sum of integer infeasibilities associated with the "lower bound-solution" is obviously zero. Consequently, an *empirical* measure of the *rate of decrease* in the optimum objective value from z^0 (the starting node) to $\underline{z}$ may be given by

$$\lambda = (z^0 - \underline{z})/s_0$$

The value of λ is updated every time an improved $\underline{z}$ is encountered.

Now, if z^k is the *optimum* objective value at node k, and s_k is its associated sum of integer infeasibilities, then the *projected* objective value of an integer solution generated from this node is estimated by

$$z_p^{\ k} = z^k - \lambda s_k$$

Thus the node having the largest $z_p^{\ k}$ is the next to be examined.

The slope λ as defined above may be viewed as the degradation in the value of z per unit decrease in the sum of integer feasibilities from s_0 to zero. If $\underline{z}$ is a good lower bound (relative to the true optimum), then the straight-line relationship for estimating $z_p^{\ k}$ may be reasonably acceptable. Computational experience indicates that the best projection criterion has two major advantages:

  (i)   It is extremely effective in locating a near optimal solution quickly; and

  (ii) It provides information to judge the quality of the best available solution by comparing $\max_k\{z_p^{\ k}\}$ with $\underline{z}$.

  (2) *Pseudocosts criterion*   The best projection criterion can be written as

$$z_p^{\ k} = z^k - \sum_{i \in I} \min\{\lambda f_i, \lambda(1 - f_i)\}$$

This implies that a variable $x_i = x_i^*$ can be forced to an integer value ($[x_i^*]$ or $[x_i^*] + 1$) at the same cost λ per unit change. This assumption may be acceptable if all the activities of the model have approximately equal costs provided that the cost per unit of forcing x_j from x_j^* to $[x_j^*]$ is approximately the same as forcing it from x_j^* to $[x_j^*] + 1$. But in many practical situations there may be a wide discrepancy between the cost coefficients; and, in this case, the best projection criterion is not expected to perform adequately.

To alleviate this problem, *individual* pseudocosts are estimated not only for each integer variable x_i but also for increasing and decreasing a variable x_i from its present *fractional* value x_i^* to the integer values $[x_i^*] + 1$ and $[x_i^*]$. Let u_i = estimated per unit cost for increasing x_i to $[x_i^*] + 1$, and d_i = estimated per unit cost for decreasing x_i to $[x_i^*]$. Then $z_p^{\ k}$ in the best projection criterion is replaced by

$$z_e^{\ k} = z^k - \sum_{i \in I} \min\{u_i\, f_i,\, d_i(1 - f_i)\}$$

Initially, the values of u_i and d_i are estimated based on the experience of the model formulator with the physical problem. For example, if $x_i = 0$ when plant i is built and $x_i = 1$ if it is not built, then given $0 < x_i{}^* < 1$, it is relatively easy to estimate d_i (for *not* building). The estimation of u_i (for building) may be much more difficult since this will bring into the picture the interaction with the other variables due to the effect of the output from plant i on the output from other plants.

Regardless of how crude the initial estimates of pseudocosts are, these estimates can be revised as soon as an integer feasible solution is attained. Namely, suppose node k is on the path (of branches) leading to a feasible integer solution. Let x_i be the branching variable at node k and assume that $(k + 1)$ and $(k + 2)$ are the successor nodes from k generated by branches $x_i \leqslant [x_i{}^*]$ and $x_i \geqslant [x_k{}^*] + 1$, respectively. Then the revised pseudocosts are

$$d_i = (z^k - z^{k+1})/f_i$$
$$u_i = (z^k - z^{k+2})/(1 - f_i)$$

In order to determine d_i and u_i for all $i \in I$ as given above, it is necessary that each x_i be used as a branching variable on the path to the feasible integer solution. Experience shows that this is generally the case. If not, then the absent variables are assigned zero pseudocosts. On the other hand, if the same variable is used for branching more than once, then it may be beneficial to average all its pseudocosts.

(3) *"Percentage-error" criterion*   At node k, the percentage-error criterion is measured by

$$\eta_k = 100(\underline{z} - z_e^{\ k})/(z^k - \underline{z})$$

where z_e is as defined by the pseudocosts above and $\underline{z}$ is a known lower bound.

The idea is that if η_k is a very large *negative* error, then there is a good chance that an improved solution exists between z^k and $\underline{z}$. If it is a very large *positive* error, then the probability of such occurrence diminishes. Thus, the next node is selected as the one with the most negative η.

C.   *Heuristics for Selection of Branching Variable*

After a node is selected for examination, an unsatisfied integer variable must be chosen at the node to effect branching. The selection based on the largest fraction among all eligible variables is generally ineffective since it concentrates only on the "local" conditions at the examined node. The use of penalties could also be futile, especially if they do not reflect the true decrease in the objective function.

The heuristic for selecting a branching variable calls for assigning priorities to the integer variables. These priorities can be based on the modeler's knowledge of the physical problem and hence the degree of "importance" of each integer variable. For example, a variable controlling a (large) number of logical relationships should be given higher priority than a variable with less complex effect on the system. Another way is to rank the variables according to their objective coefficients. Priorities based on this ranking can be acceptable if the "utilization" of the variables from the respective constraints is approximately equal.

The pseudocost developed in Section B may also be used to assign priorities to the variables, namely, select the variable corresponding to

$$\max_i(\min\{d_i f_i, u_i(1 - f_i)\})$$

This condition can be refined further by using the penalties. Thus the branching variable is associated with

$$\max_i[\min(\max\{P_d', P^*, d_i f_i\}, \max\{P_u', P^*, u_i(1 - f_i)\})]$$

where P_d', and P_u', and P^* are as defined in Section 4.3.2-B.

Reported computational experience indicates that a proper priority list, when available, could have crucial effects on solving the problem. This underscores the importance of the relevant information that the modeler could provide about the problem.

D. *Solution Strategies*

The availability of a variety of heuristics suggests the importance of manual intervention during the course of solving the problem. This means that an attempt to formulate, generate, and then solve a large integer problem immediately may prove to be a discouraging exercise in futility. Perhaps at the beginning one must aim for studying a simplified formulation of the problem. By experimenting with different heuristics, one may then accumulate sufficient information to warrant a change in the original formulation and/or develop different priority lists and pseudocosts for the problem. After the user gains sufficient confidence about the problem he may then specify a strategy for tackling the more complex model, but he should also remain prepared to change his strategy should it become evident that no progress is being achieved.

The UMPIRE code is designed to allow the user to alter his strategy depending on some feedback information. This flexibility makes it possible to develop a "dynamic" strategy which aims at:

(i) finding an (good) initial solution quickly.
(ii) deciding whether the available best solution is good enough so that, by the law of diminishing returns, further search for better solutions may not be rewarding.

This is where the different options that the above heuristics provide would be beneficial. Perhaps also as a result of this experimentation, the user may be able to develop other heuristics that suit his problem most.

4.4 Solution of Nonlinear Integer Programs by Branch-and-Bound

As shown in discussing the branch-and bound principle (Section 4.2), there are no restrictions that will prevent the use of the principle with any well-defined nonlinear problem. Perhaps the only reason for concentrating on the linear problem is that linear programming, by its simple method of solution, offers the best opportunities for manipulating the problem in a relatively efficient way.

The techniques developed in the preceding section are directly applicable, or can be readily extended, to the nonlinear problem except insofar as using those properties that are based primarily on the assumption of linearity. These typically include the use of the penalties in estimating the bounds on the objective value and the development of certain branching rules.

Following the above discussion, the original Land–Doig algorithm (Section 4.3.1) is not suitable, in general, to solve the nonlinear integer problem, primarily because the validity of the branching rules is tied with the assumption of linearity. (Why?) However, Dakin's modification of the Land–Doig algorithm (Section 4.3.2) makes the branching rule independent of the linearity condition. Specifically, given that $x_j = x_j^*$ is the optimum value of x_j, which is fractional, then in order for x_j to assume an integer value it must satisfy one of the two conditions: $x_j \geq [x_j^*] + 1$ or $x_j \leq [x_j^*]$. This clearly has nothing to do with linearity.

The major complication in solving a nonlinear problem lies generally in the difficulty of finding a solution method that guarantees the determination of the *global* optimum at *each node*. In particular, if such methods can find local optima only, then the branch-and-bound principle becomes inapplicable since the bounding rule cannot be satisfied. One notices, however, that the indicated complication does not arise as a result of using the branch-and-bound principle. Rather, the techniques of nonlinear programming still have limited capabilities for solving the continuous problems. It appears, however, that one can have reasonably good results with nonlinear algorithms if the constraint and objective functions of the problem satisfy certain conditions of convexity and concavity so that a local optimum becomes the global optimum. In these situations, Dakin's method can be applied directly in the exact manner prescribed for the linear problem.

4.5   Concluding Remarks

The major conclusion of this chapter is that, from the practical standpoint, branch-and-bound methods can be effective in solving large integer problems provided they are equipped with the types of heuristics presented in Section 4.3.3. Also, with such problems manual intervention based on the user's information and on the progress of the solution is almost a necessity if a "good" solution is to be expected in a reasonable amount of time. Based on this result, a successful solution of a problem would not only depend on the proper formulation of the model but also on one's (accumulated) experience in solving other (possibly similar) problems by available computer codes.

A notable remark about the behavior of the integer algorithms is that there seem to be potential advantages in "cross-fertilizing" the methods of implicit enumeration, cutting plane, and branch-and-bound. This is evident from the use of imbedded zero–one problems to provide a "good" lower bound on the promising feasible points. Also, the utilization of the information in Gomory's cuts does result in stronger penalties and hence tighter estimates of the upper bounds. This point will be taken again in Chapter 5 to show how the cutting plane method can benefit from the idea of cross-fertilization.

There are other branch-and-bound methods that have been developed for the integer problems, notably, Thompson's (1964) and Hillier's (1969b). But there are no indications that these methods are superior to the ones presented in this chapter. In fact, Thompson's method appears to be rigid in the sense that it is not amenable to further improvements along the lines presented in Section 4.3.2. Also, although Hillier's method assumes the availability of favorable starting conditions (that is, an initial nearly optimal feasible solution) its overall behavior does not show appreciable improvements over the improved Land–Doig algorithm under the same initial conditions.

Although the development of new branch-and-bound algorithms for the integer problems continues to be of theoretical interest, it appears that the solution of practical problems will benefit most from the development of more efficient heuristics to be superimposed on the basic structure of the branch-and-bound algorithm. Nonetheless, these theoretical developments should not be regarded as useless, since they may eventually lead to potentially good ideas. For example, the concept of penalties is actually a forerunner to the development of the very successful heuristics of Section 4.3.3.

Problems

4-1 *Assignment problem.* Consider the following problem:

minimize

$$z = \sum_{i=1}^{n} \sum_{j=1}^{n} c_{ij} x_{ij}$$

subject to

$$\sum_{i=1}^{n} x_{ij} = 1, \qquad j = 1, 2, \ldots, n$$

$$\sum_{j=1}^{n} x_{ij} = 1, \qquad i = 1, 2, \ldots, n$$

$$x_{ij} = 0 \text{ or } 1 \text{ for all } i \text{ and } j$$

Let $Q^t = \{(i, j) | x_{ij} = 1 \text{ at node } t\}$, where $Q^0 = \varnothing$. Define $I^t = \{i | (i, j) \notin Q^t\}$, $J^t = \{j | (i, j) \notin Q^t\}$, where $|I^t| = |J^t| \equiv k_t$.

(a) Prove that a proper lower bound on all feasible solutions from node t is

$$z^t = \sum_{(i,j) \in Q^t} c_{ij} + \sum_{j \in J^t} \min\{c_{ij} | i \in I^t\}.$$

(b) Show that a proper branching rule at node t is to create k_t branches, where the rth branch is defined by

$$x_{i_r j_r} = 1, \quad \begin{cases} i_r \in I^t - \{i_1, i_2, \ldots, i_{r-1}\} \\ j_r \in J^t - \{j_1, j_2, \ldots, j_{r-1}\} \end{cases} \quad r = 1, 2, \ldots, k_t$$

and

$$x_{i_p j_p} = 0, \qquad i_p \neq i_r, \quad j_p \neq j_r, \quad (i_p, j_p) \notin Q^t$$

4-2 In terms of the branch-and-bound principle (Section 4.2), define for Problem 4-1 the sets S and T and the return function w associated with T.

4-3 Solve the following assignment problem by the branch-and-bound rules in Problem 4-1. The table below gives c_{ij} for $i = 1, 2, 3, 4$ and $j = 1, 2, 3, 4$.

	1	2	3	4
1	45	∞	5	56
2	27	2	82	74
3	19	55	3	∞
4	3	10	4	84

4-4 In many integer programming applications (see Chapter 1) the constraint

$$\sum_{j=1}^{n} x_j = 1, \qquad x_j = (0 \text{ or } 1) \quad \text{for all} \quad j$$

is imbedded. If at any node of the Land–Doig algorithm, x_p is a branching variable (that is, $0 < x_p{}^* < 1$), then the algorithm creates two branches associated with $x_p = 0$ and $x_p = 1$ at this node. This is exactly equivalent to the two branches

$$\sum_{\substack{j=1 \\ j \neq p}}^{n} x_j = 1, \qquad \text{and} \qquad x_p = 1$$

The disadvantage is that the first branch may be "too loose" especially for large n. Show that a proper branching rule alleviating this problem is to create the two branches

$$\sum_{j \in R_1} x_j = 1, \qquad \text{and} \qquad \sum_{j \in R_2} x_j = 1$$

where $|R_1| \simeq |R_2|$, $R_1 \cup R_2 = \{1, 2, \ldots, n\}$, and $R_1 \cap R_2 = \emptyset$. Devise proper rules for determining R_1 and R_2.

4-5 Suppose in Problem 4-4 that the *multiple choice* constraint

$$\sum_{j=1}^{n} x_j = k, \qquad 1 < k < n, \quad x_j = (0, 1)$$

is used instead. Show that if an idea similar to that of Problem 4-4 is used, then at least $k + 1$ branches associated with

$$\sum_{j \in R_i} x_j = k, \qquad i = 1, 2, \ldots, k + 1, \ldots$$

must be created at the node, where $\bigcup_{\text{all } i} R_i = \{1, 2, \ldots, n\}$, and that necessarily $\bigcap_{\text{all } i} R_i \neq \emptyset$. Comment on the effectiveness of this branching rule. (*Note:* This is an example of a branching rule where the sets R_i do *not* "partition" $\{1, 2, \ldots, n\}$; that is, the sets R_i are not mutually exclusive.)

4-6 In Example 1.3-8 (Chapter 1), the approximation of a nonlinear function $f(x)$ by linear segments utilizes the weights t_i, $i = 1, 2, \ldots,$ to define each linear segment, where

$$\sum_{i=1}^{K} t_i = 1, \qquad t_i \geqslant 0, \quad \text{for all} \quad i$$

provided that *no more* than two t_i's are positive with the requirement that they must be for two *consecutive* values of i, that is, q and $q + 1$. Show that proper branches are given by

$$\sum_{i=1}^{q-1} t_i = 1, \qquad \text{and} \qquad \sum_{i=q+1}^{K} t_i = 1$$

4-7 Solve by the Land–Doig algorithm:

maximize

$$z = x_1 + 2x_2$$

subject to

$$x_1 + x_2 \leqslant 0.9$$
$$-2x_1 - x_2 \leqslant 0.2$$
$$x_1, x_2 \geqslant 0 \text{ and integers}$$

4-8 Solve by the Land–Doig algorithm:

maximize

$$z = 3x_1 + 3x_2 + x_3$$

subject to

$$x_1 - x_2 + 2x_3 \leqslant 4$$
$$-3x_1 + 4x_2 \qquad \leqslant 2$$
$$2x_1 + x_2 - 3x_3 \leqslant 3$$
$$x_1, x_2 \geqslant 0 \text{ and integer}$$
$$x_3 \geqslant 0$$

4-9 Solve Problem 4-8 by using Dakin's branching rule and compare the method of solution with that of the Land–Doig algorithm.

4-10 Show that the two subproblems (branches) created at a node by Dakin's branching rule (Section 4.3.2) may be replaced by a single problem, and hence one node, by augmenting a zero–one variable and proper constraints. (*Hint:* Dakin's branching rule results in either–or constraints.)

4-11   Solve Problem 4-9 by first converting it according to the procedure of Problem 4-10. Compare the two methods of solution. Are the computations enhanced if the either–or constraints are imposed on both x_1 and x_2 simultaneously?

4-12   Compute the penalties associated with node 3 in Fig. 4-12 from the simplex tableau.

4-13   Compute P^* for nodes 1 and 2 in Fig. 4-13 from the corresponding simplex tableaus.

4-14   Solve Problem 4-8 by the method of penalties.

4-15   Solve Example 4.3-4 by applying each of the heuristic rules for node selection (Section 4.3.3-B). Compare the effectiveness of the heuristics as compared with the use of penalties only. Whenever possible, use the pseudocost criterion to select the branching variable.

CHAPTER 5

Cutting Methods

5.1 Introduction

In Chapter 4, the basic idea of the branch-and-bound methods is to shift the objective function parallel to itself in the direction where its value is deteriorating starting from the continuous optimum point. These shifts are designed so that it is never possible to bypass a superior feasible point. The cutting methods, on the other hand, seek a restructuring of the feasible solution space by imposing some (specially designed) constraints on the original space such that the required optimum feasible point is expressed as a proper extreme point of the modified solution space. The general idea is that these additional constraints systematically cut off portions of the solution space such that no feasible points are ever excluded.

Because the development of the cutting methods is associated with the linear problem, the presentation in this chapter will be restricted to linear cases, unless otherwise stated.

To illustrate the application of the cutting methods, consider the graphical example in Fig. 5-1. The lattice points within the (continuous) convex solution space are the feasible integer points. The continuous optimum point is given by A, which does not satisfy the integrality condition. The cutting "planes" (I) and (II) serve as typical illustrations of how the continuous solution space is restructured so that the optimum extreme point satisfies the integrality conditions. In general, however, several (indeed, numerous)

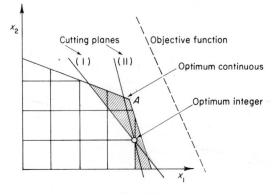

Figure 5-1

cuts may have to be applied before the required extreme point is singled out. It is obvious that there is an *infinite* number of ways in which a cut may be designed without violating any feasible points. But it will be shown later that in the cutting algorithms, which are known to converge, the number of different cuts that satisfy the *necessary* integrality condition is only finite.

Cutting methods are actually the first systematic techniques to be developed for the integer (linear) problem.† The early works of Dantzig *et al.* (1954) and Markowitz and Manne (1957) directed the attention of researchers to the importance of solving linear programs in integers, but Dantzig (1959) was the first to propose a cut for solving such problems. His idea is first to solve the linear program ignoring the integer conditions. If the resulting basic solution is noninteger, then a new set of values for the nonbasic variables must be secured. This is achieved by realizing that the sum of the nonbasic *integer* variables must at least be equal to one; that is, if $x_j, j \in NBI$, are the nonbasic integer variables, then

$$\sum_{j \in NBI} x_j \geqslant 1, \qquad x_j \geqslant 0, \quad j \in NBI$$

must be satisfied. By augmenting this cut to the current tableau, feasibility can be restored by applying the dual simplex method. Although the indicated cut represents a *necessary* condition for integrality, and indeed it was utilized to solve some problems successfully, there is no guarantee that the successive application of the cut will yield the integer solution in a finite number of iterations. [See V. J. Bowman and Nemhauser (1970) for a refinement of this cut.]

† Land and Doig (1960) report that their branch-and-bound method was developed about the same time the first systematic cutting method by Gomory (1958) became known. Their work was not available in the literature until 1960, however.

The first finite cutting algorithm was developed by Gomory (1958), for the pure integer problem. Gomory showed how the cuts can be constructed systematically from the simplex tableau. Although the algorithm is proved to converge in a finite number of iteration, it has the drawback that the machine round-off error represents a prominent difficulty. This led to the development of a new algorithm by Gomory (1960a), which improves directly on this drawback. Later, Gomory (1960b) extended his theory to cover the mixed integer problem. Other types of cuts were introduced by Glover (1965a) in what is known as the bound escalation method, and by Young (1971), Balas (1971a), Balas *et al.* (1971), and Glover (1973) in what is known as "intersection" or "convexity" cuts.

A common characteristic of the above cuts is that the associated algorithms are all of the dual type; that is, the solution to the problem is not available until the algorithm terminates. This is a major disadvantage, particularly if calculations are stopped prematurely. A rudimentary primal algorithm was first introduced by Ben-Israel and Charnes (1962), but Young (1965) was the first to develop a finite primal algorithm. Subsequent developments were given by Young (1968) and Glover (1968a).

This chapter is organized into two major sections, which categorize the cutting methods as dual and primal. In addition to giving the details of each cut, the presentation emphasizes the relationships among these cuts. These relationships may sometimes prove useful in enhancing the efficiency of the cutting methods.

5.2   Dual Cutting Methods

As mentioned in Section 5.1, dual cutting methods differ mainly in the way the cut is constructed. Consequently, the following subsections are titled according to the different cuts. The adjective mixed is used to describe those cuts that are especially designed for the mixed integer problem. The absence of this adjective signifies that the cut is designed for the pure integer problem.

5.2.1   Fractional Cut†

Suppose that the *optimum* solution to the linear program (ignoring integrality) is given by

maximize

$$z = \bar{c}_0 - \sum_{j \in NB} \bar{c}_j x_j$$

† Gomory (1958).

subject to

$$x_i = x_i^* - \sum_{j \in NB} \alpha_{ij} x_j, \qquad i \in M$$

$$x_i, x_j \geq 0, \qquad\qquad i \in M, \quad j \in NB$$

where $\bar{c}_j = z_j - c_j$, the reduced cost, as defined in Section 2.3.3. The continuous solution is given by $x_i = x_i^* \geq 0, i \in M, x_j = 0, j \in NB$, and $z = \bar{c}_0$. Because the above solution is optimal, it must be true that $\bar{c}_j \geq 0, j \in NB$. It is required that *all* x be integer.

Consider any of the constraint equations for which $x_i^* \not\equiv 0 \pmod 1$, $i \in M$; that is, x_i does not have an integer value. Let the selected equation be associated with x_k. This will be referred to as the *source row*.† Now, the x_k-equation can be written as

$$x_k = [x_k^*] + f_k - \sum_{j \in NB} ([\alpha_{kj}] + f_{kj}) x_j \qquad (source\ row)$$

or

$$x_k - [x_k^*] + \sum_{j \in NB} [\alpha_{kj}] x_j = f_k - \sum_{j \in NB} f_{kj} x_j$$

Now in order for x_k and $x_j, j \in NB$, to be integer, the right-hand side of the last equation must also be integer. This means that a necessary condition for the variables x_k and $x_j, j \in NB$, to be integral is that

$$f_k - \sum_{j \in NB} f_{kj} x_j \equiv 0 \pmod 1$$

But since

$$f_k - \sum_{j \in NB} f_{kj} x_j \leq f_k < 1$$

the necessary condition for integrality becomes

$$f_k - \sum_{j \in NB} f_{kj} x_j \leq 0$$

or

$$S - \sum_{j \in NB} f_{kj} x_j = -f_k \qquad (Gomory's\ f\text{-}cut)$$

where $S \geq 0$ is a (nonnegative) slack variable. This is the required cut, which now should be augmented to the simplex tableau from which it is derived. Since all $x_j = 0, j \in NB$, it follows that $S = -f_k$, which is infeasible. Thus by applying the dual simplex method a portion of the solution is cut

---

† Actually, the objective row may also be used as a source row. Indeed, the convergence proof at the end of this section requires that the objective variable z be restricted to integer values.

off. If the resulting optimum solution is integer the process ends. Otherwise, a new cut is constructed from the new simplex tableau and the process is repeated. If it is impossible to recover feasibility after the cut is applied, this immediately means that the original problem has no feasible (integer) solution.

The derived constraint is called the *fractial* cut (or *f*-cut) because all the coefficients f_k and f_{kj} are fractions.

A. *Properties of the f-Cut*

1. *Coefficients of the starting tableau.*   The development of the *f*-cut shows that all the variables (including the slacks) must be integer. Thus the integer problem must be formulated so that all the slacks are allowed to be integer. This is guaranteed by converting the starting tableau to an all-integer tableau by multiplying each equation by a proper multiple (and rounding the resulting coefficients if necessary). The all-integer initial tableau is thus a mandatory requirement for the *f*-cut.

   To illustrate the above point, suppose x_1 and x_2 are integer variables that are related by the inequality

   $$3x_1 + \tfrac{1}{2} x_2 \leqslant \tfrac{13}{3}$$

After adding the slack, it is seen that the equation

   $$3x_1 + \tfrac{1}{2} x_2 + y = \tfrac{13}{3}$$

cannot be satisfied for any integer y, and consequently the use of the *f*-cut will show that no feasible (integer) solution exists. This difficulty can be eliminated by multiplying the inequality by 6.

2. *Slack variable of the f-cut.*   The slack variable of the *f*-cut is necessarily integer. This is seen directly from

   $$S = -\left(f_k - \sum_{j \in NB} f_{kj} x_j\right)$$

Since the right-hand side is integer, S must also be integer. This means that the new problem resulting from augmenting the *f*-cut is a pure integer problem.

   The above point also shows that every *f*-cut necessarily passes through at least one integer point. This point may be infeasible, however.

3. *Number of different cuts.*   Different *f*-cuts can be generated from the current continuous optimum tableau. This is achieved by using as a source row: (1) any of the tableau equations, (2) a multiple

of any of these equations, or (3) a multiple combination of all the equations. From the computational standpoint, the "strongest" of these cuts is the one that cuts the deepest in the solution space without eliminating any feasible integer point.

As an illustration, consider the source row

$$x_1 - 5.4x_2 + 4.6x_3 = 7.4$$

where x_2 and x_3 are the nonbasic variables. Thus

$$x_1 - 6x_2 + 4x_3 - 7 = 0.4 - 0.6x_2 - 0.6x_3$$

which gives the cut $0.6x_2 + 0.6x_3 \geqslant 0.4$. Now, if the source row is multiplied by 2, this will yield

$$2x_1 - 11x_2 + 9x_3 - 14 = 0.8 - 0.2x_2 - 0.2x_3$$

which gives the cut $0.2x_2 + 0.2x_3 \geqslant 0.8$. By the above definition, this cut is stronger than the preceding one.

This procedure may be generalized for any integer† multiple λ of the source row. Let the source row be

$$x_k = x_k{}^* - \sum_{j \in NB} \alpha_{kj} x_j$$

Multiplying both sides by λ gives

$$\lambda x_k = \lambda x_k{}^* - \sum_{j \in NB} \lambda \alpha_{kj} x_j$$

$$= \lambda([x_k{}^*] + f_k) - \sum_{j \in NB} \lambda([\alpha_{kj}] + f_{kj}) x_j$$

or

$$\lambda x_k - \lambda[x_k{}^*] + \sum_{j \in NB} \lambda[\alpha_{kj}] x_j = \lambda f_k - \sum_{j \in NB} \lambda f_{kj} x_j$$

The left-hand side is integer, and so is the right-hand side. Thus the integer condition is maintained if the right-hand side is changed to

$$\lambda f_k - [\lambda f_k] - \sum_{j \in NB} (\lambda f_{kj} - [\lambda f_{kj}]) x_j$$

which must also be integer. Since $\lambda f_k - [\lambda f_k] < 1$, then the generalized cut (called the f^λ-cut) is

$$\lambda f_k - [\lambda f_k] - \sum_{j \in NB} (\lambda f_{kj} - [\lambda f_{kj}]) x_j \leqslant 0 \qquad (f^\lambda\text{-}cut)$$

Notice that $\lambda = 1$ yields the original f-cut.

---

† This material is covered in Garfinkel and Nemhauser (1972b), p. 193–194. Glover (1966) considers the case where λ is noninteger.

Given that λ is integer, it is now shown that the number of different f^{λ}-cuts that can be generated from any source row is finite. Indeed, if D is the absolute value of the determinant of the current basis, then *all* f^{λ}-cuts may be generated by integer λ satisfying $0 \leqslant \lambda \leqslant D - 1$ (notice that D is integer since the initial tableau is integer). To show this, let $\lambda' = \lambda + rD$, where r is an arbitrary integer and λ is an integer satisfying $0 \leqslant \lambda \leqslant D - 1$. Since $f_{kj} = d_{kj}/D$, where d_{kj} is integer and $0 \leqslant d_{kj} \leqslant D$,† then for any $j \in NB$

$$\lambda' f_{kj} - [\lambda' f_{kj}] = (\lambda + rD)\frac{d_{kj}}{D} - \left[(\lambda + rD)\frac{d_{kj}}{D}\right]$$

$$= \frac{\lambda d_{kj}}{D} + rd_{kj} - \left[\frac{\lambda d_{kj}}{D}\right] - rd_{kj}$$

$$= \lambda f_{kj} - [\lambda f_{kj}]$$

The same type of result holds for $\lambda f_k - [\lambda f_k]$.

It is important to observe that the *total* number of f^{λ}-cuts associated with a given simplex tableau is *at most* equal to D regardless of the source row (or even an integer combination of a number of source rows) used. This result is justified in Chapter 6. It is possible, however, that one source row may generate all the cuts (by using nonnegative integer $\lambda \leqslant D - 1$), while another source row will generate some of these cuts only.

4. *Strength of the f-cut.*   In (3) above, it is shown that by using different integer values of λ, a finite number of different cuts can be generated from the simplex tableau. The important point, however, is that some of these cuts are stronger than others. A cut is said to be stronger if it cuts more deeply in the solution space. One way of expressing this mathematically is as follows: The f^*-cut is stronger than the f^{**}-cut if $f_{kj}^* \leqslant f_{kj}^{**}$ for all j and $f_k^* \geqslant f_k^{**}$, with the *strict* inequality holding at least once. It is uniformly stronger if the strict inequality holds throughout.

Naturally, the above definition of cut strength is not of direct computational value. Probably the only way to implement it is to compare

---

† This result follows since if B is the current basis, then $B^{-1} = \text{Adj } B/\det B$. Because all the elements of B are integers, then so are those of the adjoint matrix, Adj B. The elements α_{kj}, $k = 1, 2, \ldots, m$, are given by $B^{-1}P_j$, where P_j defines the constraint coefficients associated with the variable x_j. Again, because all the elements of P_j are integers, then $\alpha_{kj} = I_{kj}/D$, where I_{kj} is an arbitrary integer. It then follows that $f_{kj} = d_{kj}/D$, where $0 \leqslant d_{kj} < D$.

all cuts, but this may not be feasible for large D. Consequently, there is a need for a simple procedure to be used for choosing a "strong" cut.

One way to determine a strong cut is by judiciously selecting a source row from among all the candidate rows in the tableau. Empirical measures reflecting the above definition of cut strength are to select the source row that has (a) $\max_i\{f_i | i \in M\}$, or (b) $\max_i\{f_i / \sum_{j \in NB} f_{ij} | i \in M\}$.

Although the above two conditions reflect primarily the depth of the cut, such a choice may be slightly misleading since a cut may be the deepest but unfortunately in the ineffective direction, that is, from the standpoint of reducing the value of the objective function. (To the author's knowledge, none of the available integer codes implements the idea of accounting for the objective value.) It thus appears that a more sophisticated method can be devised by applying the concept of penalties, which takes the objective value into consideration (cf. Section 4.3.2-B). If the f_i-cut for the ith basic variable is

$$- \sum_{j \in NB} f_{ij} x_j + s_i = -f_i, \qquad i \in M$$

the associated penalty P_i obtained as a result of applying the f_i-cut to the current tableau is

$$P_i = f_i \min_{j \in NB} \{\bar{c}_j / f_{ij}\}, \qquad i \in M$$

which, as shown in Section 4.3.2, is developed under the assumption of no change in basis. Consequently the best source row k is the one associated with

$$P_k = \max_{i \in M} P_i = \max_{i \in M} f_i \left(\min_{j \in NB} \{\bar{c}_j / f_{ij}\} \right)$$

This selection may yield the largest decrease in the objective value, but this result is not certain since P_i is a lower bound on the exact amount of decrease. However, it is interesting to inspect the expression for P_k. Let $r \in NB$ be the subscript corresponding to $\min\{\bar{c}_j / f_{ij} | j \in NB\}$. Then P_k may be written as

$$P_k = \bar{c}_r \max_{i \in M} \{f_i / f_{ir}\}$$

This shows that the selection of k by using the penalties may also be consistent with the selection of a deep cut. This is seen since the expression $\max\{f_i / f_{ir}\}$, over $i \in M$, may itself be taken as an empirical measure for selecting a strong cut.

Another interesting method for improving the strength of the f-cut is due to V. J. Bowman and Nemhauser (1971). The f-cut may be written as

$$\sum_{j \in NB} f_{kj} x_j \geqslant f_k$$

This cut is strongest if its right-hand side is as large as possible provided the integer requirement on all the variables is not violated and that none of the feasible integer points is excluded. But this is equivalent to making $\sum_{j \in NB} f_{kj} x_j$ as small as possible subject to the feasibility restrictions. Thus the optimum value of f_k may be determined as

$$f_k^* = \min\left\{ \sum_{j \in NB} f_{kj} x_j \,\middle|\, x_j \in R, j \in NB \right\}$$

where $R = \{x \,|\, x \text{ is integer and feasible with respect to the original problem}\}$.

The difficulty here is that the determination of f_k^* entails solving another integer problem that is not any simpler than the original one. But observe that the set R need not be as restrictive as the original solution space. Indeed, any superset of R should yield a valid value of f_k^*. For example, it may be determined with the only restriction being the source row (from which the cut is generated) and all the variables being nonnegative integers. Of course, the resulting cut would not be as strong, but the associated problem becomes computationally manageable. It is not clear, however, that the additional computations associated with this method can be justified.

▶**EXAMPLE 5.2-1**   Maximize

$$z = x_1 - 3x_2 + 3x_3$$

subject to

$$
\begin{aligned}
2x_1 + x_2 - x_3 + x_4 &&&= 4 \\
4x_1 - 3x_2 &&+ x_5 &&= 2 \\
-3x_1 + 2x_2 + x_3 &&&+ x_6 = 3 \\
x_j \geq 0 \text{ and integer,} \quad j = 1, 2, \ldots, 6
\end{aligned}
$$

The continuous optimum will be computed in this example (and also throughout the chapter) by using the column tableau format presented in Section 2.5.4. The main reason is that every variable in the original integer problem is expressed as a basic variable in the column tableau. Thus, after the f-cut is pivoted on, its row will be of the form $S - S = 0$, where S is its slack variable. This is a redundant constraint and hence can be deleted after pivoting is effected. As a result, the number of basic variables (and hence the number of rows) always remains equal to the original number of variables.

It will be assumed that the continuous optimum tableau is l-dual feasible (see Section 2.5.5). Otherwise, the optimum tableau is modified to satisfy this condition as follows: (i) In the minimization case, the rows are rearranged so that the matrix $-I$ constitutes the rows immediately below the objective row. (ii) In the maximization case, a simple rearrangement of the rows may give the required result. Otherwise, augment the constraint $\sum_{j \in NB} x_j \leqslant M$ immediately below the objective row. l-dual feasibility is important because it eliminates dual degeneracy, an essential condition for proving convergence (see Section B).

The optimum continuous tableau is given in Table 5-1 and it is already l-dual feasible. The f-cut generated from the tableau is added at the bottom and then deleted after pivoting is effected. The source row will be selected based on the basic variable with the largest fraction with ties broken arbitrarily. Later, the same example is solved by selecting the source row based on the method of penalties.

Table 5-1

		x_5	x_2	x_6	
z	14	$\frac{10}{4}$	$\frac{6}{4}$	3	
x_1	$\frac{1}{2}$	$\frac{1}{4}$	$-\frac{3}{4}$	0	
x_2	0	0	-1	0	
x_3	$4\frac{1}{2}$	$\frac{3}{4}$	$-\frac{1}{4}$	1	$\leftarrow$ Source row
x_4	$7\frac{1}{2}$	$\frac{1}{4}$	$\frac{9}{4}$	1	
x_5	0	-1	0	0	
x_6	0	0	0	-1	
S_3	$-\frac{1}{2}$	$-\frac{3}{4}$	$\boxed{-\frac{3}{4}}$	0	$\leftarrow f_3\text{-cut}$

Applying the l-dual simplex method to Table 5-1, this yields Table 5-2, from which a new cut is generated. The f_2-cut is generated from the x_2-row and is augmented as shown in Table 5-2. Table 5-3 yields the optimum integer solution.

Table 5-2

		x_5	S_3	x_6	
z	13	1	2	3	
x_1	1	1	-1	0	
x_2	$\frac{2}{3}$	1	$-\frac{4}{3}$	0	$\leftarrow$ Source row
x_3	$4\frac{2}{3}$	1	$-\frac{1}{3}$	1	
x_4	6	-2	3	1	
x_5	0	-1	0	0	
x_6	0	0	0	-1	
S_2	$-\frac{2}{3}$	0	$\boxed{-\frac{2}{3}}$	0	$\leftarrow f_2\text{-cut}$

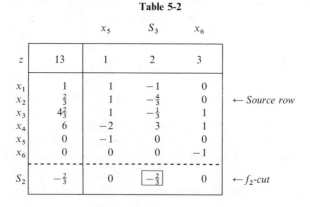

Table 5-3

		x_5	S_2	x_6	
z	11	1	3	3	
x_1	2	1	$-\frac{3}{2}$	0	
x_2	2	1	-2	0	
x_3	5	1	$-\frac{1}{2}$	1	
x_4	3	-2	$\frac{9}{2}$	1	
x_5	0	-1	0	0	
x_6	0	0	0	-1	(*Optimum*)

The coefficients of the last tableau can be made all integer by continuing to generate and add cuts. These cuts are generated from any row having fractional coefficients. Thus, in Table 5-3, any of the rows 1, 2, and 3 may be used as a source row. Actually, each of the three rows (accidently) yields the same cut,

$$-\tfrac{1}{2}S_2 + S_4 = 0$$

where S_4 is the slack variable of the cut. The augmentation of this cut to Table 5-3 will yield an all-integer tableau provided the cut is used as a pivot row. Notice that the cut will not change the optimal solution as given in Table 5-3.

It is now shown that the use of the penalties method to select the source row will lead to the optimum solution of the above example by using a smaller number of cuts. Table 5-4, which is computed from Table 5-1, summarizes the information needed to compute the penalties for x_1, x_3, and x_4. The asterisk ($*$) under f_{ij} defines j associated with P_i. The tie between $i = 1$ and $i = 4$ is broken arbitrarily. Fortunately in this case, the two rows yield the same cut.

Table 5-4

i	f_i	f_{ij}			P_i	
		$j = 5$	$j = 2$	$j = 6$		
1	$\frac{1}{2}$	$\frac{1}{4}$	$\frac{1}{4}*$	0	3	$\leftarrow k = 1$
3	$\frac{1}{2}$	$\frac{3}{4}$	$\frac{3}{4}*$	0	1	
4	$\frac{1}{2}$	$\frac{1}{4}$	$\frac{1}{4}*$	0	3	
$\bar{c}_j \rightarrow$		$\frac{10}{4}$	$\frac{6}{4}$	3		

The cut associated with $k = 1$ is given by

$$-\tfrac{1}{4}x_2 - \tfrac{1}{4}x_5 + S_1 = -\tfrac{1}{2} \qquad (f_1\text{-}cut)$$

Augmenting this cut to Table 5-1 and reoptimizing, the results are shown in Table 5-5.

Table 5-5

		x_5	S_1	x_6	
z	11	1	6	3	
x_1	2	1	-3	0	
x_2	2	1	-4	0	
x_3	5	1	-1	1	
x_4	3	-2	9	1	
x_5	0	-1	0	0	
x_6	0	0	0	-1	(*Optimum*)

It is interesting to notice that the use of the penalty method requires only one cut, which, in addition to yielding the optimum integer solution, results in an all-integer tableau. This is compared with three cuts in the other method.◀

B. *Convergence Proof under the f-Cut*

The conditions under which Gomory's algorithm is proved to converge are:

1. The value of the objective function z is bounded from below for all feasible (integer) points.
2. Only *one* cut is added at a time; that is, a single equation is added to the immediately preceding tableau from which the cut is generated.
3. The source row generating a cut is selected as the *first* equation in the tableau having $f_k \neq 0$, starting with the objective equation. This means that z must also be an integer variable.

For the current tableau, let α_0 be the right-hand side column and let α_j be the column associated with the jth nonbasic variable, where $\alpha_j \succ 0$ for all $j \in NB$ in accordance with the *l*-dual method. Let the superscript t specifically denote the tth cut,

$$-\sum_{j \in NB} f_{kj}{}^t x_j + S_k = -f_k{}^t$$

The augmentation of the $f_k{}^t$-cut to the current tableau should change its columns $\alpha_0{}^t$ and $\alpha_j{}^t$, $j \in NB$, to $\alpha_0{}^{t+1}$ and $\alpha_j{}^{t+1}$, $j \in NB$ respectively. The change is effected by using the *l*-dual method, which then guarantees that $\alpha_0{}^t > \alpha_0{}^{t+1}$. However, this together with the fact that the objective value is bounded from below are not enough to ensure finiteness, for it is possible that the decrease each time is infinitesimally small or that the objective value remains fixed above its lower bound while the values of some basic variables α_{i0} decrease indefinitely. To see that this is not the case, the following argument is introduced. Assume that the problem has a feasible integer solution.

Suppose $\alpha_{00}{}^t$ is fractional, so that a cut

$$- \sum_{j \in NB} f_{0j}{}^t x_j + S_0 = -f_0{}^t$$

is generated from it. Let $\alpha_r{}^t$ be the pivot column, then

$$\alpha_{00}{}^{t+1} = \alpha_{00}{}^t - \frac{\alpha_{0r}{}^t}{f_{0r}{}^t} f_0{}^t$$

Because $\alpha_r{}^t > 0$, it follows that $\alpha_{0r}{}^t \geqslant f_{0r}{}^t$, which shows that

$$\alpha_{00}{}^{t+1} \leqslant \alpha_{00}{}^t - f_0{}^t = [\alpha_{00}{}^t]$$

This means that $\alpha_{00}{}^{t+1}$ should decrease at least to $[\alpha_{00}{}^t]$, which is not an infinitesimal value. However, since there is a lower bound on $\alpha_{00}{}^t$, after a finite number of decrements α_{00} must remain at some fixed *integer* value for all $t > \bar{t}$.

Suppose now that $\alpha_{10}{}^{\bar{t}}$ is fractional. The associated cut is given by

$$- \sum_{j \in NB} f_{1j}{}^{\bar{t}} x_j + S_1 = -f_1{}^{\bar{t}}$$

Again, let $\alpha_{10}{}^{\bar{t}+1}$ be the pivot column; then

$$\alpha_{10}{}^{\bar{t}+1} = \alpha_{10}{}^{\bar{t}} - \frac{\alpha_{1r}{}^{\bar{t}}}{f_{1r}{}^{\bar{t}}} f_1{}^{\bar{t}}$$

Now $\alpha_{1r}{}^{\bar{t}} > 0$, for otherwise $\alpha_{0r}{}^{\bar{t}} > 0$ in order for $\alpha_r{}^{\bar{t}}$ to be *l*-positive. But this would cause $\alpha_{00}{}^{\bar{t}}$ to decrease, which contradicts the hypothesis that $\alpha_{00}{}^{\bar{t}}$ is fixed. Thus $f_{1r}{}^{\bar{t}} \leqslant \alpha_{1r}{}^{\bar{t}}$, and it follows that

$$\alpha_{10}{}^{\bar{t}+1} \leqslant \alpha_{10}{}^{\bar{t}} - f_1{}^{\bar{t}} = [\alpha_{10}{}^{\bar{t}}]$$

which shows that $\alpha_{10}{}^{\bar{t}}$ will decrease successively. Now if it becomes negative, then the α_{10}-equation must be the pivot row and the pivot element must be selected from among the *negative* $\alpha_{1j}, j \in NB$. If all $\alpha_{1j} > 0, j \in NB$, then the program must be infeasible (no integer solution), which contradicts the assumption that a feasible integer solution exists. If some $\alpha_{1j} < 0, j \in NB$, the pivoting must decrease $\alpha_{00}{}^{\bar{t}}$ since the associated α_j being l-positive necessitates that the objective equation coefficient be positive. But this contradicts the hypothesis that $\alpha_{00}{}^{\bar{t}}$ is fixed at an integer value. It thus follows that for some $t > \bar{\bar{t}} > \bar{t}$, α_{10} must be fixed at an integer value.

Continuing the same process, it can be shown that a similar argument holds for all the remaining rows. This completes the finiteness proof.

If the above proof is to be followed literally, it will be necessary always to generate the cut from the topmost noninteger row. This appears restrictive, especially in relationship to the rules given previously for the selection of a source row. In practice, however, these restrictions may actually be much stronger than needed to guarantee convergence and thus are not usually imposed. However, it may be a good idea to select the source row periodically as the topmost fractional row.

C.   *Accelerating the Effectiveness of the f-Cut*

In the preceding analysis, the successive applications of the f-cut must eventually reduce the simplex tableau into an all-integer tableau. This remark is the basis for the development of an accelerated algorithm proposed by Gomory (1963). The main idea of the algorithm is to condition the tableau through successive applications of f-cuts without actually having to achieve the optimum associated with each cut. This means that the Gaussian elimination is applied only once with each cut regardless of whether the resulting basis is infeasible. At the point where the tableau becomes all-integer, the l-dual method is applied in the normal manner to recover feasibility. If the resulting *feasible* solution is all integer, the process is terminated. Otherwise, the f-cuts are applied again. The motivation behind this procedure is to reduce the absolute value of the determinant D to one first, since this is a favorable condition for producing an integer solution.

There is a similar procedure by Martin (1963), which apparently has yielded excellent results for a special class of scheduling problems. His idea is slightly different, however, since he utilizes the f-cut only to determine the dual pivot element. He then applies a series of Gaussian eliminations that reduce the source row *only* to an all-integer equation. Using the "reverse inversion" step, he then develops from the new source row a cut that is then utilized as a regular f-cut. The development of Martin's cut is outlined in Problem 5-7.

▶EXAMPLE 5.2-2   Maximize

$$z = -x_1 + 3x_2$$

subject to

$$-2x_1 + 3x_2 \leqslant 3$$
$$4x_1 + 5x_2 \geqslant 10$$
$$x_1 + 2x_2 \leqslant 5$$
$$x_1, x_2 \geqslant 0 \quad \text{and integers}$$

The optimal linear program is given in Table 5-6. The f_2-cut is then developed and, by using it as a pivot row, *one* application of the *l*-dual method then generates Table 5-7. Although Table 5-7 is primal infeasible, feasibility is not recovered until the tableau is reduced to all-integer coefficients. Thus the f_0-cut is added to Table 5-7 to produce Table 5-8 (after one application of the *l*-dual method). Since the tableau is all integer now, the *l*-dual method is used to recover feasibility. This leads to Table 5-9, which is not a feasible integer solution. The application of the f_0'-cut yields the optimal integer solution in Table 5-10.

TABLE 5-6

		x_3	x_5	
z	$\frac{30}{7}$	$\frac{5}{7}$	$\frac{3}{7}$	
x_1	$\frac{9}{7}$	$-\frac{2}{7}$	$\frac{3}{7}$	
x_2	$\frac{13}{7}$	$\frac{1}{7}$	$\frac{2}{7}$	$\leftarrow$ *Source row*
x_3	0	-1	0	
x_4	$\frac{34}{7}$	$-\frac{3}{7}$	$\frac{22}{7}$	
x_5	0	0	-1	
S_2	$-\frac{6}{7}$	$-\frac{1}{7}$	$-\frac{2}{7}$	$\leftarrow f_2$-*cut*

TABLE 5-7

		S_2	x_3	
z	3	$\frac{3}{2}$	$\frac{1}{2}$	$\leftarrow$ *Source row*
x_1	0	$\frac{3}{2}$	$-\frac{1}{2}$	
x_2	1	1	0	
x_3	0	0	-1	
x_4	-5	11	-2	
x_5	3	$-\frac{7}{2}$	$\frac{1}{2}$	
S_0	0	$-\frac{1}{2}$	$-\frac{1}{2}$	$\leftarrow f_0$-*cut*

TABLE 5-8[a]

		S_2	S_0
z	3	1	1
x_1	0	2	-1
x_2	1	1	0
x_3	0	1	-2
x_4	-5	13	-4
x_5	3	-4	1

[a] Tableau is all integer but infeasible. Apply *l*-dual simplex to recover feasibility.

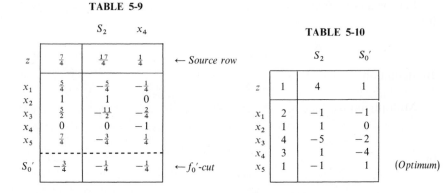

TABLE 5-9

	S_2	x_4		
z	$\frac{7}{4}$	$\frac{17}{4}$	$\frac{1}{4}$	← Source row
x_1	$\frac{5}{4}$	$-\frac{5}{4}$	$-\frac{1}{4}$	
x_2	1	1	0	
x_3	$\frac{5}{2}$	$-\frac{11}{2}$	$-\frac{2}{4}$	
x_4	0	0	-1	
x_5	$\frac{7}{4}$	$-\frac{3}{4}$	$\frac{1}{4}$	
S_0'	$-\frac{3}{4}$	$-\frac{1}{4}$	$-\frac{1}{4}$	← f_0'-cut

TABLE 5-10

	S_2	S_0'	
z	1	4	1	
x_1	2	-1	-1	
x_2	1	1	0	
x_3	4	-5	-2	
x_4	3	1	-4	
x_5	1	-1	1	(*Optimum*)

The only deviation from the regular use of the f-cut is that in Table 5-7 no immediate attempt was made to clear the infeasibility ($x_4 = 5$) since the coefficients are not all integer.◄

It is interesting to notice that, although the idea behind accelerating the effectiveness of the f-cut seems justified on a theoretical basis, there is no guarantee that this procedure will always lead to a faster convergence. A case in point is the preceding example, where the application of the l-dual method to Table 5-7 (that is, without the use of the f_0-cut) would have yielded Table 5-9 directly. This shows that the computation in Table 5-8 could have been eliminated. Nevertheless, the merit of the new idea cannot be based on the experience of one problem.

5.2.2   All-Integer Cut†

In Section 5.2.1 the use of the f-cut has the basic disadvantages that it gives rise to severe machine round-off error. Although this difficulty may be accounted for by storing numerators and denominators separately, thus avoiding floating point calculations, the resulting integer numbers may easily exceed the computer capacity. Gomory rectified this difficulty by developing a new type of cut, which consists of all-integer coefficients with its pivot element necessarily equal to -1. Thus, if the original tableau consists of all-integer coefficients, the integer property will be maintained all the time and the round-off error becomes nonexistent.

The idea of the new cut is to start with a *dual* feasible tableau with all-integer coefficients. Unlike the algorithm in Section 5.2.1, it will not be necessary to solve an initial continuous linear program. Rather, the new Gomory cut is derived directly from the initial dual-feasible tableau.

† Gomory (1960a).

Let the source row be associated with the basic x_k variable and define it as

$$x_k = b_k - \sum_{j \in NB} a_{kj} x_j \quad (source\ row)$$

By definition, $b_k < 0$. It is assumed that the source is selected such that $b_k = \min_i \{b_i \,|\, b_i < 0,\ i \in M\}$ and the problem is a maximization type.

Multiplying the source row by $1/\lambda$, $\lambda \neq 0$, this gives

$$(x_k/\lambda) + \sum_{j \in NB} (a_{kj}/\lambda) x_j = b_k/\lambda$$

or

$$([1/\lambda] + f_\lambda) x_k + \sum_{j \in NB} ([a_{kj}/\lambda] + f_{kj}) x_j = b_k/\lambda$$

Since all the variables are nonnegative, it follows that

$$[1/\lambda] x_k + \sum_{j \in NB} [a_{kj}/\lambda] x_j \leqslant b_k/\lambda$$

Because x_k and $x_j, j \in NB$, are required to be integers, then

$$[1/\lambda] x_k + \sum_{j \in NB} [a_{kj}/\lambda] x_j \leqslant [b_k/\lambda]$$

Subtracting this inequality from $[1/\lambda]$ multiple of the source row, this would give

$$\sum_{j \in NB} ([1/\lambda] a_{kj} - [a_{kj}/\lambda]) x_j \geqslant [1/\lambda] b_k - [b_k/\lambda]$$

Now if $\lambda > 1$, then the required Gomory cut is given by

$$\sum_{j \in NB} [a_{kj}/\lambda] x_j + S_k = [b_k/\lambda] \quad (\lambda\text{-}cut)$$

where S_k is a nonnegative slack variable. Since $x_j = 0$ for all $j \in NB$ and $b_k < 0$, it follows that $S_k < 0$, which is infeasible.

In order to maintain the integer property of the tableau, the pivot element in the λ-cut must be equal to -1. Clearly, this is satisfied if every negative coefficient in the left-hand side of the λ-cut is equal to -1. This can be achieved by selecting λ (>1) such that

$$\lambda \geqslant \max_{j \in NB} \{-a_{kj} \,|\, a_{kj} < -1\}$$

However, one can choose λ more selectively to produce the most decrease in the objective value. In fact, if b_0 is the current objective value and a_r the pivot column, then the new value of b_0 is given by (notice that the pivot element is assumed equal to -1):

$$b_0' = b_0 + [b_k/\lambda] a_{0r}$$

where $b_k < 0$. Thus unless $a_{0r} = 0$, b_0' is decreased the most if λ is the smallest. This means that λ should be selected as small as possible while maintaining the property that the pivot element is equal to -1.

By the lexicographic property of the (maximization) dual tableau, a_r is determined such that

$$\frac{-1}{[a_{kr}/\lambda]} a_r = l\text{-min}\left\{\frac{-1}{[a_{kj}/\lambda]} a_j \,\middle|\, j \in J^-\right\}, \qquad J^- = \{j \,|\, a_{kj} < 0, j \in NB\}$$

Then a necessary condition for $[a_{kr}/\lambda]$ to be equal to -1 is

$$\frac{-1}{(-1)} (a_r) = a_r \prec \frac{-1}{[a_{kj}/\lambda]} a_j \prec \frac{-1}{(-1)} a_j = a_j, \qquad j \in J^-$$

which means that

$$a_r = l\text{-}\min_{j \in J^-} \{a_j\}$$

Let μ_j be a positive integer. Using the necessary condition for the pivot element to be -1, determine μ_j^* as the *largest* integer satisfying

$$a_r \prec \frac{1}{\mu_j^*} a_j, \qquad j \in J^-$$

(Notice that $\mu_r^* = 1$). Thus $\mu_j^*, j \in J^-$, define the maximum ranges that guarantee a_r to be the pivot column in the lexicographic sense, which means that

$$-a_{kj}/\lambda \leqslant -[a_{kj}/\lambda] \leqslant \mu_j^*, \qquad j \in J^-$$

or

$$\lambda \geqslant -a_{kj}/\mu_j^*, \qquad j \in J^-$$

Thus, the *smallest* λ satisfying the lexicographic condition is

$$\lambda = \max_{j \in J^-}\{-a_{kj}/\mu_j^*\}$$

As specified by the λ-cut, the condition $\lambda > 1$ is always satisfied by the above selection except when the natural pivot element of the source row is equal to -1. In this case, λ is equal to 1. When this situation occurs, no λ-cut is generated, and instead the source row is used directly as a pivot row. (See Problem 5-11.)

The preceding λ-cut can be strengthened further by following an observation due to Wilson (1967). Consider the λ-cut in the form

$$S_k = [b_k/\lambda] - \sum_{j \in NB} [a_{kj}/\lambda]x_j \geqslant 0$$

or

$$[b_k/\lambda] - \sum_{j \in J^+} [a_{kj}/\lambda]x_j - \sum_{j \in J^-} [a_{kj}/\lambda] x_j \geq 0$$

where $J^- \ (J^+) = \{j \,|\, a_{kj} < 0 \ (a_{kj} > 0)\}$. Then the cut is stronger if the left-hand side is closest to zero value. This can occur by choosing larger values of λ such that each $(-[a_{kj}/\lambda]), j \in J^-$, becomes as small as possible, but provided $(-[b_k/\lambda])$ and $[a_{kj}/\lambda], j \in J^+$, remain unchanged. (Notice that these latter values can only decrease with the increase in λ, which weakens the cut.) Wilson shows that the largest λ^* satisfying the above condition is

$$\lambda^* = \max\{\lambda, \min_{j \in J_o^+} \lambda_j^*\}$$

where

$$J_0^+ = J^+ \cup \{0\}$$

$$\lambda_0^* = \left(\frac{b_k}{1 + [b_k/\lambda]}\right) - \varepsilon, \qquad \varepsilon > 0 \quad \text{and sufficiently small}$$

$$\lambda_j^* = \frac{a_{kj}}{[a_{kj}/\lambda]}, \qquad j \in J^+$$

where λ is the value obtained from the previous analysis. Notice that the choice of λ^* still keeps the pivot element equal to -1.

To illustrate the effectiveness of Wilson's idea, let the objective equation and source row, respectively, be given by

$$z = 15 - (x_2 + 2x_3 + 3x_4 + x_5)$$
$$x_1 = -20 - (-7x_2 - 8x_3 - 15x_4 + 18x_5)$$

Thus $\lambda = 7$, which yields the cut

$$x_2 + 2x_3 + 3x_4 - 2x_5 + S = -3$$

Now

$$\lambda_0^* = \frac{-20}{1 + [-20/7]} - \varepsilon = 10 - \varepsilon$$

$$\lambda_5^* = \frac{18}{[18/7]} = 9$$

Thus $\lambda^* = \max\{7, \min(10 - \varepsilon, 9)\} = 9$, yielding the cut

$$x_2 + x_3 + 2x_4 - 2x_5 + S^* = -3$$

which is stronger since x_3 and x_4 have smaller coefficients.

▶**EXAMPLE 5.2-3**   Maximize

$$z = -2x_1 - 5x_2 - 2x_3$$

subject to

$$
\begin{aligned}
x_1 - 4x_2 - \ x_3 + x_4 \qquad\qquad &= -7 \\
-x_1 - 2x_2 + \ x_3 \qquad + x_5 \qquad &= -4 \\
-3x_1 - \ x_2 - 2x_3 \qquad\qquad .\ + x_6 &= -5 \\
x_1, x_2, x_3 &\geqslant 0 \quad \text{and integer}
\end{aligned}
$$

Using the *l*-dual simplex method, the initial tableau is given in Table 5-11. The source rows will be identified throughout by the largest negative values.

TABLE 5-11

		x_1	x_2	x_3
	z	0	2	5	2
	x_1	0	-1	0	0
	x_2	0	0	-1	0
	x_3	0	0	0	-1
Source row →	x_4	-7	1	-4	-1
	x_5	-4	-1	-2	1
	x_6	-5	-3	-1	-2
λ_4-cut →	S_4	-4	0	-2	$\boxed{-1}$

The λ_4-cut is constructed from the x_4-row. Thus

$$\mu_3^* = 1, \qquad \mu_2^* = [\tfrac{5}{2}] = 2, \qquad \mu_1^* = -$$

$$\lambda_1 = -, \qquad \lambda_2 = \frac{-(-4)}{2} = 2, \qquad \lambda_3 = \frac{-(-1)}{1} = 1$$

Hence $\lambda = \max\{2, 1\} = 2$, and the λ_4-cut is given by

$$\left[\frac{1}{2}\right]x_1 + \left[\frac{-4}{2}\right]x_2 + \left[\frac{-1}{2}\right]x_3 + S_4 = \frac{-7}{2}$$

or

$$0x_1 - 2x_2 - x_3 + S_4 = -4$$

The result of applying the dual method is given in Table 5-12. The successive tableaus are shown in Tables 5-13 to 5-15.

TABLE 5-12[a]

			x_1	x_2	S_4
	z	-8	2	1	2
	x_1	0	-1	0	0
	x_2	0	0	-1	0
	x_3	4	0	2	-1
	x_4	-3	1	-2	-1
Source row →	x_5	-8	-1	-4	1
	x_6	3	-3	3	-2
λ_5-cut →	S_5	-2	-1	$\boxed{-1}$	0

[a] Pivot column: x_2. $\mu_1 = 2$, $\mu_2 = 1$, $\mu_4 = $ —, $\lambda_1 = \frac{1}{2}$, $\lambda_2 = 4$, $\lambda_4 = $ —, $\lambda = \max \{\frac{1}{2}, 4\} = 4$.

TABLE 5-13[a]

			x_1	S_5	S_4
	z	-10	1	1	2
	x_1	0	-1	0	0
	x_2	2	1	-1	0
	x_3	0	-2	2	-1
	x_4	1	3	-2	-1
	x_5	0	3	-4	1
Source row →	x_6	-3	-6	3	-2
λ_6-cut →	S_6	-1	$\boxed{-1}$	0	-1

[a] Pivot column: x_1. $\mu_1 = 1$, $\mu_5 = $ —, $\mu_4 = 2$, $\lambda_1 = 6$, $\lambda_5 = $ —, $\lambda_4 = 1$, $\lambda = \max \{6, 1\} = 6$.

It is important to notice that the convenience of dealing with integer coefficients only is attained at the expense of designing a weaker cut. Specifically, in selecting the smallest value of λ that yields the most decrease in the objective value, it was conditional that the pivot element remain equal to -1. As one might suspect, the computational efficiency of this algorithm has generally proven to be inferior to that of the fractional algorithms (Section 5.2.1).◄

The proof of convergence for the preceding algorithm is very much the same as in the fractional algorithm and hence is not discussed here.

TABLE 5-14[a]

		S_6	S_5	S_4
z	-11	1	1	1
x_1	1	-1	0	1
x_2	1	1	-1	-1
x_3	2	-2	2	1
x_4	-2	3	-2	-4
Source row → x_5	-3	3	-4	-2
x_6	3	-6	3	4
λ_5'-cut → S_5'	-1	0	$\boxed{-1}$	-1

[a] Pivot column: S_5. $\mu_6 = -$, $\mu_5 = 1$, $\mu_6 = 1$, $\lambda_6 = -$, $\lambda_5 = 4$, $\lambda_6 = 2$, $\lambda = \max\{4, 2\} = 4$.

TABLE 5-15[a]

		S_6	S_5'	S_4
z	-12	1	1	0
x_1	1	-1	0	1
x_2	2	1	-1	0
x_3	0	-2	2	-1
x_4	0	3	-2	-6
x_5	1	3	-4	2
x_6	0	-6	3	1

[a] Optimal and feasible: $z = -12$, $x_1 = 1$, $x_2 = 2$, $x_3 = 0$, $x_4 = 0$, $x_5 = 1$, $x_6 = 0$.

5.2.3 Mixed Cut†

In the previous two sections, the cuts are derived under the assumption that all the variables are integer. In the mixed integer problem, it is necessary to develop a different cut. The basis for constructing the mixed cut will thus be presented here.

Let x_k be an integer variable in the optimum (continuous) linear program and let its equation be given by

$$x_k = \beta_k - \sum_{j \in NB} \alpha_{kj} w_j \qquad \text{(source row)}$$

† Gomory (1960b).

where $w_j, j \in NB$, are the associated nonbasic variables. It was indicated in Section 4.3.2 in connection with the branch-and-bound algorithm that a necessary condition for x_k to be integer is that one of the two constraints $x_k \geqslant [\beta_k] + 1$ and $x_k \leqslant [\beta_k]$ be satisfied. This necessary condition will be the basis for developing the mixed cut.

From the x_k-equation,

$$x_k - [\beta_k] = f_k - \sum_{j \in NB} \alpha_{kj} w_j$$

Now if $x_k \leqslant [\beta_k]$, this is equivalent to

$$- \sum_{j \in NB} \alpha_{kj} w_j \leqslant -f_k \qquad (1)$$

which would be the required cut if it is known a priori that the integer value of x_k does not exceed $[\beta_k]$.

On the other hand, if integer $x_k \geqslant [\beta_k] + 1$, then this is equivalent to

$$\sum_{j \in NB} \alpha_{kj} w_j \leqslant f_k - 1 \qquad (2)$$

Since one does not know in advance which of the two conditions (1) and (2) must hold, it is necessary to develop one constraint that combines (1) and (2). Let

$$J^+ (J^-) = \{j \mid \alpha_{kj} > 0 \ (\alpha_{kj} < 0)\}$$

Then (1) implies

$$- \sum_{j \in J^+} \alpha_{kj} w_j \leqslant - \sum_{j \in NB} \alpha_{kj} w_j \leqslant -f_k \qquad (1')$$

Similarly, (2) implies

$$\sum_{j \in J^-} \alpha_{kj} w_j \leqslant \sum_{j \in NB} \alpha_{kj} w_j \leqslant f_k - 1$$

or, multiplying both sides by $f_k/(1 - f_k) > 0$, then

$$\sum_{j \in J^-} \left(\frac{f_k \alpha_{kj}}{1 - f_k} \right) w_j \leqslant -f_k \qquad (2')$$

Now since by definition (1') and (2') are mutually exclusive, then the necessary conditions for integrality can be combined into one constraint as

$$\sum_{j \in J^-} \left(\frac{f_k \alpha_{kj}}{1 - f_k} \right) w_j - \sum_{j \in J^+} \alpha_{kj} w_j \leqslant -f_k \qquad (m\text{-}cut)$$

which is the required mixed cut.

The above m-cut can be strengthened further if one takes into account the fact that some nonbasic variables w_j are also integers. The original

x_k-equation shows that the integer property of x_k will not be disturbed if α_{kj} associated with an integer w_j is increased or decreased by any integer quantity, which means that such α_{kj}'s can be made positive or negative regardless of their values in the x_k-equation. This again means that for integer w_j the dichotomization of its α_{kj} according to J^+ and J^- is artificial.

One can make use of the above result to reduce the coefficients of w_j in the m-cut in the manner that will make it the strongest. Thus, for an integer w_j, the smallest value of $\alpha_{kj} < 0$ $(\alpha_{kj} > 0)$ is $f_{kj} - 1$ (f_{kj}), where $\alpha_{kj} = [\alpha_{kj}] + f_{kj}$, $0 \leqslant f_{kj} \leqslant 1$. This means that in the m-cut the smallest *absolute* values for the coefficients must be $f_k(1 - f_{kj})/(1 - f_k)$ for $j \in J^-$ and f_{kj} for $j \in J^+$. But since it was shown above that the dichotomization $j \in J^+$ and $j \in J^-$ is insignificant if w_j is integer, then the smallest coefficients for an integer w_j, $j \in I$, are given by

$$\lambda_{kj} = \min\left\{ \frac{f_k(1 - f_{kj})}{1 - f_k}, f_{kj} \right\}, \qquad j \in I$$

which yields

$$\lambda_{kj} = \begin{cases} f_{kj}, & \text{if } f_{kj} < f_k \\ \dfrac{f_k(1 - f_{kj})}{1 - f_k}, & \text{if } f_{kj} \geqslant f_k \end{cases}$$

and the stronger m-cut is given by

$$\sum_{j \in J_n^-} \frac{f_k \alpha_{kj}}{1 - f_k} w_j - \sum_{j \in J_n^+} \alpha_{kj} w_j - \sum_{j \in I} \lambda_{kj} w_j \leqslant -f_k \qquad (m^*\text{-cut})$$

where

$$J_n^- \ (J_n^+) = \{ j \mid j \in J^- \cap \sim I \ (j \in J^+ \cap \sim I) \}.$$

(The notation $\sim I$ means not I.)

It is interesting to notice that if $J_n^- = J_n^+ = \varnothing$, that is, if all the variables are integers (pure problem), then the m^*-cut will yield a stronger cut than the f-cut (Section 5.2.1). This is easily seen, since for $f_{kj} \geqslant f_k$

$$f_k(1 - f_{kj})/(1 - f_k) < f_{kj}$$

which follows directly from the definition of λ_{kj}. This result was recognized by Scheurmann (1971) and independently by Salkin (1971).

It is important to notice, however, that if in the pure problem the m^*-cut is used in place of the f-cut, the associated slack variable will no longer be restricted to be integer. Thus, following the *first* application of the m^*-cut *as a pure integer cut*, it will be necessary to resort back to the m^*-cut in its *mixed* mode, that is, J_n^+ and J_n^- may no longer be empty and, as a result, the m^*-cut will not be comparable with the f-cut. At this point, it is not clear, in general, which procedure should lead to a rapid convergence to the optimal integer solution.

The convergence proof in the mixed case is very much the same as the one in Section 5.2.1. It must be stated, however, that the proof relies on the fact that the objective variable z must be integer, a situation which may not be satisfied in general due the mixed nature of the variables.

▶**EXAMPLE 5.2-4** Maximize

$$z = -5x_2 - 10x_4 + 20$$

subject to

$$x_1 - \tfrac{5}{3}x_2 \quad - \tfrac{1}{3}x_4 = \tfrac{5}{3}$$
$$- \tfrac{4}{3}x_2 + x_3 + \tfrac{11}{3}x_4 = \tfrac{7}{3}$$
$$x_1, x_2 \geqslant 0$$
$$x_3, x_4 \geqslant 0 \quad \text{and integer}$$

This problem happens to yield the optimum continuous solution directly. The associated l-dual tableau is given in Table 5-16.

TABLE 5-16

		x_2	x_4	
z	20	5	10	
x_1	$1\tfrac{2}{3}$	$-\tfrac{5}{3}$	$-\tfrac{1}{3}$	
x_2	0	-1	0	
x_3	$2\tfrac{1}{3}$	$-\tfrac{4}{3}$	$\tfrac{11}{3}$	← *Source row*
x_4	0	0	-1	
S_3	$-\tfrac{1}{3}$	$-\tfrac{2}{3}$	$-\tfrac{1}{6}$	← m_3^*-*cut*

The cut is developed from the x_3-equation, which can be written as

$$x_3 + \frac{-4}{3}x_2 + \left(3 + \frac{2}{3}\right)x_4 = 2 + \frac{1}{3}$$

(Notice that x_2 is continuous and x_4 is integer.) Thus the mixed cut is given by

$$S_3 + ((\tfrac{1}{3})(-\tfrac{4}{3})/(1 - \tfrac{1}{3}))x_2 - ((\tfrac{1}{3})(1 - \tfrac{2}{3})/(1 - \tfrac{1}{3}))x_4 = -\tfrac{1}{3}$$

or

$$S_3 - \tfrac{2}{3}x_2 - \tfrac{1}{6}x_4 = -\tfrac{1}{3} \qquad (m_3^*\text{-}cut)$$

This yields the results in Table 5-17. Since $x_3 = 3$ and $x_4 = 0$ are integers, the optimum solution is at hand.

TABLE 5-17

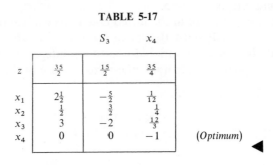

	S_3		x_4	
z	$\frac{35}{2}$	$\frac{15}{2}$	$\frac{35}{4}$	
x_1	$2\frac{1}{2}$	$-\frac{5}{2}$	$\frac{1}{12}$	
x_2	$\frac{1}{2}$	$\frac{3}{2}$	$\frac{1}{4}$	
x_3	3	-2	$\frac{12}{3}$	
x_4	0	0	-1	*(Optimum)* ◀

5.2.4   Bound-Escalation Cut†

Glover (1965a) proposed a dual algorithm for solving a pure integer problem of the form:

maximize

$$z = \sum_{j \in NB} c_j w_j$$

subject to

$$\sum_{j \in NB} a_{ij} w_j \geq b_i, \qquad i \in M$$

$$w_j \geq 0 \quad \text{and integer}$$

It is assumed that all c_j, a_{ij}, and b_i are integers. Moreover, the starting tableau associated with the problem is dual feasible (that is, $c_j \leq 0$ for all j and $b_i < 0$ for at least one $i \in M$) with all its columns (except possibly the right-hand side column) being *l*-positive.

The proposed method has the general properties of Gomory's all-integer algorithm (Section 5.2.3) but differs, at least in the apparent sense, in that no cuts are developed explicitly nor is it necessary to carry out the pivoting process of the *l*-dual method in the usual manner.

Basically, the general idea of Glover's method is to find the optimal integer solution by determining lower bounds on each w_j in such a way as to satisfy a necessary integrality condition. For example, if in the constraint $\sum_{j \in NB} a_{ij} w_j \geq b_i > 0$, $a_{ik} > 0$ while all $a_{ij} \leq 0$, $j \in NB$ and $j \neq k$, then w_k, being integer, must satisfy the lower bound

$$w_k \geq \langle b_i / a_{ik} \rangle$$

† Glover (1965a).

where $\langle b_i/a_{ik}\rangle$ is the *smallest* integer $\geqslant b_i/a_{ik}$. Glover, however, does not consider the constraints one at a time, but uses a number of constraints simultaneously, which allows the determination of lower bounds on several variables. This requires that each such constraint contain but one positive a_{ij}. Since this is not generally the case, the desired result can be effected by systematically adding or subtracting the appropriate left columns of the tableau. This naturally leads to a transformation of the original integer variables from w_j to say w_j', where w_j' are also integers. Glover proves that such transformation is unique so that there is always a one-to-one correspondence between the original and transformed variables and that the optimum of the transformed problem implies the optimum for the original problem. This means that the optimal values of the original problem can always be recovered from the current transformed problem.

Realizing now the uniqueness of the problem transformation, Glover develops rules for successively escalating the lower bounds on w_j'. As is known from the theory of lower bounding technique, such lower bounds can only affect the column on the right of the tableau in the sense that $\alpha_0' = \alpha_0 - \beta_k \alpha_k$, where α_0 and α_0' are the current and updated right columns, β_k is defined as the integer lower bound on w_k, and α_k is the left column associated with w_k. Now if α_k is always selected *l*-positive and if $\beta_k > 0$, then eventually α_0 should become nonpositive. At this point, the process ends, since the optimum solution is given by $w_j' = 0$. (Notice that all $c_j \leqslant 0$, which according to the rules of the algorithm will remain to satisfy dual feasibility.)

The above discussion provides only the general idea of Glover's bound-escalation method. Rather than presenting the details, a special case is considered where the lower bound of only one variable at a time is computed. The objective here is to give some insights into Glover's method and to show its basic relationship to Gomory's all-integer method.

Define a source constraint such that

$$\sum_{j\in J^+} a_{ij}w_j + \sum_{j\in J^-} a_{ij}w_j \geqslant b_i > 0 \qquad (source\ constraint)$$

where $J^+\,(J^-) = \{j\in NB\,|\,a_{ij} > 0\,(a_{ij} < 0)\}$. Designate $w_k, k\in J^+$, as the (integer) variable for which a lower bound is to be determined. Let $m_{ij} = \langle a_{ij}/a_{ik}\rangle, j\in J^+$, and define y_k such that

$$y_k = w_k + \sum_{\substack{j\in J^+ \\ j\neq k}} m_{ij}w_j \geqslant 0 \qquad (y_k\text{-}equation)$$

By definition, y_k is a nonnegative integer variable. Substituting for w_k in the source constraint, this gives

$$\sum_{\substack{j \in J^+ \\ j \neq k}} (a_{ij} - a_{ik} m_{ij}) w_j + a_{ik} y_k + \sum_{j \in J^-} a_{ij} w_j \geq b_i \qquad (\textit{modified source constraint})$$

Since $a_{ij} - a_{ik} m_{ij} \leq 0$, then y_k is the only variable having a positive coefficient a_{ik}, which means that

$$y_k \geq \langle b_i / a_{ik} \rangle \equiv \beta_k \qquad (\textit{bounding form})$$

The variable w_k is now transformed to y_k in *all* the constraints, with its (unique) relationship to w_k specified by the y-equation. The lower bound $y_k \geq \beta_k$ is then used to update the right column α_0 as indicated previously. Also the creation of y_k has changed the coefficients of $w_j, j \in J^+$ and $j \neq k$, as shown above. This is what Glover accomplishes by using what he calls "elemental transformation."

It is important to notice that the sequence of α_0 vectors is *l*-decreasing only if $\alpha_k \succ 0$, assuming $\beta_k > 0$. But the choice of m_{ij} as given above does not generally guarantee that $\alpha_k \succ 0$. This can be accomplished, however, by following two rules:

(i)   Select k such that $\alpha_k = l\text{-min}_{j \in J^+}\{\alpha_j\}$.
(ii) Replace m_{ij} by $t_{kj} = \min(m_{ij}, r_{ij})$, where r_{ij} is the largest integer that keeps $\alpha_j' = \alpha_j - r_{ij}\alpha_k \succ 0$, for all $j \in J^+, j \neq k$.

Rule (ii) has the drawback that if $r_{ij} < m_{ij}$ for *at least one* $j \in J^+$, then it is no longer true that the modified source constraint will have all but one of its left-side coefficients nonpositive. This means that β_k cannot be taken equal to $\langle b_i / a_{ik} \rangle$, and indeed the only explicit value β_k can have is zero. It follows that α_0 will not be modified. But since $\alpha_j, j \in J^+, j \neq k$, will be changed, a succeeding trial must eventually lead to the determination of $\beta_k > 0$. Actually, when $\beta_k = 0$, the vectors $\alpha_j, j \in J^+$, are conditioned through elemental transformation so as to preserve the *l*-positive property of the tableau.

To show that the preceding discussion can be expressed in the form of a legitimate cut, the necessary condition $y_k \geq \beta_k$ can actually be written as

$$y_k = w_k + \sum_{\substack{j \in J^+ \\ j \neq k}} m_{ij} w_j \geq \beta_k \qquad (\textit{b-cut})$$

This will be referred to as the bounding cut. Rather than substituting for w_k, the *b*-cut is applied directly to the *l*-dual tableau in the regular manner, with α_k being the appropriate pivot column. Since the coefficient of w_k is $+1$, this preserves the integer property of the tableau. It is clear that this cut must produce the same changes in α_0 and $\alpha_j, j \in J^+, j \neq k$, as $y_k \geq \beta_k$. But since y_k is not used explicitly, the successive application of the *b*-cut will terminate with the optimal values of w_j directly. Notice again that maintaining $\alpha_k \succ 0$ is important to guarantee convergence. The use of m_{ij} as defined in the b-cut may not allow this so that the cut must be modified to accomodate this point. This is best explained by considering a different interpretation of the b-cut.

It is clear that the source row implies

$$\sum_{j \in J^+} a_{ij} w_j \geqslant \sum_{j \in NB} a_{ij} w_j \geqslant b_i$$

Letting w_k be a designated variable, then the last inequality yields

$$w_k + \sum_{\substack{j \in J^+ \\ j \neq k}} \langle a_{ij}/a_{ik} \rangle w_j \geqslant \langle b_i/a_{ik} \rangle \equiv \beta_k$$

which is the b-cut. Notice now that if some m_{ij} are replaced by $r_{ij} < m_{ij}$ to preserve $\alpha_j \succ 0, j \in J^+$, then the immediately preceding inequality will no longer hold in general. But since m_{ij} and $w_j, j \in J^+$, are nonnegative, the b-cut reduces to

$$w_k + \sum_{\substack{j \in J^+ \\ j \neq k}} \min(r_{ij}, m_{ij}) w_j \geqslant 0$$

The application of this cut will not affect the α_0-vector but will cause changes in $\alpha_j, j \in J^+, j \neq k$. This is equivalent to carrying out the elemental transformation proposed by Glover.

It is interesting to show that the b-cut is a special case of the λ-cut (Section 5.2.2). The relationship is seen if the cut as determined by the all-integer algorithm is such that

$$a_{ik} = \lambda \geqslant \max_{j \in J^-}\{-a_{ij}\}, \qquad k \in J^+$$

Then the b- and λ-cuts become precisely the same. (Notice that in Section 5.2.2 the set J^+ is the same as J^- used in this section.) But it can be seen from the construction of the λ-cut that the cut will be weakened by such a choice, since this will eliminate all the terms associated with $j \in J^-$ ($j \in J^+$ in Section 5.2.2). This hints at the point that the b-cut may be inferior to the λ-cut. Although this appears to be true if a single source constraint is used to generate the lower bound, Glover indicates that this need not be the case when several bounds are acquired simultaneously.

5.2.5   Convexity Cut†

The convexity cut is a general form of legitimate integer cut that was initially started under the name "hypercylindrical" cut by Young (1971) for

---

† Young (1971), Balas (1971a), Glover (1973).

solving a special class of mixed zero–one problems in which the zero–one variables x_j are related by constraints of the type $\sum_{j=1}^{n} x_j = k$ and $0 \leqslant x_j \leqslant 1$, where $0 < k < n$. The hypercylindrical property stems from converting the special constraints to $\sum_{j=1}^{n} x_j^2 = k$. Independently of Young's work, Balas (1971a) developed what he called the "intersection" cut for solving general integer (mixed or pure) linear problems. His idea is to utilize the local property of the integer points enclosing the current optimum continuous solution to develop a legitimate integer cut. Actually, this cut is determined to pass through the intersection points of the half-lines emanating from the current optimum (continuous) extreme point with a specially defined hypersphere. The sphere passes through all the vertices of the unit hypercube enclosing the continuous extreme point.

Later, Glover (1973) developed a general theory for constructing legitimate cuts, which extends Young's and Balas' ideas to a more general class of mathematical programs including, of course, integer programming. His idea is that the hypersphere (or hypercylinder) can be replaced by any *convex* set provided that none of the feasible points lie in its *interior*. In this case, the intersection of the half-lines emanating from the continuous vertex with the convex set defines what Glover calls the "convexity" cut.

Because Glover's theory encompasses Young's and Balas' ideas (indeed, Glover shows that Gomory's m-cut can be interpreted as a convexity cut), this section starts by discussing Glover's results and then specializes to the specific cases. In order to appreciate the idea of the new cut, the graphical example in Fig. 5-2 is used for illustration. It is assumed that the two

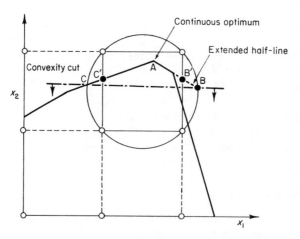

Figure 5-2

variables x_1 and x_2 are integers. The current optimum continuous vertex is shown as A. The square enclosing A replaces the hypercube referred to in Balas' intersection cut and the encompassing circle represents the convex set. Notice that no feasible integer point lies in the interior of the circle. One can see that a legitimate convex set can be constructed in a variety of ways but that the specific choice of the convex set has a direct bearing on the strength or weakness of the resulting cut. For example, the circle can be replaced legitimately by the square it encloses. In this case, the cut passes through B′ and C′ rather than B and C, which is a weaker cut. It is obvious, however, that a cut is strongest if a legitimate convex set is chosen such that the (extended) half-lines emanating from the continuous point are the longest. In Fig. 5-2, AB > AB′, AC > AC′, and thus the cut BC is stronger than B′C′. This observation is what makes Glover's theory interesting since it allows flexibility in selecting the best set that exploits the structure of the problem.

Glover's general results will now be explained. Let the mathematical programming problem be such that the current solution point is defined or implied by the constraints

$$x = x^* - \sum_{j \in NB} \alpha_j w_j$$

which are expressed in matrix form with x, x^*, and α_j column vectors. In terms of an equivalent linear programming problem, x represents the basic variable, while $w_j, j \in NB$, are the associated zero nonbasic variables. Glover's cut is described by the following lemma:

▶GLOVER'S CONVEXITY CUT LEMMA   Let S be a given set of feasible points. If R is a convex set whose interior contains no points in S and if $x = x^*$ (possibly a boundary point of R) has a deleted feasible neighborhood that lies in the interior of R, then for any constants $w_j^* > 0, j \in NB$, such that the points

$$\xi_j^* = (x^* - \alpha_j w_j^*) \in R, \qquad \text{for all} \quad j \in NB$$

the convexity cut

$$\sum_{j \in NB} w_j / w_j^* \geqslant 1 \qquad (c\text{-}cut)$$

excludes the point $x = x^*$, but never any point in S.

Proof   Assume for simplicity that $x = x^*$ is an interior point of R. Consider

$$x = r_0 x^* + \sum_{j \in NB} r_j (x^* - \alpha_j w_j^*)$$

where $r_0 + \sum_{j \in NB} r_j = 1, r_j \geq 0$ for $j \in \{0\} \cup NB$, which means that x is a convex combination of the points $x^*, \{x^* - \alpha_j w_j^*\}, j \in NB$. From the assumptions, it follows that $\sum_{j \in NB} r_j < 1$ implies that x is an interior point of R. But the above convex combination can be written as

$$x = x^* - \sum_{j \in NB} \alpha_j (r_j w_j^*)$$

This shows that the set of convex combinations satisfies the solution space associated with $x = x^*$ if $r_j w_j^* = w_j$, or $r_j = w_j/w_j^*$. It follows that the condition $\sum_{j \in NB} r_j < 1$ can be written as

$$\sum_{j \in NB} w_j/w_j^* < 1$$

which is the half-space complementing the c-cut. This means that all non-interior points $x \in S$ that satisfy $x = x^* - \sum_{j \in NB} \alpha_j w_j$ cannot be excluded by the c-cut. ◄

Although the convexity cut is developed here in the context of integer programming, the same idea is applicable to other mathematical programming problems. Notable among these is concave minimization over a convex polyhedron in which the optimum solution is characterized by its occurrence at an extreme point [see Taha (1973)]. Section 7.3.2-C will show how the fixed-charge problem, which is a special case of concave minimization over a convex polyhedron, can be solved by using the convexity cut.

The use of the convexity cut with integer linear problems is now presented. The presentation is based on Balas (1971a). This is followed by an interpretation of Gomory's m-cut in terms of Glover's theory.

A.   *Balas' Intersection Cut*

It is assumed as usual that the l-dual method is used to obtain the solution, so that the *constraint* equations associated with a current continuous tableau can be written as

$$x = x^* - \sum_{j \in NB} \alpha_j w_j, \qquad S = S^* - \sum_{j \in NB} \delta_j w_j$$

where x is the integer vector and S the (continuous) slack vector. Notice that, unlike the fractional algorithm (Section 5.2.1), S is not restricted to integer values. However, if S or a subset thereof is integer, this can be taken care of in the succeeding analysis by incorporating the appropriate elements of δ_j as part of $\alpha_j, j \in NB$.

For the purpose of illustrating the method, we further assume that the problem is of the pure integer type. The treatment of the mixed case has also been developed but will not be presented here.

Balas' development is interpreted in terms of Glover's theory as follows: Let S be the set of feasible integer points defined by the feasible vertices of the unit hypercube, whose $2n$ closed halfspaces are given by

$$[x^*] \leqslant x \leqslant [x^*] + e$$

where each element of the vector e is 1. Assume temporarily that x^* has no integer elements, so that the unit hypercube is uniquely defined. The definition of S shows that the immediate concern concentrates only on the local integer points surrounding x^*. As mentioned previously, since the above unit hypercube contains no interior feasible points, it can be used directly to represent the convex set R. A stronger cut may be obtained if this unit hypercube is replaced by a (enclosing) hypersphere whose center is $[x^*] + e/2$ with a radius $n^{1/2}/2$. The equation of this hypersphere is given by

$$x^{\mathrm{T}}(e + 2[x^*] - x) = [x^*]^{\mathrm{T}}(e + [x^*])$$

Clearly, the new set R (hypersphere) contains no interior integer feasible points.

According to Glover's theory, the values of w_j^* defining the cut are determined from the intersection of the half-lines $\xi_j = x^* - \alpha_j w_j$ with the sphere $x^{\mathrm{T}}(e + 2[x^*] - x) = [x^*]^{\mathrm{T}}(e + [x^*])$. Again, assume that $x = x^*$ is nondegenerate, so that there are exactly n distinct halflines associated with $x = x^*$. (The important case of degeneracy will be treated later.) Thus the points of intersection defined by w_j^* are determined by letting $x = \xi_j$ in the equation of the sphere and then solving for w_j, that is,

$$(x^* - w_j^*\alpha_j)^{\mathrm{T}}(e + 2[x^*] - x^* + w_j^*\alpha_j) = [x^*]^{\mathrm{T}}(e + [x^*])$$

After some algebraic manipulations, this gives

$$w_j^* = \frac{1}{p_j}\{h_j + |[(h_j^2 + f^{\mathrm{T}}(e - f)p_j)]^{1/2}|\}, \qquad j \in NB$$

where

$$f = x^* - [x^*]$$
$$p_j = \alpha_j^{\mathrm{T}}\alpha_j \quad (\text{inner product of } \alpha_j)$$
$$h_j = (f - \tfrac{1}{2}e)^{\mathrm{T}}\alpha_j$$

The absolute value of the square root is consistent with the requirement that $w_j^* > 0$. It follows that Balas' intersection cut is

$$\sum_{j \in NB} w_j/w_j^* \geqslant 1 \qquad (\textit{Balas' i-cut})$$

Notice that the expression for w_j^* is always defined.

The basis for the preceding development is that n distinct half-lines exist. However, if $x = x^*$ is degenerate, this means that the current extreme point is "overdetermined," that is, the vertex is the intersection of $n + k$ hyperplanes, $k > 0$, while in reality only n hyperplanes are necessary to determine the point.

Balas overcomes the above difficulty by dropping each constraint equation in which the associated basic variable is at zero level. These are the constraints that create the degenerate situation. He then proves that this must yield a set of n distinct half-lines that emanate from x^*. The proof depends primarily on the fact that the column vectors α_j are by construction independent. Although this method overcomes the degeneracy problem, the resulting i-cut may be weaker depending on the deleted set of constraints.

▶**Example 5.2-5** Consider Example 5.2-1. Assume that the only integer variables are x_1, x_2, and x_3. The slacks x_4, x_5, and x_6 are assumed continuous.

From the continuous optimum solution in Table 5-1 one gets

$$x^* = \begin{pmatrix} \frac{1}{2} \\ 0 \\ 4\frac{1}{2} \end{pmatrix} \quad \alpha_5 = \begin{pmatrix} \frac{1}{4} \\ 0 \\ \frac{3}{4} \end{pmatrix} \quad \alpha_2 = \begin{pmatrix} -\frac{3}{4} \\ -1 \\ -\frac{1}{4} \end{pmatrix} \quad \alpha_6 = \begin{pmatrix} 0 \\ 0 \\ 1 \end{pmatrix}$$

Thus $f^{\mathrm{T}} = (\frac{1}{2}, 0, \frac{1}{2})$, and $p_5 = \frac{5}{8}$, $p_2 = \frac{13}{8}$, $p_6 = 1$, $h_5 = 0$, $h_2 = \frac{1}{2}$, $h_6 = 0$, and $f^{\mathrm{T}}(e - f) = \frac{1}{2}$. This yields

$$w_5^* = \tfrac{8}{5}(0 + [0 + \tfrac{1}{2}(\tfrac{5}{8})]^{1/2}) = 0.894$$
$$w_2^* = \tfrac{8}{13}(\tfrac{1}{2} + [\tfrac{1}{2} + \tfrac{1}{2}(\tfrac{13}{8})]^{1/2}) = 1.01$$
$$w_6^* = 1(0 + [0 + \tfrac{1}{2}(1)]^{1/2}) = 0.707$$

and the associated i-cut is given by

$$\frac{x_5}{0.894} + \frac{x_2}{1.01} + \frac{x_6}{0.707} \geqslant 1$$

or

$$-1.118x_5 - 0.99x_2 - 1.414x_6 \leqslant -1$$

This cut is now augmented to the current tableau and the *l*-dual simplex algorithm applied. If the resulting optimum solution is still noninteger, the process is repeated to generate new cuts.◀

Notice that the values of w^* are generally irrational. However, from Glover's convexity cut lemma, any rational w' such that $0 < w' < w^*$ can replace w^*.

Several remarks will now be made concerning the strength and convergence of Balas' algorithm. Balas observes that the i-cut may be strengthened in two ways. The first is to replace the i-cut by its "integerized" equivalence by employing Gomory's theory for developing the λ-cut (Section 5.2.2). The integerization will be attractive only if it yields a stronger cut, in the sense of making the appropriate coefficients smaller. One notices also that it is unnecessary to require the pivot element to be equal to -1 unless the current tableau is (nearly) all integer. This follows because this condition generally weakens the cut.

The second way is to replace the hypersphere by a dual unit hypercube, which is defined as a "rotated" cube in which a vertex of the unit cube $[x^*] \leqslant x \leqslant [x^*] + e$ touches the center point of a face of the dual cube. A graphical example of this situation is shown in Fig. 5-3 for a two-dimensional problem, which indicates that the new cut is stronger. A general

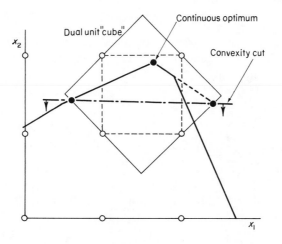

Figure 5-3

development by Balas *et al.* (1971) proves this result. Indeed, it must be obvious that the set R, as defined in Glover's theory, can be constructed in various ways to yield a strongest cut [see Balas (1971a)]. However, the difficulty here is that it may involve excessive computation.

Balas also observes that when $x = x^*$ is degenerate, the unit cube $[x^*] \leqslant x \leqslant [x^*] + e$ is not unique. In this case the specific choice of a cube may affect the strength of the cut. Balas shows how the best cube can be selected but, unfortunately, at the expense of additional computations.

Finally, the question of convergence is settled. Balas develops an argument that ensures the finiteness of the algorithm, based primarily on

(1) development of rules ensuring that the determinant of the current basis remains finite as the number of pivots tends to infinity, (2) the objective function is bounded from below, (3) cuts are added in integer form, and (4) the l-dual method is used to solve the problem.

B.   Gomory's m-Cut Viewed as a Convexity Cut

Let x_k be an integer variable whose value $x_k{}^*$ is currently noninteger. The set $[x_k{}^*] \leqslant x_k \leqslant [x_k{}^*] + 1$, which is obviously convex, can be used to replace R in Glover's lemma. The boundary planes $x_k = [x_k{}^*]$ and $x_k = [x_k{}^*] + 1$ must include all the vertices of the unit cube

$$[x^*] \leqslant x \leqslant [x^*] + e,$$

and the new R has no feasible interior point.

Now each of the half-lines from x^* will intersect at most one of the two planes $x_k = [x_k{}^*]$ and $x_k = [x_k{}^*] + 1$, since, by definition,

$$[x_k{}^*] < x_k{}^* < [x_k{}^*] + 1.$$

The intersecting plane is easily determined as the one yielding $w_j{}^* > 0$. Thus the half-lines associated with $x_k{}^*$ are $x_k{}^* - \alpha_{kj} w_j$, and their intersection points are determined by

$$x_k{}^* - \alpha_{kj} w_j{}^* = [x_k{}^*] \qquad \text{or} \qquad [x_k{}^*] + 1$$

which yields

$$w_j{}^* = \begin{cases} f_k/\alpha_{kj}, & \text{if} \quad \alpha_{kj} > 0 \\ (f_k - 1)/\alpha_{kj}, & \text{if} \quad \alpha_{kj} < 0 \\ \infty, & \text{if} \quad \alpha_{kj} = 0 \end{cases}$$

where $f_k = x_k{}^* - [x_k{}^*]$. Substituting for $w_j{}^*$ in the c-cut, the m-cut is obtained directly.

C.   Strength of the Intersection Cut

It is interesting to compute Gomory's m^*-cut (Section 5.2.3) for Example 5.2-5. (The m^*-cut is the stronger version of the m-cut). Using the first equation as the source row, the resulting m^*-cut is

$$\frac{x_5}{2} + \frac{x_2}{1} + \frac{x_6}{\infty} \geqslant 1$$

This cut is (almost) uniformly stronger than the one produced by the i-cut, a result that should be expected in general.

Although no computational experience has been reported about the intersection cut, it is not unreasonable to conclude that the new cut will not result in a real breakthrough in the computational difficulty that characterizes all integer cutting methods. However, viewed differently, the convexity cut (as introduced by Glover) has already shown promise in shedding light on new methods for solving difficult mathematical programming problems. A typical example is the fixed-charge problem, which can be formulated as a mixed 0–1 model (see Example 1.3-7). But observe that the 0–1 variables are extraneous and are used only to allow a manageable formulation of the problem. Section 7.3.2-B will show that this problem can be solved directly by using the convexity cut without having to introduce the auxiliary zero–one variables.

Glover (1973) has also originated a new type of cut, called "cut search," for integer programming problems, which allows testing "solution points" for feasibility. The new procedure, although certainly in the spirit of convexity cut theory, does not require constructing a specific convex set R. However, it can be viewed as a search method that decomposes the solution space into properly selected unit hypercubes and then tests their associated vertices for feasibility.

5.3 Primal Cutting Methods

One of the basic drawbacks of dual cutting methods is that the solution remains dual feasible until the optimal integer solution is attained. This means that if the computation is terminated prematurely, no information about the optimum solution is available. This difficulty is remedied by developing primal methods, which, by definition, can provide primal feasible solutions.

In this section, two methods are introduced: the first may be considered as "almost" dual since it modifies the dual cutting method so a primal feasible solution may become attainable during the course of the calculations. The second is a straightforward primal cutting method in which the cuts are applied so that the initial primal feasibility is maintained throughout the calculations.

5.3.1 Almost Dual Method

This method is basically a dual cutting algorithm in which a primal feasible solution may be determined by solving a zero–one problem (to be defined later) associated with the current continuous solution point. It is assumed that the *fractional* cut (Section 5.2.1) is used to solve the problem. The idea was introduced previously in connection with the branch-and-bound algorithm in order to find a good initial bound (see Section 4.3.2-C).

Suppose that after the tth (fractional) cut the equivalent continuous problem is given by

$$\max_{x \geq 0}\{x_0{}^t = CX \mid A^t X \leq b^t\} \qquad (\text{P}^t)$$

and let $X = X^t$ define the associated optimal solution. Let

$$X = [X^t] + Y, \qquad Y \in (0, 1)$$

where Y is a vector of zero–one variables. The zero–one problem whose (feasible and infeasible) solution points are given by the vertices of the unit hypercube

$$[X^t] \leq X \leq [X^t] + e, \qquad e = (1, 1, \ldots, 1)^{\text{T}}$$

is defined as

$$\max_{Y}\{z_0{}^t = CY + C[X^t] \mid A^t Y \leq b^t - A^t[X^t], Y \in (0, 1)\} \qquad (\text{Z}^t)$$

The feasible space of (Z) identifies the feasible vertices of the unit hypercube enclosing $X = X^t$.

The algorithm for solving (P^0), that is, the original integer problem is as follows: Assume for simplicity that (P^0) is the continuous version of the integer problem. Assume further that all the coefficients in the objective function are integers. Let $x_0{}^t$ be the optimum objective value of (P^t), and $x_u{}^t$, $x_l{}^t$ the upper and lower bounds associated with the optimum objective value of (P^t).

Step 0   Solve (P^0), that is, without cuts added. Let $x_u{}^0 = x_0{}^0$. If there is no obvious solution to the integer problem, set $x_l{}^0 = -\infty$. Let $t = 0$ and go to Step 1.

Step 1   Define (Z^t) and find its solution satisfying $x_l{}^t \leq z_0{}^t \leq x_u{}^t$. If (Z^t) has no feasible solution, go to Step 2. Otherwise, define Y^t as its optimum solution and set $x_l{}^{t+1} = z_0{}^t$ and $x_u{}^{t+1} = x_u{}^t$. Go to Step 3.

Step 2   Develop the $(t + 1)$th cut and solve (P^{t+1}). Let $x_u{}^{t+1} = x_0{}^{t+1}$ and $x_l{}^{t+1} = x_l{}^t$, then go to Step 3. If (P^{t+1}) has no feasible solution, terminate; no integer feasible solution exists.

Step 3   If $x_u{}^{t+1} - x_l{}^{t+1} < 1$, stop; $X = Y^t + [X^t]$ is the optimum integer solution. Otherwise, set $t = t + 1$; then go to Step 1.

The preceding algorithm is actually a cutting plane method with a superimposed branch-and-bound capability. It is this latter capability that may yield primal feasible solutions.

The proposed method actually takes advantage of the desirable properties of both cutting and branch-and-bound methods. Specifically, the effects of the following drawbacks are either minimized or eliminated:

(i) The machine round-off error associated with the cutting method is more controllable since, in general, the integer solution is determined by the zero–one subproblems.

(ii) The branch-and-bound algorithm is generally characterized by its severe taxation of the computer memory. However, in the present algorithm this is not the case since branching is characterized by the zero–one subproblem, which is automatically determined from the current continuous solution.

▶**EXAMPLE 5.3-1** Maximize

$$x_0 = 2x_1 + x_2$$

subject to

$$12x_1 + 7x_2 \leqslant 55$$
$$5x_1 + 2x_2 \leqslant 18$$
$$x_1, x_2 \geqslant 0 \quad \text{and integers}$$

The solution of (P^0) is given by Table 5-18.

TABLE 5-18[a]

		x_3	x_4
x_0	$8\frac{3}{11}$	$\frac{1}{11}$	$\frac{2}{11}$
x_1	$1\frac{5}{11}$	$-\frac{2}{11}$	$\frac{7}{11}$
x_2	$5\frac{4}{11}$	$\frac{5}{11}$	$-\frac{12}{11}$
x_3	0	-1	0
x_4	0	0	-1

[a] $x_0{}^0 = 8\frac{3}{11}$, $x_u{}^0 = 8\frac{3}{11}$, $x_l{}^0 = -\infty$.

Now $X^0 = (1\frac{5}{11}, 5\frac{4}{11})$ and $[X^0] = (1, 5)$, so that (Z^0) is obtained by substituting $x_1 = y_1 + 1$ and $x_2 = y_2 + 5$ in (P^0), yielding

maximize

$$z^0 = 2y_1 + y_2 + 7 \qquad\qquad (Z^0)$$

subject to

$$12y_1 + 7y_2 \leqslant 8$$
$$5y_1 + 2y_2 \leqslant 3$$
$$y_1, y_2 = (0, 1)$$

The solution to (Z^0) is $(y_1{}^0, y_1{}^0) = (0, 1)$, which yields $x_l{}^0 = 8$. (This solution is supposedly obtained by using Balas' additive algorithm, Section 3.5.1.) Since $x_u{}^0 - x_l{}^0 = 8\frac{3}{11} - 8 = \frac{3}{11} < 1$, it follows that the solution $x_1 = 1 + 0 = 1$ and $x_2 = 5 + 1 = 6$ is optimal and the process is terminated.

It is interesting to notice that the same solution is obtained by applying three Gomory's fractional cuts.◀

5.3.2   Primal Cuts†

Ben-Isreal and Charnes (1962) were the first to suggest a primal cutting algorithm along the same ideas of Gomory's all-integer λ-cut method (Section 5.2.2). The basic difference is that primal (integer) feasibility is maintained at all stages of calculations. However, no finiteness proof is provided for the algorithm.

Later, Young (1965) introduced some original changes in the Ben-Isreal and Charnes algorithm that are sufficient to support finiteness. This algorithm was later improved by Young (1968), who stated that the improvements were greatly influenced by the concurrent work of Glover (1968).

In this section, the Ben-Isreal and Charnes rudimentary primal algorithm is first presented. The weak points of this algorithm leading to the work of Young are then considered. Young's (simplified) primal algorithm is stated and its finiteness discussed.

Consider the starting linear program in the form:

maximize

$$z = \sum_{j \in N} c_j x_j$$

subject to

$$\sum_{j \in N} a_{ij} x_j \leqslant b_i, \qquad i \in M$$

$$x_j \text{ nonnegative integers}, \qquad j \in N$$

It is assumed that c_j, a_{ij}, and b_i are integers and that the starting program is primal feasible.

The proposed cut must satisfy two basic conditions: (1) It must preserve the integrality of the tableau. (2) It must maintain the primal feasibility property. These conditions are satisfied by constructing the cut so it acts as a *natural* pivot row with its pivot element equal to $+1$.

To show how this cut can be constructed, let x_k be the leaving variable at the current iteration, which is determined by applying the regular feasibility condition of the (primal) simplex method before the current cut

---

† Ben-Israel and Charnes (1962), Glover (1968a), Young (1965, 1968).

is adjoined. Then define the source row as

$$x_k = x_k^* - \sum_{j \in NB} \alpha_{kj} w_j$$

where x_k^* is the current value of x_k and w_j, $j \in NB$, are the current nonbasic variables. Let $\alpha_{kr} > 1$ be the pivot element (that is, α_r is the pivot column). Then following the same argument of Gomory's λ-cut (Section 5.2.2) and taking $\lambda = \alpha_{kr} > 1$, this yields the cut

$$S_k = [x_k^*/\alpha_{kr}] - \sum_{j \in NB} [\alpha_{kj}/\alpha_{kr}]w_j \qquad (p\text{-}cut)$$

where S_k is a nonnegative (integer) slack variable. (If $\alpha_{kr} = 1$, the source row is used directly for pivoting.)

The p-cut satisfies the above stated conditions since, when augmented to the current tableau, its row will act as the natural pivot row for the primal simplex method with its pivot element equal to $+1$. To show that this is true, the rth (pivot) element in the p-cut is given by $[\alpha_{kr}/\alpha_{kr}] = 1$ as desired. Moreover, S_k will become the natural leaving variable in the tableau, since

$$\frac{[x_k^*/\alpha_{kr}]}{1} \leqslant \frac{x_k^*}{\alpha_{kr}} = \min_{i \in M} \left\{ \frac{x_i^*}{\alpha_{ir}} \middle| \alpha_{ir} > 0 \right\}$$

As long as $x_k^*/\alpha_{kr} \geqslant 1$, that is, $[x_k^*/\alpha_{kr}] \geqslant 1$, there is a guarantee that the augmentation of the p-cut would lead to an improvement in the objective value. However, if $x_k^*/\alpha_{kr} < 1$, then $[x_k^*/\alpha_{kr}] = 0$ and the objective value remains unchanged. This situation is similar to degeneracy in linear programming. Unfortunately, the methods for resolving such a problem in linear programming cannot be implemented in integer programming because cutting methods create new constraints and variables, thus destroying the main property that the maximum number of possible basic solutions associated with a degenerate point is finite. This is where the Ben-Israel and Charnes procedure breaks down, since there is no assurance that the algorithm is finite.

Two important modifications were introduced by Young to guarantee the finiteness of the algorithm. These include the new rules for selecting the pivot column and the source row. First, consider the rules for determining the pivot column.

The procedure for selecting the pivot column calls for adjoining an additional constraint to the initial tableau, which limits the sum of the original nonbasic variables. This row will be referred to as the L-row and is given by

$$\sum_{j \in NB} w_j + S_L = b_L \qquad (L\text{-}row)$$

where b_L is an integer value selected large enough so that none of the feasible integer points is excluded. Now define the column vector

$$R_j = \left(\frac{\bar{c}_j}{\alpha_{Lj}}, \frac{\alpha_{1j}}{\alpha_{Lj}}, \dots, \frac{\alpha_{mj}}{\alpha_{Lj}} \right)^{\mathrm{T}}, \qquad j \in P \equiv \{ j \in NB \,|\, \alpha_{Lj} > 0 \}$$

where $\bar{c}_j$ is the *current* objective coefficient in the jth column and $\alpha_{1j}, \dots, \alpha_{mj}, \alpha_{Lj}$ are the constraint coefficients associated with the same column. The columns $R_j, j \in P$, are thus determined from among the current nonbasic columns such that $\alpha_{Lj} > 0$. The pivot column α_r is now selected from among $\alpha_j, j \in P$ such that R_r is the lexicographically smallest among all $R_j, j \in P$.

The selection of the source row is now considered. Given the pivot column r, if $[x_k^*/\alpha_{kr}] \geq 1$, the associated simplex iteration is referred to as a *transition* cycle, since this guarantees an improvement in the objective value. Otherwise, if $[x_k^*/\alpha_{kr}] = 0$, this will lead to a *stationary* cycle implying no change in the objective value. If finiteness is to be established, only a finite number of stationary cycles should occur before the optimum is achieved. Young shows that any rule compatible with the following condition should work:

For any index row k in which $[x_k^*/\alpha_{kr}] = 0$, a succeeding tableau must occur after a finite number of iterations such that $[\hat{x}_k^*/\hat{\alpha}_{kr}] \geq 1$.

What this condition implies is that a source row currently responsible for producing a stationary cycle must eventually be conditioned (its coefficients properly changed) so that, after a finite number of iterations, it either ceases to be a source row or produces a transition cycle. This means that in some succeeding iteration, a transition cycle must be realized.

Young indicates that several rules for selecting the source row can be developed according to the given condition. One of these rules is as follows: Let $V(r)$ be the set of all rows satisfying

$$0 \leq [x_t^*/\alpha_{tr}] \leq \theta_r \equiv \min_{\text{all } i} \{ x_i^*/\alpha_{ir} \,|\, \alpha_{ir} > 0 \}$$

Thus $V(r)$ defines the set of legitimate rows for generating a cut given the pivot column α_r.

If $\theta_r \geq 1$, then *any* $k \in V(r)$ can produce a transition cycle and the selection of the source row is arbitrary among all the elements in $V(r)$. However, if $\theta_r < 1$, then every $k \in V(r)$ will produce a stationary cycle, and a procedure is needed to force the condition $x_k^* > \alpha_{kr}$ for all $k \in V(r)$. This is shown to occur by the following theorem.

▶**THEOREM** Given the pivot column α_r and the source row $k \in V(r)$, let $\alpha_{kj}, j \in NB$, and $\hat{\alpha}_{kj}, j \in \widehat{NB}$, be typical coefficients of the source row before and after the associated p-cut is applied to the current tableau, respectively. Then, $\hat{\alpha}_{kj} \geqslant 0$ for all $j \in \widehat{NB} - \{v\}$, where v is the index of the slack variable associated with the cut, and

$$\hat{\alpha}_{kj} < \alpha_{kr}, \qquad \text{for all} \quad j \in \widehat{NB}$$

Proof The use of the cut as a pivot row reduces the elements of the kth row to

$$\hat{\alpha}_{kv} = -\alpha_{kr},$$

and

$$\hat{\alpha}_{kj} = \alpha_{kj} - [\alpha_{kj}/\alpha_{kr}]\alpha_{kr}, \qquad j \in \widehat{NB} - \{v\}$$
$$= N\alpha_{kr} + e_{kj} - [\alpha_{kj}/\alpha_{kr}]\alpha_{kr}$$

where N is an integer and $0 \leqslant e_{kj} < \alpha_{kr}$. Thus, taking $N = [\alpha_{kj}/\alpha_{kr}]$, this yields

$$0 \leqslant \hat{\alpha}_{kj} = e_{kj} < \alpha_{kr}, \qquad \text{for all} \quad j \in \widehat{NB} - \{v\}$$

Since $\hat{\alpha}_{kv} = -\alpha_{kr}$, it follows

$$\hat{\alpha}_{kj} < \alpha_{kr}, \qquad \text{for all} \quad j \in \widehat{NB} ◀$$

The implication of the above theorem is as follows: In a stationary cycle, x_k^* remains unchanged, but the constants α_{kj} always decrease each time a stationary cycle occurs. Thus by always selecting the same source row, eventually a future iteration is reached at which $x_k^* \geqslant \alpha_{kr}$, where r is the pivot column. At this point, the kth row will no longer produce stationary cycles. Thus a new source row is chosen, and if it produces a stationary cycle, the same procedure is repeated.

It may first appear that even though a given source row will eventually cease to produce stationary cycles within a finite interval, it is possible that, at a later iteration, this same source will produce stationary cycles again with the result that no future transition cycle is ever possible. Although it is possible that an excluded row can become a source row again before a transition cycle occurs, Young proves that this process cannot be repeated indefinitely and that after a finite number of iterations either a transition cycle must occur or an optimum is attained. The proof ensures that the finiteness of the algorithm is closely related to the rule for selecting the pivot column. Indeed, the finiteness proof is based on two points: (1) The vector R_r (associated with the pivot column) must strictly increase lexicographically from one tableau to the next over any sequence of stationary cycles; and (2) the first component of R_r (associated with the objective

equation) must increase at finite intervals, and a finite number of such
increases should render R_r with its first element nonnegative. Because of
the way R_r is constructed, it is obvious that this condition is sufficient for
optimality. The details of the proof can be found in the work of Young (1968).

▶EXAMPLE 5.3-2   Maximize

$$z = 2x_1 + x_2 + x_3$$

subject to

$$-4x_1 + 5x_2 + 2x_3 \leqslant 4$$
$$4x_1 - x_2 + x_3 \leqslant 3$$
$$2x_1 - 3x_2 + x_3 \leqslant 1$$

x_1, x_2, x_3 nonnegative integer

The successive tableaus are given in Tables 5-19 to 5-25.

TABLE 5-19[a]

		↓			
		x_1	x_2	x_3	
z	0	-2	-1	-1	
x_1	0	-1	0	0	
x_2	0	0	-1	0	
x_3	0	0	0	-1	
x_4	4	-4	5	2	
x_5	3	4	-1	1	← Source row
x_6	1	2	-3	1	
S_L	10	1	1	1	
S_1	0	$\boxed{1}$	-1	0	Cut

[a] $V(x_1) = \{5, 6\}$.

It is interesting to observe that the optimum solution is obtained after
one iteration (Table 5-21), but this optimality is not verified until Table 5-25.
This illustrates one of the computational difficulties of the algorithm.◀

The performance of the primal algorithm can be improved if a "good"
initial solution can be secured. (The method in Section 5.3.1 can be used
to determine such a starting solution.) However, this solution must be
expressed as a basic solution. The following method, due to Ben-Israel and

TABLE 5-20[a]

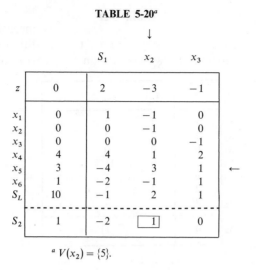

	$\downarrow$		
	S_1	x_2	x_3
z	0	2	-3	-1
x_1	0	1	-1	0
x_2	0	0	-1	0
x_3	0	0	0	-1
x_4	4	4	1	2
x_5	3	-4	3	1	$\leftarrow$
x_6	1	-2	-1	1
S_L	10	-1	2	1
S_2	1	-2	$\boxed{1}$	0

[a] $V(x_2) = \{5\}.$

Charnes (1962), can be used to convert an interior solution into a basic solution.

Suppose the initial problem is given as:

maximize
$$z = c_N X_N$$
subject to
$$IX_B + DX_N = b$$
$$X_B, X_N \geqslant 0 \quad \text{and integer}$$

TABLE 5-21[a]

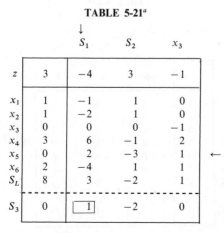

	$\downarrow$		
	S_1	S_2	x_3
z	3	-4	3	-1
x_1	1	-1	1	0
x_2	1	-2	1	0
x_3	0	0	0	-1
x_4	3	6	-1	2
x_5	0	2	-3	1	$\leftarrow$
x_6	2	-4	1	1
S_L	8	3	-2	1
S_3	0	$\boxed{1}$	-2	0

[a] $V(S_1) = \{4, 5\}.$

TABLE 5-22[a]

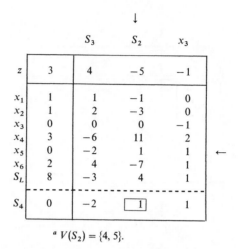

		S_3	S_2	x_3	
z	3	4	-5	-1	
x_1	1	1	-1	0	
x_2	1	2	-3	0	
x_3	0	0	0	-1	
x_4	3	-6	11	2	
x_5	0	-2	1	1	←
x_6	2	4	-7	1	
S_L	8	-3	4	1	
S_4	0	-2	$\boxed{1}$	1	

[a] $V(S_2) = \{4, 5\}$.

where X_B and X_N are the starting basic and nonbasic vectors, respectively. Let $(X_B, X_N) = (\xi_B, \xi_N)$ be the available good integer solution. Define $d_\xi = D\xi_N$ and $c_\xi = c_N \xi_N$.

Let x_ξ be a new nonbasic nonnegative integer variable. The new starting tableau that expresses (ξ_B, ξ_N) as a basic solution is given by:

maximize

$$z = c_N X_N + c_\xi x_\xi$$

TABLE 5-23[a]

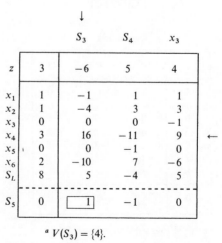

		S_3	S_4	x_3	
z	3	-6	5	4	
x_1	1	-1	1	1	
x_2	1	-4	3	3	
x_3	0	0	0	-1	
x_4	3	16	-11	9	←
x_5	0	0	-1	0	
x_6	2	-10	7	-6	
S_L	8	5	-4	5	
S_5	0	$\boxed{1}$	-1	0	

[a] $V(S_3) = \{4\}$.

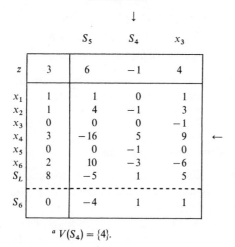

TABLE 5-24^a

↓

		S_5	S_4	x_3
z	3	6	-1	4
x_1	1	1	0	1
x_2	1	4	-1	3
x_3	0	0	0	-1
x_4	3	-16	5	9 ←
x_5	0	0	-1	0
x_6	2	10	-3	-6
S_L	8	-5	1	5
S_6	0	-4	1	1

^{a} $V(S_4) = \{4\}$.

subject to

$$IX_B + DX_N + d_\xi x_\xi = b$$
$$X_B, X_N, x_\xi \geqslant 0 \quad \text{and integer}$$

It is evident that $d_\xi \leqslant b$ since $0 \leqslant \xi_B = b - d_\xi$, which follows from the fact that (ξ_B, ξ_N) is a feasible solution. Thus, by using x_ξ as the pivot column, a transition cycle (that is, an improved initial value of z) is possible. The primal algorithm can then be applied directly to the resulting tableau.

In order to recover the optimal values of the original variables, let

TABLE 5-25^a

		S_5	S_6	x_3
z	3	2	1	5
x_1	1	1	0	1
x_2	1	0	1	4
x_3	0	0	0	-1
x_4	3	4	-5	4
x_5	0	-4	1	-1
x_6	2	2	3	-3
S_L	8	-1	-1	4

^{a} *Optimum*: $z = 3$, $x_1 = x_2 = 1$, $x_3 = 0$.

(X_B^0, X_N^0, x_ξ^0) be the optimal solution to the transformed problem. Then

$$X_B^0 + DX_N^0 + d_\xi x_\xi^0 = b$$
$$X_B^0 + DX_N^0 + D\xi_N x_\xi^0 = b$$
$$X_B^0 + D(X_N^0 + \xi_N x_\xi^0) = b$$

Comparing the last equation with the original problem, if follows that the optimal values for the original variables are given by

$$X_B^* = X_B^0, \qquad X_N^* = X_N^0 + \xi_N x_\xi^0$$

5.4 Comments on Computational Experience

The cutting plane methods presented in this chapter have not been subjected to extensive computational experience. In particular, there appear to be no reported computational results for (i) the dual methods including mixed cuts, intersection cuts, and bound escalation cuts, and (ii) primal methods. However, a paper by Trauth and Woolsey (1969) compares the use of different codes for solving twenty-nine pure integer problems with sizes varying from 1 by 10 to 15 by 50. The codes are based on either Gomory's f-cuts (Section 5.2.1) or λ-cuts (Section 5.2.2).

The general conclusions drawn from this experience may be summarized as:

1. Computation time is code dependent. While a problem may be easily solved by one code, another code may prove a complete failure. Unfortunately, no one code is uniformly better for all the problems.
2. Except for a class of allocation problems, the codes based on the λ-cut are generally inferior to those based on the f-cut.
3. The round-off error presents a serious difficulty in all the codes.
4. The solution of integer programs by cutting methods is generally difficult and uncertain. Based on the reported experience, it is impractical to attempt solving problems larger than 100 by 100 by cutting methods. The computation time and/or round-off error would, in general, present acute difficulties. This, however, does not mean that problems smaller than 100 by 100 are solvable. Rather, there is an improved chance of reaching the optimum solution. But in view of conclusion (1), it may be necessary to experiment with a number of codes before one can decide whether a solution is attainable with the obvious drawback of increasing the computation time.

Interestingly, Trauth and Woolsey report that in some problems, a simple rearrangement of the constraints has converted a relatively easy problem (from the computational standpoint) to an extremely difficult one. There are no rules, however, as to how the data can be conditioned before the solution is attempted (cf. implicit enumeration methods, Chapter 3).

Of course, there are reports of very large problems that have been solved by a cutting method [see Martin (1963)]. But it appears that the solved problems have special structures, and the integer codes were developed to exploit these structures.

5.5 Concluding Remarks

In this chapter different cuts for the integer problem have been developed. Each type of cut provides a certain computational advantage, at least from the theoretical viewpoint.

Although some discussion is given about how stronger cuts can be developed, these (sometimes elaborate) ramifications practically contribute little to the computational efficiency of the cutting algorithms. The experiences reported to date about these methods have been discouraging. Indeed, it can almost be concluded that not much hope should be expected (as far as computation is concerned) from improving the theoretical properties of the cuts. The disappointing fact is that, in cutting plane methods, only the first "few" cuts seem to produce significant improvements toward reaching the optimum integer solution. The "cutting power" of the succeeding cuts is usually hampered by the massive degeneracy typical of integer programming algorithms and/or the severe effects of machine round-off errors.

Problems

5-1 (Bowman and Nemhauser, 1970). Show that Dantzig's cut

$$\sum_{j \in NB} x_j \geq 1$$

can be strengthened as follows. Let x_i be an integer basic variable whose continuous optimum value is fractional and let x_i represent the ith source row. Define $NB_i = \{j \in NB \mid f_{ij} > 0\}$. Then a stronger cut is

$$\sum_{j \in NB_i} x_j \geq 1$$

In this sense, the strongest cut generated from the entire simplex tableau is derived from the source row yielding $\min_i\{|NB_i|\}$.

5-2   Prove that for the ith source row (for which basic $x_i = [x_i^*] + f_i$), the constraint

$$\sum_{j \in NB_i} (1 - f_{ij})x_j \geq 1 - f_i$$

is a legitimate cut with respect to the integer problem, where NB_i is defined in Problem 5-1.

5-3   Show that the results of Problems 5-1 and 5-2 can be combined to produce the f-cut (Section 5.2-1).

5-4   Consider the following integer problem:

maximize

$$z = 2x_1 + 4x_2$$

subject to

$$2x_1 + 6x_2 \leqslant 23$$
$$x_1 - x_2 \leqslant 1$$
$$x_1 + x_2 \leqslant 6$$
$$x_1, x_2 \geqslant 0 \quad \text{and integer}$$

Solve by the f-cut based on selecting the source row by (a) the largest fractional basic variable, and (b) the method of penalties.

5-5   Solve Problem 5-4 by augmenting all f-cuts from every eligible source row simultaneously. Compare the computations with those of Problem 5-4.

5-6   Solve Problem 5-4 by using Gomory's accelerated f-cut.

5-7   *Martin's f-cut* (Martin, 1963; cf. Garfinkel and Nemhauser, 1972b, p. 185).   Suppose a legitimate source row for the f-cut is given by

$$x_i = \beta_i^1 - \sum_{j \in NB} \alpha_j^1 x_j \qquad \text{(source row 1)}$$

The f-cut obtained from (1) is

$$S_1 = -f_i^1 - \sum_{j \in NB} f_{ij}^1 x_j \qquad (f^1\text{-cut})$$

By using the f^1-cut as a pivot row with its pivot element determined from the l-dual simplex optimality condition, let x_k, $k \in NB$, be the selected pivot element. Then applying the Gaussian elimination (row operations) to source row 1 only, this yields the following new source row:

$$x_i = \beta_i^2 - \sum_{j \in NB - \{k\}} \alpha_{ij}^2 x_j - \alpha_{is_1}^2 S_1 \qquad \text{(source row 2)}$$

Suppose at least one of the coefficients $\beta_i^2, \alpha_{ij}^2, j \in NB - \{k\}$, and α_{is_1} is noninteger. Then a new f-cut can be generated from source row 2. Let this cut be given by

$$S_2 = -f_i^2 - \sum_{j \in NB - \{k\}} f_{ij}^2 x_j - f_{is_1}^2 S_1 \qquad (f^2\text{-cut})$$

The idea now is to use the f^2-cut as a pivot row with the pivot element now given by S_1 (note S_1 now occupies the column of x_k in the f^1-cut). This leads to source row 3:

$$x_i = \beta_i^3 - \sum_{j \in NB - \{k\}} \alpha_{ij}^3 x_j - \alpha_{is_2}^3 S_2 \qquad (\text{source row 3})$$

The process is repeated, always pivoting on the slack variable of the cut, until a source row results with all its coefficients integers. At this point, no further f-cuts can be generated. The different f-cuts may now be used to develop the composite cut. Let the f^r-cut be the last cut generated. Then the r cuts are composed into one cut as follows. The f^1-cut is used to substitute out the slack S_1 in the f^2-cut; then the resulting f^2-cut, now in terms of S_2 and $x_j, j \in NB$, is used to substitute S_2 in the f^3-cut, and so on. At the point when the *last* cut (f^r-cut) is expressed in terms of S_r and $x_j, j \in NB$, the resulting cut is the required composite cut. This cut may now be used as a pivot row for the entire tableau with x_k being the pivot column.

(a) By using the above scheme, show that source row r can be written as

$$x_i = \beta_i^r - \sum_{j \in NB - \{k\}} \alpha_{ij}^r x_j - \alpha_{is_r}^r S_r \qquad (\textit{source row r})$$

(b) If the (rth) *composite* cut is defined as

$$S_r = a - \sum_{j \in NB} b_j x_j \qquad (\textit{Martin's cut})$$

prove that

$$a = -(\beta_i^1 - \beta_i^r)/\alpha_{is_r}^r$$
$$b_j = -(\alpha_{ij}^1 - \alpha_{ij}^r)/\alpha_{is_r}^r, \qquad j \neq k$$
$$b_k = \alpha_{ij}^r/\alpha_{is_r}^r$$

5-8 Solve Problem 5-4 by using Martin's composite cuts. Compare the resulting computations with those in Problem 5-4.

5-9 Show that the selection of the pivot element of the *composite* cut equal to the pivot element of the f^1-cut may destroy the l-dual feasibility when the composite cut is pivoted on. Hence, according to the convergence proof under the f-cut, the composite cut does not guarantee convergence.

5-10 Since, as indicated in Problem 5-9, the use of the composite cut may destroy the l-dual feasibility, it is possible that an all-integer iteration is encountered where the solution is neither primal nor dual feasible. Show that primal feasibility when it exists, can be recovered by applying the dual simplex feasibility condition without regard to the optimality condition.

5-11   Consider the following problem:

minimize

$$z = 3x_1 + 2x_2 + 4x_3$$

subject to

$$-x_1 + \ x_2 + 3x_3 \geqslant 8$$
$$3x_1 - 4x_2 + \ x_3 \geqslant 9$$
$$2x_1 + \ x_2 - \ x_3 \geqslant 6$$
$$x_1, x_2, x_3 \geqslant 0 \quad \text{and integer}$$

(a)   Solve by using the λ-cut.
(b)   Solve by using Wilson's improved λ-cut.

5-12   Solve Problem 5-11 with the objective function changed to

minimize $z = 2x_1 - x_2 + 3x_3$

5-13   Solve the following problem by using the m-cut:

maximize

$$z = x_1 + 1.6x_2 + 1.2x_3$$

subject to

$$0.4x_1 + 1.3x_2 + 0.2x_3 \leqslant 2.2$$
$$0.9x_1 + 0.6x_2 + \ x_3 \leqslant 2.8$$
$$x_1, x_2, x_3 \geqslant 0$$
$$x_1, x_3 \text{ integer}$$

5-14   Solve Problem 5-4 by using the m-cut. Note that the problem is of the pure integer type. Thus the first m-cut contains all integer variables except for its slack variables. In the succeeding m-cuts, these slacks are treated as noninteger variables. How does this method of solution compare with the use of the f-cut?

5-15   Solve Problem 5-11 by using Glover's bound escalation cut.

5-16   Compare the first m-cut in Problem 5-13 with a corresponding Balas' intersection cut.

5-17   Solve the following problems by the primal integer algorithm:

(a) maximize

$$z = 4x_1 + 3x_2$$

subject to

$$4x_1 + x_2 \leqslant 10$$
$$2x_1 + 3x_2 \leqslant 8$$
$$x_1, x_2 \geqslant 0 \quad \text{and integer}$$

(b) minimize

$$z = 2x_1 + 5x_2$$

subject to

$$x_1 + x_2 \geqslant 4.5$$
$$2x_1 + 6x_2 \geqslant 22$$
$$x_1, x_2 \geqslant 0 \quad \text{and integer}$$

CHAPTER 6

The Asymptotic Integer Algorithm

6.1 Introduction

In integer linear programs, given that the continuous optimum is non-integer, the integer optimum is determined by seeking an interior (or possibly a boundary) point of the continuous solution space. In the enumeration and branch-and-bound methods (Chapters 3 and 4), this is achieved by partitioning the solution space into continuous subproblems in such a way that their optimum solutions actually characterize those promising interior (or boundary) points. In the cutting methods, the promising points are sought by modifying the continuous space so that the optimum integer point is properly identified with the optimum extreme point of the modified continuous space.

Another idea for effecting the same result is to recognize that the determination of the optimum integer point is equivalent to assigning positive values to some (or all) of the *nonbasic* variables associated with the current continuous optimum solution. Indeed, this idea is used (on a limited scale) in connection with the rounding procedure introduced in Section 1.2. But in view of the elementary application of the idea, only *slack* variables are allowed to assume positive values; otherwise, the simple rounding procedure becomes incapable of coping with the interaction that may result from elevating other nonbasic variables above zero values.

To alleviate the problem, a more convenient representation of the problem becomes necessary. This is what the so-called "asymptotic" algorithm seeks to accomplish. It appears more appealing at this point to introduce the basic idea of the method in terms of graphical representations. These will also serve to justify the name of the algorithm as well as the importance of the method. The mathematical treatment that follows should then provide a rigorous foundation for the technique.

6.2   The Idea of the Asymptotic Algorithm

Consider the following numerical example:

maximize

$$z = x_1 + 2x_2$$

subject to

$$x_1 + 2x_2 \leqslant 8$$
$$2x_1 \qquad \leqslant 7$$
$$-2x_1 + 4x_2 \leqslant 9$$
$$x_1, x_2 \geqslant 0 \quad \text{and integer}$$

It is assumed, without loss of generality, that all the coefficients of the problem are integers. This means that all the slack variables must also be integers.

The optimal continuous solution of the example is given in Table 6-1 (for convenience, the simplex compact tableau, Example 2.3-2, is used in the solution). The basic variables are $x_1 = \frac{7}{2}$, $x_2 = \frac{9}{4}$, and $S_3 = 7$.

TABLE 6-1

		S_1	S_2
z	8	0	1
x_1	$\frac{7}{2}$	0	$\frac{1}{2}$
x_2	$\frac{9}{4}$	$\frac{1}{2}$	$-\frac{1}{4}$
S_3	7	-2	2

This solution is shown in Fig. 6-1 as the optimum extreme point of the continuous solution space. The idea now is to allow the nonbasic variables S_1 and S_2 to assume positive values. But in order not to be in the same difficult situation created in the rounding procedure of Section 1.2,

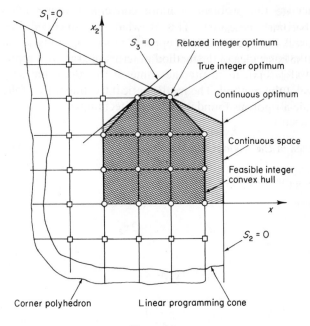

Figure 6-1

the nonnegativity condition on the basic variables x_1, x_2, and S_3 is dropped so that the values of the nonbasic variables may now be adjusted independently. In terms of Fig. 6-1, the relaxed solution space is given by what is known as the *linear programming cone*. The boundaries of the cone are given by $S_1 = S_2 = 0$, and its interior is created by $S_1 > 0$ and $S_2 > 0$. Notice that dropping the nonnegativity condition on the (basic) slack variable S_3 makes the corresponding constraint trivial. This means that the cone is determined only by the binding constraints at the continuous optimum.

Since the prime interest is in the integer points, one must concentrate on the convex hull of the integer solutions in the cone. This is usually known as the (unbounded) *corner polyhedron*. In the following discussion a *relaxed problem* is defined such that its solution space is given by the integer points in the corner polyhedron.

The relaxed space obviously includes all the feasible integer points of the integer problem. Thus, if the optimum solution of the relaxed problem is feasible with respect to the original problem (that is, if all the basic variables are nonnegative), then it is also optimal for the original problem. In other words, if the optimum solution given the corner polyhedron coincides with an extreme point of the convex hull of all feasible integer points, then this also gives optimum integer solution. Referring to Fig. 6-1, one can see that this property is satisfied and hence the point ($x_1 = 2$, $x_2 = 3$) yields the optimum integer solution.

The question now is: What are the advantages of this procedure? The main answer is that there exist reasonably efficient algorithms for solving the relaxed integer problem, and if the solution happens to satisfy the property illustrated in Fig. 6-1, then the original integer problem may be solved efficiently. Indeed, the usefulness of the method is enhanced by the fact that a sufficient (but unfortunately not necessary) condition can be established so that (some of) the problems satisfying the indicated property can be recognized prior to solving the relaxed problem. This condition is given in Section 6.3.1.

Regrettably, many integer problems exist that are not *directly* solved by the relaxed problem. Indeed, trivial changes in the data of an otherwise well-behaved problem may give rise to a computationally tedious program. To illustrate this point, if the right-hand side of the third constraint in Fig. 6-1 is changed to 7 (instead of 9) then, as shown in Fig. 6-2, the

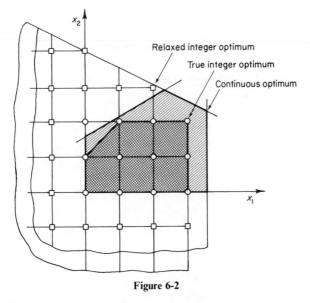

Figure 6-2

relaxed integer optimum does not coincide with the "true" integer optimum. In this case, further (possibly voluminous) computations are needed to determine the true integer optimum. This point is discussed in Section 6.5.

Perhaps it is possible now to justify the name "asymptotic" problem. From the example given above, it is noticed that when the right-hand side of the third constraint *increases* from 7 (Fig. 6-2) to 9 (Fig. 6-1), the relaxed problem automatically solves the integer problem. The general observation is that as the right-hand sides of the constraints of the integer problem tend to be sufficiently large, the relaxed problem *asymptotically* solves the (true) integer problem. For obvious reasons, the asymptotic problem is also called *integer programming over cones*.

The above discussion indicates that the application of the asymptotic algorithm requires two basic elements:

(i) an algebraic representation of the corner polyhedron (that is, the convex hull of all the integer points in the linear programming cone), and

(ii) a method for determining the optimum integer point in the corner polyhedron.

The first point is discussed in Section 6.3. This also gives a sufficient condition for the relaxed problem to solve the integer problem. Section 6.4 presents two methods for solving the relaxed problem. These are based on the use of the shortest route and dynamic programming techniques.

6.3 Development of the Asymptotic Algorithm†

Suppose that the integer linear program is given by:

maximize

$$z = \{cx \mid Ax = b, x \geq 0, \text{ and integer}\}$$

It is assumed that all the elements of c, A, and b are integers. In the absence of the integrality condition on x, the optimal solution to the equivalent linear program is given by:

maximize

$$z = c_B B^{-1}b - (c_B B^{-1}N - c_N)x_N$$

subject to

$$x_B = B^{-1}b - B^{-1}Nx_N$$

$$x_B, x_N \geq 0$$

where x_B and x_N are the (optimal) basic and nonbasic variables, and B is the basic matrix, so that $(B, N) = A$. Similarly, c_B and c_N are the basic and nonbasic components of c.

† Gomory (1965).

Since at the optimum continuous solution $x_N = 0$, then the optimum linear program is given by $x_B = B^{-1}b$. As illustrated in Section 6.2, the corner polyhedron is obtained from the optimum linear program by imposing the integrality condition on x_B and x_N while maintaining the nonnegativity condition on x_N only. This means that the corner polyhedron is given by

$$x_B = B^{-1}b - B^{-1}Nx_N$$
$$x_B \text{ integer}$$
$$x_N \geq 0 \text{ and integer}$$

Since x_B is unrestricted in sign, the corner polyhedron becomes equivalent to

$$B^{-1}b - B^{-1}Nx_N \equiv 0 \quad (\text{mod } 1)$$
$$x_N \geq 0 \quad \text{and integer}$$

Essentially, this says that only the difference $B^{-1}b - B^{-1}Nx_N$ must be integer for nonnegative integer x_N. Thus by adding or subtracting integer values to $B^{-1}b$ and $B^{-1}Nx_N$ the relationship remains unchanged. This means that the corner polyhedron is given by

$$f(B^{-1}b) - f(B^{-1}Nx_N) \equiv 0 \quad (\text{mod } 1)$$
$$x_N \geq 0 \text{ and integer}$$

where f operates on each element of the column vector and is defined such that $f(a) = a - [a]$, and $[a]$ is the largest integer included in a. It then follows that the corner polyhedron is defined by Gomory's fractional cuts (f-cuts) (see Section 5.2.1) derived for each of the rows in the continuous optimal tableau.

The optimal solution over the corner polyhedron (that is, the optimal solution of the *relaxed problem*) is now expressed by the problem:

minimize

$$z_r = (c_B B^{-1}N - c_N)x_N$$

subject to

$$f(B^{-1}b) - f(B^{-1}Nx_N) \equiv 0 \quad (\text{mod } 1)$$
$$x_N \geq 0 \text{ and integer} \qquad (relaxed\ problem)$$

The minimization of z_r follows from the maximization of $z = c_B B^{-1}b - z_r$ after neglecting the constant. Notice that from the *optimality* of the linear program $c_B B^{-1}N - c_N \geq 0$ and hence the relaxed problem must have a bounded solution.

The methods for solving the relaxed problem are developed based on some special properties, which will be explained by using the linear program whose solution is given in Table 6-1. The continuous optimum solution is given by:

maximize

$$z = 8 - S_2$$

subject to

$$\begin{pmatrix} x_1 \\ x_2 \\ S_3 \end{pmatrix} = \begin{pmatrix} \frac{7}{2} \\ \frac{9}{4} \\ 7 \end{pmatrix} - \begin{pmatrix} 0 & \frac{1}{2} \\ \frac{1}{2} & -\frac{1}{4} \\ -2 & 2 \end{pmatrix} \begin{pmatrix} S_1 \\ S_2 \end{pmatrix}$$

$$x_1, x_2, S_1, S_2, S_3 \geqslant 0$$

Thus the relaxed problem is given by:

minimize

$$z_r = S_2$$

subject to

$$\begin{pmatrix} \frac{1}{2} \\ \frac{1}{4} \\ 0 \end{pmatrix} - \left[\begin{pmatrix} 0 \\ \frac{1}{2} \\ 0 \end{pmatrix} S_1 + \begin{pmatrix} \frac{1}{2} \\ \frac{3}{4} \\ 0 \end{pmatrix} S_2 \right] \equiv \begin{pmatrix} 0 \\ 0 \\ 0 \end{pmatrix} \pmod 1$$

$$S_1, S_2 \geqslant 0 \quad \text{and integer}$$

Note that the row associated with basic S_3 has all zero coefficients. This should be expected since the associated constraint $(-2x_1 + 4x_2 \leqslant 9)$ is not binding at the continuous optimum and hence does not define the corner polyhedron (see Fig. 6-1). This means that the constraints of the relaxed problem are described by

$$\begin{pmatrix} \frac{1}{2} \\ \frac{1}{4} \end{pmatrix} - \left[\begin{pmatrix} 0 \\ \frac{1}{2} \end{pmatrix} S_1 + \begin{pmatrix} \frac{1}{2} \\ \frac{3}{4} \end{pmatrix} S_2 \right] \equiv \begin{pmatrix} 0 \\ 0 \end{pmatrix} \pmod 1$$

$$S_1, S_2 \geqslant 0 \quad \text{and integer}$$

The idea of these constraints is that the difference between the two left-hand side elements must be an integer vector. It can be shown that regardless of the (positive integer) values that may be assumed by S_1 and S_2, the resulting vector

$$\left[\begin{pmatrix} 0 \\ \frac{1}{2} \end{pmatrix} S_1 + \begin{pmatrix} \frac{1}{2} \\ \frac{3}{4} \end{pmatrix} S_2 \right]$$

after applying the f-operator, can only assume one of a *finite* number of possibilities (exactly four in the example). One is thus interested in the (S_1, S_2) combination, which yields $f(B^{-1}b) = (\frac{1}{2}, \frac{1}{4})^T$ while minimizing z_r.

To illustrate the above point, a tabulation of $(0, \frac{1}{2})^T S_1$ and $(\frac{1}{2}, \frac{3}{4})^T S_2$ is given in Table 6-2. The table reveals the important observation that, for all (positive integer) values of S_1 and S_2, there are only a finite number of vectors for $f\{(0, \frac{1}{2})S_1\}$ and $f\{(\frac{1}{2}, \frac{3}{4})S_2\}$, namely, $f\{(0, \frac{1}{2})S_1\} = \{(0, 0) \text{ or } (0, \frac{1}{2})\}$ and $f\{(\frac{1}{2}, \frac{3}{4})S_2\} = \{(0, 0), (\frac{1}{2}, \frac{3}{4}), (0, \frac{1}{2}), \text{ or } (\frac{1}{2}, \frac{1}{4})\}$. Thus the only effect of making $S_1 \geqslant 2$ and $S_2 \geqslant 4$ is simply to change $\{(B^{-1}b) - (B^{-1}Nx_N)\}$ by an integer vector. But since all the coefficients of the objective function in the relaxed problem are nonnegative by definition, one must be interested only in the integer values of S_1 and S_2 satisfying $S_1 = 0$ or 1, and $S_2 = 0$, 1, 2, or 3. In other words, although values outside these ranges may render $\{f(B^{-1}b) - f(B^{-1}Nx_N)\}$ an integer vector, they cannot minimize

$$z_r = (c_B B^{-1}N - c_N)x_N$$

since $c_B B^{-1}N - c_N \geqslant 0$.

TABLE 6-2

S_1	0	1	2	3	4	5	6	$\cdots$
$(0, \frac{1}{2})S_1$	$(0, 0)$	$(0, \frac{1}{2})$	$(0, 1)$	$(0, \frac{3}{2})$	$(0, \frac{4}{2})$	$(0, \frac{5}{2})$	$(0, \frac{6}{2})$	$\cdots$
$f\{(0, \frac{1}{2})S_1\}$	$(0, 0)$	$(0, \frac{1}{2})$	$(0, 0)$	$(0, \frac{1}{2})$	$(0, 0)$	$(0, \frac{1}{2})$	$(0, 0)$	$\cdots$
S_2	0	1	2	3	4	5	6	$\cdots$
$(\frac{1}{2}, \frac{3}{4})S_2$	$(0, 0)$	$(\frac{1}{2}, \frac{3}{4})$	$(1, \frac{6}{4})$	$(\frac{3}{2}, \frac{9}{4})$	$(2, 3)$	$(\frac{5}{2}, \frac{15}{4})$	$(3, \frac{9}{2})$	$\cdots$
$f\{(\frac{1}{2}, \frac{3}{4})S_2\}$	$(0, 0)$	$(\frac{1}{2}, \frac{3}{4})$	$(0, \frac{1}{2})$	$(\frac{1}{2}, \frac{1}{4})$	$(0, 0)$	$(\frac{1}{2}, \frac{3}{4})$	$(0, \frac{1}{2})$	$\cdots$

Applying the above ideas to solve the given relaxed problem, note that an integer $\{f(B^{-1}b) - f(B^{-1}Nx_N)\}$ is accomplished for $S_1 = 1$ and $S_2 = 1$ with $z_r = 1$, namely,

$$(\tfrac{1}{2}, \tfrac{1}{4})^T - [(0, \tfrac{1}{2})^T(1) + (\tfrac{1}{2}, \tfrac{3}{4})^T(1)] = (0, -1)^T$$

which satisfies $(0, 0)^T \pmod{1}$. Notice also that $(S_1 = 1, S_2 = 5)$ satisfies the same condition but yields $z_r = 5$ and hence cannot be optimum.

It is easy now to check if the relaxed problem automatically solves the integer problem. This is done by checking whether the resulting values of the basic variables x_1, x_2, and S_3 are nonnegative. Thus

$$x_1 = \tfrac{7}{2} - \tfrac{1}{2}(1) = 3$$
$$x_2 = \tfrac{9}{4} - \tfrac{1}{2}(1) + \tfrac{1}{4}(1) = 2$$
$$S_3 = 7 + 2(1) - 2(1) = 7$$

which is feasible. Hence the optimal solution is $x_1 = 3$, $x_2 = 2$, and $z = 7$.

(Notice that the integrality of the basic variables is not an issue here, since this is guaranteed by the definition of the relaxed problem. Only non-negativity condition may be checked.) Of course, if any of the basic variables is negative, then the relaxed problem does not solve the integer problem and further calculation will be necessary to secure the true optimum. Indeed, this calculation extends the search beyond the specified (minimum) ranges for the nonbasic variables (in the above example, $S_1 \leqslant 1$ and $S_2 \leqslant 3$). As one may suspect this could lead to very laborious computations (see Section 6.5).

It is possible now to introduce the principles underlying the solution of the relaxed problem. Referring to Table 6-2, for S_1, $S_2 = 0$, 1, 2, ..., $f\{(0, \frac{1}{2})^T S_1\}$ and $f\{(\frac{1}{2}, \frac{3}{4})^T S_2\}$ necessarily reduce to the elements of what is known as a *finite* (Abelian) *group*, namely,

$$G = \{(0, 0), (\tfrac{1}{2}, \tfrac{3}{4}), (0, \tfrac{1}{2}), (\tfrac{1}{2}, \tfrac{1}{4})\}$$

In general, the corner polyhedron, as defined above, can be represented by a finite group G whose order or cardinality (number of distinct elements) is given by $|G| = D$, where D is the *absolute* value of the determinant of the basis associated with the *continuous* optimal linear program. For example, in the above numerical problem

$$B = \begin{pmatrix} 1 & 2 & 0 \\ 2 & 0 & 0 \\ -2 & 4 & 1 \end{pmatrix}$$

for which $D = |-2(2)| = 4$.

It is important to notice that in Table 6-2 the vector $(\frac{1}{2}, \frac{3}{4})^T S_2$ is capable of generating every element of the group G, while the vector $(0, \frac{1}{2})^T S_1$ can generate only two elements. Generally, if any *one* element of the group can generate the entire group, the group is said to be *cyclic*. Such a property, when it exists, allows a simpler method for solving the relaxed problem. Problems associated with noncyclic groups can also be solved but at the expense of additional computation. These points are discussed in Section 6.4.

In view of the above discussion, it is possible to write the relaxed problem in what is known as a *group problem* as follows:

minimize

$$z_g = \sum_{j \in NB} c_j(g_j)x_j$$

subject to

$$\sum_{j \in NB} g_j x_j \equiv g_0$$

$$x_j \geqslant 0 \quad \text{and integer} \qquad \text{(group problem)}$$

where $g_0 = f(B^{-1}b)$, $g_j = f(B^{-1}P_j|P_j$ is the jth column of N), and $c_j(g_j) = j$th component of $(c_B B^{-1}N - c_N)$ associated with the group element g_j. It is understood that the addition is done in modulus arithmetic.

Using the group problem definition, efficient solution methods based on network theory and dynamic programming can be developed. These methods are presented in Section 6.4.

6.3.1 A Sufficient Condition for Feasibility of Basic Variables†

This section develops a sufficient condition for the group problem to solve the integer program directly.

Given that D is the absolute value of the determinant of the optimum basis, then, as clear from the above discussion, $x_j \leqslant D - 1$. This condition can be made more restrictive by replacing it by the inequality $\sum_{j \in NB} x_j \leqslant D - 1$. An intuitive justification for its validity is as follows. (See Problem 6-4 for an outline of a rigorous proof.) Suppose $\sum_{j \in NB} x_j \geqslant D$. Then this would mean that $\sum_{j \in NB} g_j x_j$ would lead to more than D columns. But since the group G cannot include more than D distinct elements (including the zero column), it must be that some of these columns are repetitions.

Let K_B be the cone generated by the columns of the *optimal* basis B. Since $Bx_B = b$, and $x_B \geqslant 0$, then b is a nonnegative linear combination of the columns of B.

Let $x_N{}^*$ be the optimal solution for the group problems; then the associated basic variables $x_B{}^*$ are feasible if and only if

$$x_B{}^* = B^{-1}(b - Nx_N{}^*) \geqslant 0$$

This condition is thus satisfied if and only if $b - Nx_N$ is in K_B, since again $b - Nx_N$ is a nonnegative combination of the columns of B. However, since N and $x_N{}^*$ are not known a priori (otherwise there is no need for the sufficiency condition), the containment of $b - Nx_N{}^*$ in K_B cannot be checked directly. Instead, by establishing a calculable upper bound on the magnitude of the vector $Nx_N{}^*$, a (weaker) sufficiency condition can be established.

Let $\|z\|$ be the Euclidean length of the vector z. Then

$$\|Nx_N{}^*\| = \left\| \sum_{j \in NB} \alpha_j x_j \right\| \leqslant \left\| \xi \sum_{j \in NB} x_j \right\| \leqslant (D-1)\|\xi\|$$

where ξ is defined such that $\|\xi\| = \max_{j \in NB}\{\|\alpha_j\|\}$. The inequality is established from the previous result $\sum_{j \in NB} x_j \leqslant D - 1$.

Now let $d = (D-1)\|\xi\|$ and define $K_B(d)$ as a restricted cone contained in K_B whose points are at least a distance d from the boundaries

† Gomory (1965).

of K_B. Since $\|Nx_N{}^*\| \leq d$, the sufficient condition is satisfied if b is contained in $K_B(d)$.

▶**EXAMPLE 6.3-1**   The sufficient condition is applied to the examples in Figs. 6-1 and 6-2. Notice that the two problems differ only in the right-hand side vector b.

Consider the determination of K_B and $K_B(d)$ first. From the continuous optimum solution,

$$B = \begin{pmatrix} 1 & 2 & 0 \\ 2 & 0 & 0 \\ -2 & 4 & 1 \end{pmatrix}, \qquad x_B = (x_1, x_2, S_3), \qquad x_N = (S_1, S_2)$$

$$N = (\alpha_1, \alpha_2) = \begin{pmatrix} 0 & \frac{1}{2} \\ \frac{1}{2} & -\frac{1}{4} \\ -2 & 2 \end{pmatrix}$$

Now

$$\xi = (\tfrac{1}{2}, -\tfrac{1}{4}, 2)^T, \qquad \|\xi\| = ((\tfrac{1}{2})^2 + (-\tfrac{1}{4})^2 + 2^2)^{1/2} \cong 2.1$$
$$d = (4 - 1)(2.1) = 6.3$$

The cone K_B as defined by the vectors of B is illustrated graphically in Fig. 6-3. The restricted cone $K_B(d)$, $d = 6.3$, is then contained in K_B, also as shown in Fig. 6-3.

The planes defining the boundaries of K_B can be computed from the data in Fig. 6-3. These are given by

$$\begin{array}{llc} 2y_1 - y_2 & = 0 & ① \\ 2y_1 - 2y_2 - y_3 & = 0 & ② \\ y_2 & = 0 & ③ \end{array}$$

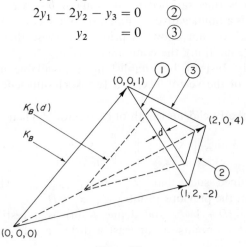

Figure 6-3

Thus the cone K_B is given by

$$
\begin{array}{rl}
2y_1 - \quad y_2 \quad\quad \geqslant 0 & \text{①} \\
-2y_1 + 2y_2 + y_3 \geqslant 0 & \text{②} \quad\quad \text{(cone } K_B) \\
y_2 \quad\quad \geqslant 0 & \text{③}
\end{array}
$$

(The direction of the inequalities is determined so that the vector not lying on the plane is feasible with respect to the inequality. For example, the vector $(2, 0, 4)$ is infeasible if the inequality in ① is reversed. See Fig. 6-3.)

The restricted cone $K_B(d)$ is then defined by the inequalities

$$
\begin{array}{rl}
2y_1 - \quad y_2 \quad\quad \geqslant 6.3 & \\
-2y_1 + 2y_2 + y_3 \geqslant 6.3 & \text{(cone } K_B(d)) \\
y_2 \quad\quad \geqslant 6.3 &
\end{array}
$$

Now for the problem in Fig. 6-1, $b = (8, 7, 9)$, which satisfies the inequalities of $K_B(d)$. This means that the sufficient condition is satisfied. On the other hand, in Fig. 6-2, $b = (8, 7, 7)$, which does not satisfy the sufficiency condition. Observe however, that in general a problem may not satisfy the sufficiency condition, yet the group problem will automatically solve the integer program.◀

It must be realized that the implementation of the above sufficiency condition may, in general, require prohibitive computations. Consequently, the practical value of the condition cannot be promising. Nevertheless, it provides important geometric insight. Other sufficient conditions are given by Hu (1969) and J. F. Shapiro (1968b) (see Problems 6-5 and 6-7). Balas (1973) proves that Gomory's condition is never satisfied for the 0-1 problem (see Problem 6-11).

6.4   Solution of the Group (Relaxed) Problem

This section presents two methods for solving the group problem. The first method is based on a "shortest-route" representation of the problem. The second method utilizes dynamic programming.

6.4.1   Shortest-Route Algorithm†

The group problem

minimize

$$
z_g = \left\{ \sum_{j \in NB} c(g_j) x_j \;\middle|\; \sum_{j \in NB} g_j x_j = g_0, \, x_j \geqslant 0 \text{ integer} \right\}
$$

† Hu (1969), pp. 348–354.

can be represented as a network by realizing that the entire group consists of D distinct elements (including the zero column Θ). Let each element of the group represent a *node*. From the definition of the group problem, a realization of a new node g_k from another node g_j is effected by incrementing the value of x_j by one. This is equivalent to saying that *directed* arcs, each associated with a *unit* value of x_j, must emanate from node g_j to another (appropriate) node. The length of the arc resulting from incrementing x_j is $c(g_j)$. Naturally, arcs emanate from a node only if they lead to another node other than the one representing the zero column. Also, there is no need to draw arcs emanating from node g_0 representing the right-hand side. The objective is to find the shortest route from Θ, where all nonbasic x_j are zero, to node g_0.

To illustrate the network representation, consider the example of Fig. 6-1. The group elements associated with the problem are given by

$$\Theta = \begin{pmatrix} 0 \\ 0 \end{pmatrix}, \qquad g_1 = \begin{pmatrix} \frac{1}{2} \\ \frac{3}{4} \end{pmatrix}, \qquad g_2 = \begin{pmatrix} 0 \\ \frac{1}{2} \end{pmatrix}, \qquad g_3 = \begin{pmatrix} \frac{1}{2} \\ \frac{1}{4} \end{pmatrix}$$

Thus the group problem may be written as

minimize

$$z_g = \{0\, S_1 + S_2 \,|\, g_2 S_1 + g_1 S_2 = g_3,\, S_1, S_2 \geqslant 0 \text{ and integer}\}$$

The network representation is shown in Fig. 6-4. At most two arcs (associated with S_1 and S_2) can emanate from any node. Notice that no arcs are directed to the Θ-node, since obviously such a route cannot be optimum. (For example, an S_1-arc from g_2 leads to Θ.) No arcs emanate from g_3 because it is associated with the right-hand side vector. The problem reduces to finding the shortest route between Θ and g_3, which obviously in this case is given by $S_1 = 1$ and $S_2 = 1$.

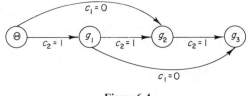

Figure 6-4

It is mportant to notice that a change in the right-hand side vector b *only* cannot change the nodes of the network. The only possible variation is that the node representing the right-hand side may change, in which case old arcs may be omitted and new ones added. For example, in the problem in Fig. 6-2, the right-hand side is $b = (8, 7, 7)^\mathrm{T}$. This gives the continuous solution $x_B = (\frac{7}{2}, \frac{9}{4}, 5)$, which yields the same right-hand element g_3. Hence the network remains essentially unchanged. On the other hand, if $b = (7, 8, 7)$, then $x_B = (4, \frac{3}{2}, 9)^\mathrm{T}$, and the right-hand side element becomes g_2. In this case, it is not necessary to draw the S_2-arc from g_2 to g_3 or the S_1-arc from g_1 to g_3 as shown in Fig. 6-4.

The implication of the above discussion is that a variation in the vector b basically changes the terminal node (together with some arcs). Consequently, in solving the group problem, it appears promising to develop all the shortest routes from Θ to every other node in the network on the assumption that such a node may represent the right-hand side element. The results should prove useful in carrying out sensitivity analysis. This necessitates that *all* proper arcs emanating from *every* node be in the tree, except those terminating at Θ.

Without loss of generality, the group problem may be written as:

minimize

$$z_g = \sum c(g)t(g)$$

subject to

$$\sum_{g \in G - \Theta} g \cdot t(g) = g_0$$

$$t(g) \geqslant 0 \quad \text{and integer}$$

This definition differs from the one given in Section 6.3 in that *all* the elements of the group [less the zero element (column) Θ] are represented in the left-hand side of the constraints. The two definitions can be made equivalent by assigning $c(g) = \infty$ to all $g \in G - \Theta$ that are not automatically defined by the relaxed problem.

As mentioned previously, the algorithm is developed by assuming that g_0 can be replaced by any element $g \in G - \Theta$, that is, it is required to find the shortest routes from the zero element Θ to every element $g \in G - \Theta$.

Rather than using the network directly to solve the problem, a more convenient representation is to have each $g \in G - \Theta$ represented by a square in a table of $D - 1$ elements, where $D = |G|$. Each square includes $c(g)$ and an index ρ to keep track of the intermediate nodes that lie on the shortest route from Θ to g. A typical *initial* table is given in Table 6-3. This assumes that the calculation starts so that an element g_k is reached directly from Θ. This means that for $g_0 = g_k$, $t(g) = 1$ for $g = g_k$ and zero otherwise.

Initially, all the indices ρ are called *temporary* labels, since one is not sure that $g_\rho = g_\rho$ is the cheapest route to reach g_ρ. A temporary label

TABLE 6-3

	g_1	g_2	g_3	$\cdots$	g_{D-1}
$c(g)$	$c(g_1)$	$c(g_2)$	$c(g_3)$	$\cdots$	$c(g_{D-1})$
$\rho(g)$	1	2	3	$\cdots$	$D-1$

becomes *permanent* if it is not altered by later calculation. Let T and P be the sets of temporary and permanent labels, respectively, at a given stage of the calculation.

The steps of the algorithm are now stated as follows:

Step 1   Mark as permanent the square having the smallest $c(g_\rho)$ among all the temporary labels $\rho \in T$. Ties are broken arbitrarily.

Step 2   Let g_r be the last group element that was labelled permanently. For all the group elements with permanent labels $i \in P$, compute

$$c(g_i) + c(g_r) \text{ and } c(g_i + g_r), \qquad \text{for all} \quad i \in P$$

If $c(g_i) + c(g_r) \geqslant c(g_i + g_r)$, the sets T and P remain unchanged, since it is not cheaper to reach the element $(g_i + g_r)$ via the elements g_i and g_r. Otherwise, the entries of the column associated with $g_i + g_r$ must be changed to $c(g_i) + c(g_r)$ and m, where $m = \rho(g_i)$ or $\rho(g_r)$.

Step 3   If all the labels are permanent, stop. Otherwise, go to Step 1.

The solution is now determined by starting from g_0 and then proceeding in the reverse order to Θ. This should give the values of $t(g)$ in increments of one.

▶**EXAMPLE 6.4-1**   Suppose that the continuous optimum problem associated with an integer program is given by:

maximize

$$z = 100 - (2x_3 + 7x_4 + 4x_5)$$

subject to

$$\binom{x_1}{x_2} = \binom{\frac{15}{8}}{\frac{9}{4}} - \left[\binom{\frac{5}{8}}{\frac{3}{4}}x_3 + \binom{\frac{5}{2}}{0}x_4 + \binom{-\frac{5}{8}}{\frac{5}{4}}x_5 \right]$$

$$x_1, x_2, x_3, x_4, x_5 \geqslant 0$$

The resulting group problem is written as:

minimize

$$z_g = 2x_3 + 7x_4 + 4x_5$$

subject to

$$\begin{pmatrix} \frac{5}{8} \\ \frac{3}{4} \end{pmatrix} x_3 + \begin{pmatrix} \frac{1}{2} \\ 0 \end{pmatrix} x_4 + \begin{pmatrix} \frac{3}{8} \\ \frac{1}{4} \end{pmatrix} x_5 = \begin{pmatrix} \frac{7}{8} \\ \frac{1}{4} \end{pmatrix}$$

$$x_3, x_4, x_5 \geq 0 \quad \text{and integer}$$

The group is of order 8 and is cyclic since all eight elements can be generated from $(\frac{5}{8}, \frac{3}{4})^{\text{T}}$.

Representing the group by using Hu's notation, one gets

$$g_1 = \begin{pmatrix} \frac{5}{8} \\ \frac{3}{4} \end{pmatrix}, \qquad g_2 = \begin{pmatrix} \frac{1}{4} \\ \frac{1}{2} \end{pmatrix}, \qquad g_3 = \begin{pmatrix} \frac{7}{8} \\ \frac{1}{4} \end{pmatrix}, \qquad g_4 = \begin{pmatrix} \frac{1}{2} \\ 0 \end{pmatrix}$$

$$g_5 = \begin{pmatrix} \frac{1}{8} \\ \frac{3}{4} \end{pmatrix}, \qquad g_6 = \begin{pmatrix} \frac{3}{4} \\ \frac{1}{2} \end{pmatrix}, \qquad g_7 = \begin{pmatrix} \frac{3}{8} \\ \frac{1}{4} \end{pmatrix}$$

The starting information is thus summarized in Table 6-4. This implies that $x_3 = t(g_1)$, $x_4 = t(g_4)$, and $x_5 = t(g_7)$. The entries $c(g) = \infty$ imply that the associated "node" cannot be reached directly from the zero element Θ. All the labels $\rho(g)$ are temporary.

TABLE 6-4

	g_1	g_2	g_3	g_4	g_5	g_6	g_7
$c(g)$	2	∞	∞	7	∞	∞	4
$\rho(g)$	1	2	3	4	5	6	7

Since $c(g_1) = 2$ is the smallest for all g, $\rho(g_1) = 1$ is permanent. (This is noted by an asterisk in the g_1-column.) Thus $P = \{\rho(g_1)\}$ and

$$T = \{\rho(g_2), \ldots, \rho(g_7)\}.$$

Now

$$c(g_1) + c(g_1) = 2 + 2 = 4 < \infty = c(g_2) = c(g_1 + g_1)$$

Thus put $c(g_2) = 4$ and $\rho(g_2) = 1$, which gives Table 6-5.

TABLE 6-5

	g_1	g_2	g_3	g_4	g_5	g_6	g_7
$c(g)$	2*	4	∞	7	∞	∞	4
$\rho(g)$	1	1	3	4	5	6	7

The elements g_2 and g_7 have the same minimal cost $c(g) = 4$. Breaking the tie arbitrarily, $\rho(g_7) = 7$ is permanent. This yields $P = \{\rho(g_1), \rho(g_7)\}$. Hence

$$c(g_1) + c(g_7) = 6 > 0 = c(\Theta) = c(g_1 + g_7)$$
$$c(g_7) + c(g_7) = 8 < \infty = c(g_6) = c(g_7 + g_7)$$

The changes are $c(g_6) = 8$, $\rho(g_6) = 7$, and this yields Table 6-6.

TABLE 6-6

	g_1	g_2	g_3	g_4	g_5	g_6	g_7
$c(g)$	2*	4	∞	7	∞	8	4*
$\rho(g)$	1	1	3	4	5	7	7

The index $\rho(g_2) = 1$ is permanent. Thus $P = \{\rho(g_1), \rho(g_7), \rho(g_2)\}$:

$$c(g_1) + c(g_2) = 6 < \infty = c(g_3) = c(g_1 + g_2), \qquad \text{change } g_3$$
$$c(g_7) + c(g_2) = 8 > 2 = c(g_1) = c(g_7 + g_2)$$
$$c(g_2) + c(g_2) = 8 > 7 = c(g_4) = c(g_2 + g_2)$$

The resulting changes are given in Table 6-7.

TABLE 6-7

	g_1	g_2	g_3	g_4	g_5	g_6	g_7
$c(g)$	2*	4*	6	7	∞	8	4*
$\rho(g)$	1	1	1	4	5	7	7

Continuing in the same manner, the successive tables are Tables 6-8, 6-9, and 6-10.

TABLE 6-8

	g_1	g_2	g_3	g_4	g_5	g_6	g_7
$c(g)$	2*	4*	6*	7	10	8	4*
$\rho(g)$	1	1	1	4	1	7	7

TABLE 6-9

	g_1	g_2	g_3	g_4	g_5	g_6	g_7
$c(g)$	2*	4*	6*	7*	9	8	4*
$\rho(g)$	1	1	1	4	1	7	7

TABLE 6-10

	g_1	g_2	g_3	g_4	g_5	g_6	g_7
$c(g)$	2*	4*	6*	7*	9*	8*	4*
$p(g)$	1	1	1	4	1	7	7

To find the values of $t(g)$, a reverse procedure is followed. Given $g_0 = g_3$, from Table 6-10 one has $p(g_3) = 1$, or $t(g_1) \geq 1$. Now $g_3 - g_1 = g_2$ and $p(g_2) = 1$, which gives $t(g_1) \geq 2$. Again $g_2 - g_1 = g_1$ and $p(g_1) = 1$, which gives $t(g_1) \geq 3$. But since $g_1 - g_1 = \Theta$, the reverse procedure is complete and $t(g_1) = 3$. This means that $x_3 = 3$, while $x_4 = x_5 = 0$ gives the optimal solution to the relaxed problem.

Notice that the calculation could be stopped at Table 6-8, that is, as soon as $g_0 = g_3$ is permanently labeled. However, by using Table 6-10, a shortest route can be determined for all $g_0 = g_i$, $i = 1, 2, \ldots, D - 1$.

The solution $x_3 = 3$, $x_4 = x_5 = 0$, yields feasible values for the basic variables x_1 and x_2, namely, $x_1 = x_2 = 0$. Consequently, the group problem solves the integer problem directly.◀

The validity proof for the algorithm may be outlined as follows:

(1) It is impossible to create a cheaper group element $g_k = \sum_{j \in Q} g \cdot t(g)$ such that Q includes at least one element from T with the remaining elements (if any) being from P; T and P are the sets of temporary and permanent labels.

(2) It suffices to consider the sums of elements in P taken two at a time only.

(3) Permanent labels represents the actual minimum cost of the associated group elements.

6.4.2   Dynamic Programming Algorithm†

The group problem defined in terms of the nonbasic variables $x_j, j \in NB$, is given by

minimize

$$z_g = \sum_{j \in NB} c(g_j) x_j$$

† Gomory (1965).

subject to

$$\sum_{j \in NB} g_j x_j = g_0$$

$$x_j \geqslant 0 \quad \text{and integer}$$

Let y be defined such that

$$f_n(y) = \min \sum_{j=1}^{n} c(g_j) x_j, \qquad n = 1, 2, \ldots, |NB|$$

subject to

$$\sum_{j=1}^{n} g_j x_j = y$$

$$x_j \geqslant 0 \quad \text{and integer}, \qquad j = 1, 2, \ldots, n$$

In other words, y defines the right-hand side group element given fixed values for $x_j, j = 1, 2, \ldots, n$. Clearly, $f_n(\Theta) = 0$.

A dynamic programming algorithm can be developed for the above problem by observing that in forming y, the group element g_n is used *at least once* or is not used at all. This is equivalent to saying that for

$$x_n \begin{cases} \geqslant 1, & f_n(y) = c(g_n) + f_n(y - g_n) \\ = 0, & f_n(y) = f_{n-1}(y) \end{cases}$$

Thus the recursive equation may be written as

$$f_n(y) = \min\{f_{n-1}(y), c(g_n) + f_n(y - g_n)\}$$
$$f_n(\Theta) = 0$$

In order for the recursive equation to be computable, one must be able to compute $f_n(y)$ for all y. This is always possible if g_n generates the entire group G so that $y - g_n$ can be formed. The elements y are thus given by $y = r g_n, r = 1, 2, \ldots, |g_n|$, where $|g_n|$ is the order of g_n.

In the case where g_n does not generate G, that is, $|g_n| < |G|$ (notice that this can happen in a cyclic group also), the following procedure may be followed: First compute $f_n(y)$ for all $y = r g_n, r = 1, 2, \ldots, |g_n|$. Let y_1 be an element of G not reached by g_n, that is, $y_1 \neq r g_n, r = 1, 2, \ldots, |g_n|$. From the recursive equation

$$f_n(y_1) = \min\{f_n(y_1 - g_n) + c(g_n), f_{n-1}(y_1)\}$$

Thus one may assume initially that $f_n(y_1) = f_{n-1}(y_1)$. Clearly, this assumption is correct only if $f_n(y_1 - g_n) + c(g_n) > f_{n-1}(y_1)$. Consequently, to reflect this point, a new function $f_n{}^1(y_1) = f_{n-1}(y_1)$ is defined. As will be shown later, $f_n{}^1$ must eventually determine the true function f_n.

Now compute

$$f_n{}^1(y_1 + rg_n) = \min\{f_n{}^1(y_1 + rg_n - g_n) + c(g_n), f_{n-1}(y_1 + rg_n)\}$$
$$r = 1, 2, \ldots, |g_n|$$

The idea is that, if $f_n{}^1(y_1 + rg_n) = f_n(y_1 + rg_n)$ for some $r = d \leqslant |g_n|$, then all subsequent $f_n(y_1 + rg_n)$, $r = d + 1$, ..., $|g_n|$ are computed directly by $f_n{}^1(y_1 + rg_n)$. In particular, this condition guarantees that the true value of $f_n(y_1)$ is $f_n{}^1(y_1 + |g_n|g_n)$. If this value is the same as the initially assumed value $f_n(y_1) = f_{n-1}(y_1)$, then there is nothing more to be done and $f_n{}^1$ yields the true values of f_n. Otherwise, the new value of $f_n(y_1)$ is now used to compute $f_n(y_1 + rg_n)$, $r = 1, 2, \ldots, d$, by simply going through the recursive equation once more. Of course, the value $r = d$ is not known a priori. But this is accounted for by computing $f_n(y_1 + rg_n)$ for $r = 1, 2, \ldots$, until for some $r < |g_n|$, $f_n(y_1 + rg_n) = f_n{}^1(y_1 + rg_n)$. This gives d. Notice, however, that there is no computational value in knowing the explicit value of d.

The validity of the above procedure reduces to proving that there exists some $r = d \leqslant |g_n|$ such that $f_n{}^1(y_1 + rg_n) = f_n(y_1 + rg_n)$. This is verified as follows. For any $r = 1, 2, \ldots, |g_n|$, $f_n(y_1 + rg_n)$ is equal to either $f_{n-1}(y_1 + rg_n)$ or $f_n(y_1 + rg_n - g_n) + c(g_n)$. Consider the first case,

$$f_n(y_1 + rg_n) = f_{n-1}(y_1 + rg_n).$$

From the recursive equation defining $f_n{}^1$,

$$f_n{}^1(y_1 + rg_n) \leqslant f_{n-1}(y_1 + rg_n)$$

Also, from the definition of $f_n{}^1(y_1)$, $f_n(y_1) \leqslant f_n{}^1(y_1)$, and consequently

$$f_n{}^1(y_1 + rg_n) \geqslant f_n(y_1 + rg_n)$$

since the original overestimate can only be propagated. Because the right-hand sides of the two inequalities are equal by assumption, it follows that

$$f_n{}^1(y_1 + rg_n) = f_n(y_1 + rg_n), \qquad \text{for some} \quad r \leqslant |g_n|$$

In the second case,

$$f_n(y_1 + rg_n) = f_n(y_1 + rg_n - g_n) + c(g_n)$$

Utilizing this condition, then

$$f_n(y_1) = f_n(y_1 + |g_n|g_n) = f_n(y_1 + (|g_n| - 1)g_n) + c(g_n)$$
$$= \cdots = f_n(y_1) + |g_n|c(g_n)$$

This is a contradiction unless $c(g_n) = 0$. But if $c(g_n) = 0$, then $f_{n-1}(y_1 + rg_n) = f_n(y_1)$ or $f_{n-1}(y_1 + rg_n) = f_n(y_1 + rg_n)$. This is the same as the first case.

Since the order of the group $|G|$ is always a multiple of $|g_n|$, the above procedure is repeated $[(|G|/|g_n|) - 1]$ times, each with a different starting y that has not been reached in previous calculation. This guarantees computing $f_n(y)$ for all $y \in G$.

In order to facilitate reading the optimal solution after the recursive computations are completed, the following scheme is used: After obtaining $f_n(y)$, an index $\rho_n(y)$ is defined such that

$$\rho_n(y) = \begin{cases} \rho_{n-1}(y), & \text{if } f_n(y) = f_{n-1}(y) \\ n, & \text{otherwise} \end{cases}$$

This index keeps track of the last variable x_j made equal to one. After $f_n(y)$ is computed for all $n = 1, 2, \ldots, |NB|$, a backtracking procedure (similar to the one used with the shortest-route algorithm) is utilized to find the optimal solution.

▶ EXAMPLE 6.4-2   The same problem in Example 6.4-1 is solved again. The group problem is:

minimize

$$z_g = 2w_1 + 7w_4 + 4w_7$$

subject to

$$\begin{pmatrix} \frac{5}{8} \\ \frac{3}{4} \end{pmatrix} w_1 + \begin{pmatrix} \frac{1}{2} \\ 0 \end{pmatrix} w_4 + \begin{pmatrix} \frac{3}{8} \\ \frac{1}{4} \end{pmatrix} w_7 = \begin{pmatrix} \frac{7}{8} \\ \frac{1}{4} \end{pmatrix}$$

$$w_3, w_4, w_5 \quad 0 \geqslant \text{and integer}$$

(For convenience, the variables x_j are redefined so that w_1 is associated with g_1, w_4 with g_4, and w_7 with g_7. The group elements g_1, g_4, and g_7 are as defined in Example 6.4-1. Notice that $g_4 = 4g_1$ and $g_7 = 7g_1$.)

Consider $g_1 = (\frac{5}{8}, \frac{3}{4})^{\mathsf{T}}$ with $|g_1| = 8$, which is equal to $|G|$. Thus

$$
\begin{aligned}
f_1(\Theta) &= 0 \\
f_1(g_1) &= c(g_1) = 2, & \rho_1(g_1) &= 1 \\
f_1(g_2) &= f_1(2g_1) = 4, & \rho_1(g_2) &= 1 \\
f_1(g_3) &= f_1(3g_1) = 6, & \rho_1(g_3) &= 1 \\
f_1(g_4) &= f_1(4g_1) = 8, & \rho_1(g_4) &= 1 \\
f_1(g_5) &= f_1(5g_1) = 10, & \rho_1(g_5) &= 1 \\
f_1(g_6) &= f_1(6g_1) = 12, & \rho_1(g_6) &= 1 \\
f_1(g_7) &= f_1(7g_1) = 14, & \rho_1(g_7) &= 1
\end{aligned}
$$

Next consider $g_4 = (\frac{1}{2}, 0)^T$ with $|g_4| = 2$. Thus

$$f_4(\Theta) = 0$$
$$f_4(g_4) = \min\{f_4(\Theta) + c(g_4), f_1(g_4)\} = \min\{0 + 7, 8\} = 7, \quad \rho_4(g_4) = 4$$
$$f_4(\Theta) = f_4(2g_4) = \min\{7 + 7, 0\} = 0$$

Let g_1 be the first starting element (there will be $\frac{8}{2} - 1 = 3$ different starting elements). Hence

$$f_4{}^1(g_1) = f_1(g_1) = 2$$
$$f_4{}^1(g_5) = f_4{}^1(g_1 + g_4) = \min\{f_4{}^1(g_1) + c(g_4), f_1(g_5)\}$$
$$= \min\{2 + 7, 10\} = 9, \quad \rho_4(g_5) = 4$$
$$f_4{}^1(g_1) = f_4{}^1(g_1 + 2g_4) = \min\{f_4{}^1(g_5) + c(g_4), f_1(g_1)\}$$
$$= \min\{9 + 7, 2\} = 2, \quad \rho_4(g_1) = 1$$

Since $f_4{}^1(g_1)$ as calculated the *second* time is equal to the assumed value $f_1(g_1)$, no further calculation associated with the starting element g_1 is needed. This means that $f_4{}^1$ yields f_4 directly for g_1 and g_5.

Consider g_3 as the second starting element. Then

$$f_4{}^1(g_3) = f_1(g_3) = 6$$
$$f_4{}^1(g_7) = f_4{}^1(g_3 + g_4) = \min\{f_4{}^1(g_3) + c(g_4), f_1(g_7)\}$$
$$= \min\{6 + 7, 14\} = 13, \quad \rho_4(g_7) = 4$$
$$f_4(g_3) = f_4(g_3 + 2g_4) = \min\{f_4{}^1(g_7) + c(g_4), f_1(g_3)\}$$
$$= \min\{13 + 7, 6\} = 6, \quad \rho_4(g_3) = 1$$

Again this means that $f_4(g_3) = f_4{}^1(g_3)$ and $f_4(g_7) = f_4{}^1(g_7)$.

Finally, consider g_6 as the third starting point. Thus

$$f_4{}^1(g_6) = f_1(g_6) = 12$$
$$f_4{}^1(g_2) = f_4{}^1(g_6 + g_4) = \min\{f_4{}^1(g_6) + c(g_4), f_1(g_2)\}$$
$$= \min\{12 + 7, 4\} = 4$$
$$f_4{}^1(g_6) = f_4{}^1(g_6 + 2g_4) = \min\{f_4{}^1(g_2) + c(g_4), f_1(g_6)\}$$
$$= \min\{4 + 7, 12\} = 11, \quad \rho_4(g_6) = 4$$

Since $f_4{}^1(g_6)$ computed the second time is different from $f_1(g_6)$, further calculation is needed this time starting with $f_4(g_6) = 11$ and $\rho_4(g_6) = 4$. Thus

$$f_4(g_2) = f_4(g_6 + g_4) = \min\{11 + 7, 4\} = 4, \quad \rho_4(g_2) = 1$$

Notice that $f_4(g_2) = f_4{}^1(g_2)$ mainly because the minimum is equal to $f_1(g_2)$.

At this point, all $f_4(g_j)$, $j = 1, 2, \ldots, 7$, are available, and a summary follows:

$$f_4(\Theta) = 0$$

$$
\begin{aligned}
f_4(g_1) &= 2, & \rho_4(g_1) &= 1 \\
f_4(g_2) &= 4, & \rho_4(g_2) &= 1 \\
f_4(g_3) &= 6, & \rho_4(g_3) &= 1 \\
f_4(g_4) &= 7, & \rho_4(g_4) &= 4 \\
f_4(g_5) &= 9, & \rho_4(g_5) &= 4 \\
f_4(g_6) &= 11, & \rho_4(g_6) &= 4 \\
f_4(g_7) &= 13, & \rho_4(g_7) &= 4
\end{aligned}
$$

Consider $g_7 = (\tfrac{3}{8}, \tfrac{1}{4})^{\mathrm{T}}$ with $|g_7| = 8$. Thus

$$
\begin{aligned}
f_7(\Theta) &= 0 \\
f_7(g_7) &= \min\{f_7(\Theta) + c(g_7), f_4(g_7)\} \\
&= \min\{0 + 4, 13\} = 4, & \rho_7(g_7) &= 7 \\
f_7(g_6) = f_7(2g_7) &= \min\{4 + 4, 11\} = 8, & \rho_7(g_6) &= 7 \\
f_7(g_5) = f_7(3g_7) &= \min\{8 + 4, 9\} = 9, & \rho_7(g_5) &= 4 \\
f_7(g_4) = f_7(4g_7) &= \min\{9 + 4, 7\} = 7, & \rho_7(g_4) &= 4
\end{aligned}
$$

$$\boxed{f_7(g_3) = f_7(5g_7) = \min\{7 + 4, 6\} = 6, \qquad \rho_7(g_3) = 1} \quad \leftarrow g_0$$

$$
\begin{aligned}
f_7(g_2) = f_7(6g_7) &= \min\{6 + 4, 4\} = 4, & \rho_7(g_2) &= 1 \\
f_7(g_1) = f_7(7g_7) &= \min\{4 + 4, 2\} = 2, & \rho_7(g_1) &= 1
\end{aligned}
$$

Notice that the calculation could be stopped at $f_7(g_3)$ since $g_0 = g_3$. Notice also that the solution is now readily available for any $g_0 = g_j$, $j = 1, 2, \ldots, 7$.

The optimal solution is obtained as follows: Since $\rho_7(g_3) = 1$, then $w_1 \geqslant 1$. Next consider $f_4(g_3 - g_1) = f_4(g_2) = 4$, and $\rho_4(g_2) = 1$ implies $w_1 \geqslant 2$. Finally, $f_1(g_2 - g_1) = f_1(g_1) = 2$ and $\rho_1(g_1) = 1$; hence $w_1 \geqslant 3$. Because there are no more "stages," backtracking is complete and $w_1 = 3$. The variables w_4 and w_7 are equal to zero. This is the same solution obtained by Hu's algorithm.◀

Hu (1968) gives a comparison between his (shortest-route) algorithm and Gomory's algorithm. He concludes that:

(i)   The number of operations in his algorithm is less than that in Gomory's ($2D^2$ operations instead of $2D^2$ to $4D^2$; D is the order of the group G);

(ii)   The steps in Hu's algorithm are independent of D or the order of its elements g_j ; and

(iii)   It requires fewer computer memory locations.

6.5   Solution of Integer Programs by the Group Problem

In Section 6.3, it is indicated that the group problem solves the integer program directly only if the corresponding values of the basic variables X_B in the optimum linear program are feasible. Figure 6-2 illustrates that this need not be the case always. In this section, two methods are presented to show how the solution of the group problem can be extended to handle such situations. The first method extends Hu's algorithm in a straightforward manner. The second is a branch-and-bound method in which each sub-problem is itself a legitimate group problem and may thus be solved by either one of the two methods in Section 6.4. There is also a third method, which develops a dynamic programming algorithm that automatically accounts for the nonnegativity of the basic variables [see Greenberg (1969b)].

6.5.1   Extension of Hu's Algorithm

In Hu's algorithm (Section 6.4.1) the shortest route is determined by ranking the nodes in ascending order of their distance (cost) from the origin (zero-column). The procedure terminates at the node for which the associated group element is equal to the right-hand side element. Because the optimum solution of the relaxed problem (defined in Section 6.3) is associated with an extreme point of the corner polyhedron, it is only necessary to consider the nodes associated with the finite group G.

When the optimum extreme point of the corner polyhedron does not coincide with a feasible solution to the integer problem, it will be necessary to search in the *interior* of the corner polyhedron for the best integer point satisfying feasibility. This is achieved by utilizing the same principle as in Hu's algorithm, that is, ranking the nodes in ascending order of their distances from the origin (zero element). However, rather than confining the network to $D = |G|$ nodes, every integer point in the corner polyhedron is represented by a node. New nodes are created from a current node by incrementing each nonbasic variable by one. Although the resulting network is infinite, there are only D different *types* of nodes, namely, those defined by the group. The ranking of the nodes starts from the zero element Θ, with all $x_j = 0$, $j \in NB$, and is carried out until a node with the proper congruence relationship is reached, which also yields feasible values for the basic variables.

There is only one small difficulty with this approach, illustrated by a numerical example. Suppose the relaxed problem is defined by two *nonbasic* variables x_1 and x_2. The node $(x_1 = 2, x_2 = 2)$ can be reached from the origin following six different routes, for example, $(x_1 = 1, x_1 = 1, x_2 = 1, x_2 = 1)$, $(x_1 = 1, x_2 = 1, x_1 = 1, x_2 = 1)$, ..., etc. Thus, for the sake of efficiency, a scheme is needed so that the same node is not encountered more than once in the network. This is achieved by using the following simple rule: Index all the nonbasic variables in ascending order. If a current node is reached by incrementing the variable x_k, then only nodes reached by incrementing $x_k, x_{k+1}, x_{k+2}, \ldots$, can be created from the current node. The rule guarantees a unique set of arcs (path) between any two nodes in the network and that eventually *every* node in the network can be reached by a path from the origin.

The extended algorithm may be regarded as a (simplified) branch-and-bound procedure in which the next node to be considered (branched) is the one having the smallest objective value z_g. If a node is reached by incrementing x_k, the branches emanating from this node correspond to $x_k, x_{k+1}, \ldots$, and x_p, where p is the highest index for the variables of the group problem. If a node gives a solution to the group problem and also yields feasible values for the basic variables, it is marked as an upper bound. This upper bound is updated if a smaller value is encountered. The procedure terminates when the node associated with the current upper bound is selected for branching. Notice that there is no optimization problem at each node. This is obviously a weak point in the procedure, which may be alleviated as shown in the next section.

▶EXAMPLE 6.5-1   Suppose in Example 6.4-1 that the right-hand side is changed so that the values of the basic variables in the continuous optimum solution are given as $(x_1, x_2) = (\frac{15}{8}, \frac{5}{4})$. This change does not affect the associated group problem, but its optimal solution $(x_3 = 3, x_4 = x_5 = 0)$ will no longer yield feasible values for the basic variables, since $x_1 = 0$ and $x_2 = -1$.

Utilizing the extended Hu's algorithm, the associated search tree is shown in Fig. 6-5. The order in which a node is considered is indicated by the top number in the circle. The incremented nonbasic variable is indicated on the branch. The choice of variables associated with the branches emanating from a node follows the rule given above.

Node 4 gives the optimum solution to the group problem, which is infeasible with respect to the basic variables. At node 7 the branch x_5 leads to node 15, which *at this point* sets an upper bound on z_g ($=11$) since it yields feasible values for the basic variables. This upper bound allows terminating the search along the branches yielding $z_g \geq 11 - 1 = 10$. (Notice that $\text{lcm}\{c_j | j \in NB\} = 1$.) As shown in Fig. 6-5, node 15 corresponds to the optimal solution of the integer problem. This gives $x_4 = x_5 = 1$, which yields $x_1 = \frac{15}{8} - \frac{20}{8} + \frac{5}{8} = 0$ and $x_2 = \frac{5}{4} - 0 - \frac{5}{4} = 0$, with $z = 100 - 11 = 89$.◀

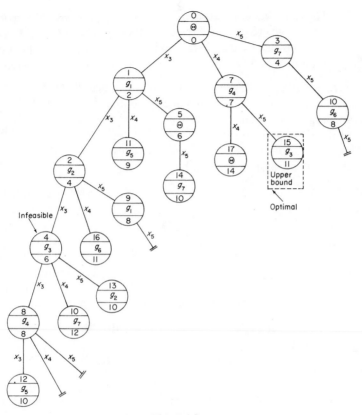

Figure 6-5

Observe that because the corner polyhedron contains an infinite number of integer points, there is the difficulty that the search may never be terminated if the integer problem has no feasible solution. This problem is taken care of by establishing an upper bound u_j on the feasible values of each variable x_j. A branch is thus terminated if its associated value of x_j exceeds u_j.

6.5.2   Branch-and-Bound Algorithm†

The idea of Shapiro's algorithm is more or less the same as the one used in the preceding section. There are two basic differences:

(1)   The group problem is first solved using one of the algorithms in Section 6.4, and the resulting solution provides a lower bound on z_g $(=\underline{z}_g)$. This will be the starting node (0) in the tree.

---

† Shapiro (1968b).

(2)   The branches emanating from node k are defined by $x_j \geqslant d_j^k + 1$, $j \in NB$, where $d_j^0 = 0$ for all j at the starting node (0). (Compare with the idea used by Dakin to improve the Land–Doig algorithm, Section 4.3.2.)

However, in order to have a unique path to each node, the same idea utilized in Section 6.5.1 is implemented, so that if the kth node is reached by branching on x_r, then only branches associated with x_r, x_{r+1}, ..., and x_p need to emanate from the kth node.

The use of a lower bound on x_j now creates an optimization problem. At node (0), this problem is given by:

minimize

$$z_g{}^0 = \sum_{j \in NB} c(g_j) x_j$$

subject to

$$\sum_{j \in NB} g_j x_j = g_0$$

$$x_j \geqslant 0 \quad \text{and integer,} \qquad j \in NB$$

Suppose the branch $x_r \geqslant 1$ emanates from (0); then the optimization problem becomes

minimize

$$z_g{}^1 = \sum_{j \in NB} c(g_j) x_j$$

subject to

$$\sum_{j \in NB} g_j x_j = g_0$$

$$x_j \geqslant 0 \quad \text{and integer,} \qquad j \in NB$$

$$x_r \geqslant 1 \quad \text{and integer}$$

This can be converted into a regular group minimization problem by using the substitution $x_r = x_r' + 1$, which yields:

minimize

$$z_g{}^1 = \sum_{j \neq r} c(g_j) x_j + c(g_r) x_r' + c(g_r)$$

subject to

$$\sum_{j \neq r} g_j x_j + g_r x_r' = g_0 - g_r$$

$$x_j, x_r' \geqslant 0 \quad \text{and integer}$$

Notice that the only difference (from the optimization viewpoint) between this problem and the one at (0) is the change in the right-hand side element from g_0 to $g_0 - g_r$. But since the methods in Section 6.4 yield the solution to the group problem for *all* possible right-hand side elements, it will not be necessary to resolve the resulting group problem anew.

The above idea can be generalized for any two succeeding nodes. If g is the right-hand side element at the current node and if $x_r \geq d$ is the branch leading to the succeeding node, then the right-hand side element for the new group problem is given by $g - dg_r$. This, of course, is accompanied by the proper change to the new variable $x_r' = x_r - d$.

One remark that may lead to efficient computation can be made here. Suppose the optimum of a group problem at a given node k yields $x_r = b$. Define the integer $a \leq b$ and let $x_r \geq a$, $x_r \geq a + 1$, ..., and $x_r \geq b$ define a path from node k. Then all the nodes on this path have the same solution obtained at k, since the additional constraints are redundant.

The rule for selecting the next node to be branched may be either the one used in Section 6.5.1 or the last-in–first-out (LIFO) rule introduced in Section 4.5.2.

►EXAMPLE 6.5-2   Example 6.5-1 will be solved by Shapiro's algorithm. The associated solution tree is shown in Fig. 6-6. The group element defining the *right-hand* side of the associated group problem is indicated in the middle of the node. Given this element, the corresponding solution is read directly from Example 6.4-1 (or 6.4-2).

Notice that because the solution at node 0 yields $x_3 = 3$, the solution at nodes 1, 2, and 3 is the same as at node 0. However, at node 4 where $x_3 \geq 4$, the solution changes.

The LIFO rule is used to select the next node to be branched. Branch $x_4 \geq 1$ from node 0 establishes the upper bound $\bar{z}_g = 11$. All the nodes between 0 and 10 are excluded because they yield $z_g > 11$.

To illustrate how the group problem at each node is defined and solved, consider node 5. The right-hand side element is

$$g = g_3 - 3g_1 - g_4 = g_4$$

The substitution $x_3 = x_3' + 3$ and $x_4 = x_4' + 1$ is used. The solution is determined as follows: From Example 6.4-2, $p_7(g_4) = 4$, or $x_4' \geq 1$. Next consider $g_4 - g_4 = \Theta$; thus $x_3' = x_5 = 0$ and $x_4' = 1$. This gives $x_3 = 3$, $x_4 = 2$, $x_5 = 0$, and $z_g = 20$.

The optimal solution at node 11 (also at node 12) is $x_3 = 0$, $x_4 = x_5 = 1$, with $z_g = 11$.◄

Work related to Shapiro's algorithm appears in Gorry *et al.* (1970) and Gorry and Shapiro (1971). A relaxation procedure that reduces the size of

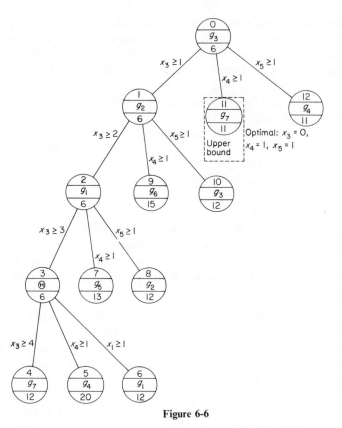

Figure 6-6

the group is given by Gorry *et al.* (1972). An extension of the group theoretic approach to the mixed integer case and other cases can be found in the work of Wolsey (1971a, b).

Other algorithms for solving integer programs by the group problem were developed by Glover and Litzler (1969) and by Greenberg (1969b). Greenberg's method is based on a dynamic programming algorithm similar to the one introduced in Section 6.4.2 except that an additional state variable is defined to account automatically for the feasibility of the basic variables. There is no evidence that these methods offer any computational advantages over the ones presented above.

6.6   Reducing the Number of Congruences

The amount of computation in the group problem is seen to increase with the order of the group as well as the associated number of congruences. The size of the group is directly related to the determinant of the basis in the optimal linear program. The number of congruences, on the other hand, may be reduced if it can be shown that the deleted congruences can be generated from the remaining one. To illustrate this point, consider the group

$$\begin{pmatrix} \frac{1}{3} \\ \frac{2}{3} \end{pmatrix} x_1 + \begin{pmatrix} \frac{2}{3} \\ \frac{1}{3} \end{pmatrix} x_2 \equiv \begin{pmatrix} \frac{1}{3} \\ \frac{2}{3} \end{pmatrix} \pmod{1}$$

Multiplying the first congruence by 2 and then taking modulo 1 of the resulting coefficients immediately yields the second congruence. This means that the given group may be equivalently reduced to

$$\tfrac{1}{3}x_1 + \tfrac{2}{3}x_2 \equiv \tfrac{1}{3} \pmod{1}$$

Notice that such a reduction does not affect the size of the group, which remains equal to 3.

The above idea is the basis for developing general procedures for reducing the number of congruences in a group, namely, by (systematically) interchanging rows (or columns), adding an integer multiple of one row (column) to another, and reversing the sign of a row (column). The detailed steps for carrying out such reductions are incorporated in what are known as the Smith normal form and the Hermite normal form. These details are tedious, and hence the associated algorithms will not be presented here. The interested reader may consult Hu (1968) for the calculation of the Smith normal form. A development by Bradley (1971a) shows how the Hermite and Smith normal forms may be computed.

6.7   Faces of the Corner Polyhedron

An interesting idea is to compute the constraints (faces) defining the corner polyhedron described in Sections 6.2 and 6.3. The idea is appealing because the solution of the relaxed problem reduces to solving an ordinary linear program with all the convenient properties that it possesses. Moreover, the constraints defining the corner polyhedron actually represent the strongest cuts with respect to the original integer problem, and thus their determination may be beneficial in enhancing the efficiency of the cutting algorithms (Chapter 5).

Unfortunately, the methods available for generating these faces are computationally inefficient, so the details of the development will not be presented here. The reader may consult Gomory (1967, 1969) for a treatment of the problem.

6.8   Concluding Remarks

This chapter presents an alternative approach to the solution of the integer problem. Although the development is theoretically interesting, it does not provide the remedy for the acute computational difficulty that characterizes integer programming. Naturally, there are some problems that can be solved more efficiently by this method (especially those yielding a small group size), but there are also many others where the method proves ineffective. Perhaps the method is more effective when the relaxed problem solves the integer problem directly, but unfortunately there exists no (necessary and sufficient) condition that can be used to identify such problems a priori.

It is interesting to notice that the method produces some useful results, which, as a byproduct, may enhance the effectiveness of other solution methods. For example, if the sufficient condition (Section 6.3.1) is satisfied, the result $\sum_{j \in NB} x_j \leqslant D - 1$ may be used in the branch-and-bound algorithms of Chapter 4 to terminate the search along branches that violate this condition. This may actually reduce the number of subproblems solved and hence increase the computational efficiency.

Problems

6-1   Define the relaxed problem for the following integer model:

maximize
$$z = 7x_1 + 9x_2 + 3x_3$$

subject to
$$2x_1 - x_2 + 3x_3 \leqslant 6$$
$$3x_1 + 7x_2 + x_3 \leqslant 35$$
$$x_1, x_2, x_3 \geqslant 0 \text{ and integers}$$

Does this problem satisfy the sufficiency condition?

6-2   Consider the following problem:

maximize
$$z = \sum_{j=1}^{n} c_j x_j$$

subject to
$$\prod_{i=1}^{n} x_j = b$$
$$x_j \text{ are positive integers}$$

where all c_j are constants and b is a positive integer constant. Prove that the optimal solution is given by

$$x_k = b$$
$$x_j = 1, \qquad j \neq k$$

where, without loss of generality, $c_k = \max\{c_1, c_2, \ldots, c_n\}$.

6-3 Consider

minimize

$$z_g = \sum_{j \in R} c_j t_j$$

subject to

$$\sum_{j \in R} g_j t_j = g_0$$

$$t_j \geq 0 \quad \text{and integer}$$

over the group G. Assume that the problem has a feasible solution. Prove that

$$\prod_{j \in R} (1 + t_j) \leq |G| = D$$

6-4 By using the result of Problems 6-2 and 6-3 prove that

$$\sum_{j \in R} t_j \leq D - 1$$

6-5 (Hu, 1969). Prove that a sufficient condition for the relaxed problem to solve the integer problem directly is

$$B^{-1}b - (D - 1)\zeta \geq 0$$

where the ith element of ζ is the maximum element in row i of $B^{-1}N$.

6-6 Consider the group problem defined in Problem 6-3. Let $r_j \leq D$ be the order of g_j in G, $j \in R$. Prove that the group problem has an optimal solution in which $t_j \leq r_j - 1$, for all $j \in R$.

6-7 (J. F. Shapiro, 1968a). Using r_j as defined in Problem 6-6 and the notation in Section 6.3, prove that a sufficient condition for the relaxed problem to solve the original problem is

$$B^{-1}b \geq |N^*|r$$

where

$$r = (r_1, r_2, \ldots, r_{|R|})^{\mathsf{T}}$$
$$N^* = (n_{ij}) = B^{-1}N$$

and $|N^*|$ is obtained from N^* by taking the absolute value of every element n_{ij}.

6-8   Apply the sufficiency conditions in Problems 6-5 and 6-7 to Problem 6-1 and compare the results.

6-9   Show that when the optimum basis B is unimodular, then all the sufficient conditions developed in Problems 6-5 and 6-7 and in Section 6.3.1 are always satisfied. Give an intuitive interpretation of the result.

6-10   Consider the following problem:

minimize

$$z = 5x_1 + 4x_2$$

subject to

$$2x_1 + 3x_2 \geqslant 7$$
$$x_1 + 3x_2 \geqslant 2$$
$$4x_1 + x_2 \geqslant 5$$
$$x_1, x_2 \geqslant 0 \quad \text{and integer}$$

Solve the relaxed problem as (a) a shortest route problem, and (b) a dynamic programming problem. Does the relaxed problem satisfy any of the sufficient conditions in Section 6.3.1, Problems 6-5 and 6-7?

6-11   (Balas, 1973). Prove that Gomory's sufficiency condition (Section 6.3.1) is never satisfied if the original integer problem is a pure zero–one problem by showing that $b \notin K_B(d)$ whenever $d > 1$.

6-12   Prove that the result in Problem 6-11 can be extended to an arbitrary integer program having a constraint of the form $a_{kj} x_j = b_k$ with $a_{kj} \geqslant 0$ for all j, provided the condition $d > 1$ is replaced by $d > b_k$.

6-13   Let $Ax = b$ be the constraint of the zero–one problem. Show that if A is an integer matrix and unless the linear programming optimum is all integer with all the nonbasic columns being unit vectors, then it is always true that $D \geqslant 2$ and $d = \|\xi\|(D - 1) > 1$.

CHAPTER 7

Algorithms for Specialized Integer Models

7.1 Introduction

This chapter presents algorithms that exploit the special properties of certain integer models, including the knapsack, the fixed-charge, the traveling salesman, and the set covering problems. General formulations of these problems are given in Section 1.3. Although the general algorithms presented in the preceding chapters can be used, at least in principle, to solve the indicated models, the specialized methods should prove efficient computationally.

7.2 Knapsack Problem

The knapsack problem is defined in Example 1.3-4 as:

minimize

$$z = \left\{ \sum_{j=1}^{n} (-c_j) x_j \,\middle|\, \sum_{j=1}^{n} a_j x_j \leq b,\, x_j \geq 0 \right\}$$

It is assumed that all the coefficients a_j and b are nonnegative. Generally, all c_j are expected to be nonnegative, but, as will be shown below, there are variations of the problem that for analytic convenience may cause some c_j to be negative.

In Example 1.3-4, it is indicated that the knapsack problem has applications in its own right, but its importance is enhanced by the fact that it can be used in an auxiliary manner to secure efficient solution methods to more complex problems. In Chapter 6, a problem related to the knapsack is used to develop the asymptotic algorithm for integer programming. Also, another important application of the knapsack model to the cutting stock problem is given in Section 7.2.1.

In order to stress the potential usefulness of the knapsack model further, Section 7.2.2 presents a method by which any general integer problem, in principle, can be reduced to a knapsack problem. Section 7.2.3 then presents a number of special algorithms for solving the knapsack model.

7.2.1   Applications of Knapsack Models in Cutting-Stock Problems†

The cutting-stock problem assumes that for a given material a stock of standard lengths $L_1, L_2, \ldots, L_r$ is maintained. There is no limitation on the number of pieces available of each length. All stocked pieces have the same width. It is required to fill m orders, where the ith order requests *at least* b_i pieces each of length l_i and having the same standard width. An order can be filled so long as l_i does not exceed $\max_k \{L_k | k = 1, 2, \ldots, r\}$.

Let the jth cutting pattern be defined such that a_{ij} pieces each of length l_i can be generated from a standard length L_k. In this case, the trim loss associated with such a pattern is equal to $c_j = L_k - \sum_{i=1}^m l_i a_{ij} \geqslant 0$. It is possible that $a_{ij} = 0$ for some $i = 1, 2, \ldots, m$. Also, the same pattern may be cut from different standard lengths thus yielding different values of c_j. However, under optimal conditions, a pattern j is cut from a standard length L_k, which yields the smallest trim loss c_j. In addition, c_j must be strictly less than $\min_i l_i$.

Suppose x_j is the number of standard pieces to be cut according to pattern j; then the problem becomes:

minimize
$$z = \sum_j c_j x_j$$

subject to
$$\sum_j a_{ij} x_j \geqslant b_i, \qquad i = 1, 2, \ldots, m$$
$$x_j \geqslant 0 \quad \text{and integer} \quad \text{for all } j$$

† Gilmore and Gomory (1961, 1963).

The obvious difficulty with this formulation is that x_j is restricted to integer values, so that for a large number of variables the problem may not be computationally tractable. Another difficulty is that it may be impossible to enumerate all the cutting patterns in advance.

Gilmore and Gomory account for these difficulties as follows: First, drop the integer condition on x_j so that the resulting problem becomes a regular linear program; then round the optimal continuous solution in some appropriate manner. Second, instead of enumerating all cutting patterns in advance, a pattern is considered only if it is promising from the viewpoint of improving the linear programming solution. This is where the knapsack model proves valuable.

At any iteration of the simplex method, let the associated basis be defined by

$$B = (P_1, \ldots, P_i, \ldots, P_m)$$

where P_i is an m-column vector, $i = 1, 2, \ldots, m$. Let $C_B = (c_1, c_2, \ldots, c_m)$ be the coefficients of the objective function associated with the vectors $P_1, P_2, \ldots, P_m$. Then from the theory of linear programming (see Section 2.3.3), a cutting pattern j is promising if its associated reduced cost

$$z_j - c_j = C_B B^{-1} P_j - c_j$$

is positive (minimization problem), where

$$P_j = (a_{1j}, a_{2j}, \ldots, a_{mj})^T$$

is the vector representing the number of pieces of length l_i, $i = 1, \ldots, m$, generated from the cutting pattern j.

At this point, the elements of P_j are not known, that is, the new cutting pattern has not been determined. From linear programming theory, the most promising pattern is the one yielding the largest $z_j - c_j$ among all possible (nonbasic) patterns. This is equivalent to solving the problem

maximize

$$w = \sum_{i=1}^{m} \pi_i \alpha_i$$

subject to

$$\sum_{i=1}^{m} l_i \alpha_i \leqslant L$$

$$\alpha_i \geqslant 0 \quad \text{and integer.}$$

The coefficient π_i is the ith element of $C_B B^{-1}$. The problem is solved for r values of the right-hand side. Namely, $L = L_1, L_2, \ldots,$ and L_r, where r is the total number of available standard lengths.

The above problem is a knapsack model. Let the optimal solution be given by $w = w^*$ and $\alpha_i = \alpha_i^*$, $i = 1, 2, \ldots, m$. Next compute $c = L - \sum_{i=1}^{m} l_i \alpha_i^*$. Then the pattern

$$P = (\alpha_1^*, \alpha_2^*, \ldots, \alpha_m^*)^T$$

is promising only if $w^* - c > 0$. In this case, the vector P can be introduced into the basis. Naturally, if more than one L yield $w^* - c > 0$, then the pattern yielding the largest $w^* - c$ is chosen. At the point where $w^* - c \leqslant 0$ for all L, the algorithm terminates with the current basis yielding the optimum solution, namely, $X_B = B^{-1}b$, where $b = (b_1, b_2, \ldots, b_m)^T$. The solution may be rounded now if necessary.

The solution of r knapsack problems in order to generate one pattern may be too costly from the computational standpoint. A possible ramification is to take the first pattern yielding $w^* - c > 0$.

Gilmore and Gomory (1961) first utilized Dantzig's dynamic programming approach for solving the knapsack problem. This is presented in Section 7.2.3-B. In a later paper (Gilmore and Gomory, 1963), they added improvements leading to computational savings. A new and more efficient dynamic programming model was developed by the same authors (Gilmore and Gomory, 1965). This model is similar to the one presented in Section 6.4.2 for the group problem. A still superior dynamic programming model was introduced by Gilmore and Gomory (1966) (see Problem 7-6), but in spite of all these improvements, the dynamic programming algorithms tend to get unwieldy as the right-hand side constant L becomes large.

7.2.2   Reduction of Integer Problems to Knapsack Models

The idea of reducing multiple constraint problems into a single constraint originated with a classic theorem by Mathews (1897), stated as follows:

▶**THEOREM 7-1**   Consider a system of two equations with *strictly positive* integer coefficients

$$w_1 \equiv \sum_{j \in N} a_{1j} x_j = b_1, \qquad w_2 \equiv \sum_{j \in N} a_{2j} x_j = b_2$$

where x_j are assumed to be nonnegative integers.

   (a)   If the two equations have a feasible solution, then $b_1 a_{2j} \leqslant b_2 a_{1j}$ for at least one j;

(b)   If λ is any positive *integer* satisfying

$$\lambda > b_2 \max_{j \in N}\{a_{1j}/a_{2j}\} > b_1$$

then

$$w_1 + \lambda w_2 = b_1 + \lambda b_2$$

has the same set of feasible solutions as the original system.

Proof   Suppose (a) is not true; then $b_1 a_{2j} > b_2 a_{1j}$ for all j. Let $\{x_j\}$ be a feasible solution to the given system. Since $x_j \geqslant 0$, then $b_1 \sum_{j \in N} a_{2j} x_j > b_2 \sum_{j \in N} a_{1j} x_j$ or $b_1 w_2 > b_2 w_1$, which contradicts the fact that both sides equal $b_1 b_2$. Part (a) now shows that the right inequality in (b) is legitimate.

To prove part (b), notice that a solution to the original system clearly implies the new equation. To show that the reverse is true, suppose $w_2 = b_2 + k$, which by the new equation gives $w_1 = b_1 - \lambda k$, where k is an integer. Then (b) is proved if one proves $k = 0$. Suppose $k > 0$; then the condition $\lambda > b_1$ implies $w_1 = b_1 - \lambda k < 0$. But this is impossible by the fact that all coefficients of the w_1-equation are positive. If $k < 0$, then $w_2 < b_2$ and $w_1 \geqslant b_1 + \lambda$. From the condition $\lambda > b_2(a_{1j}/a_{2j})$ and by multiplying both sides by x_j, one gets $\lambda \sum_{j \in N} a_{2j} x_j > b_2 \sum_{j \in N} a_{1j} x_j$, or $\lambda w_2 > b_2 w_1$. Utilizing $w_1 \geqslant b_1 + \lambda$, then $\lambda w_2 > b_2 w_1$ implies $\lambda w_2 > b_1 b_2 + \lambda b_2 \geqslant \lambda b_2$ or $w_2 > b_2$, which leads to contradiction. This means that $k = 0$ as was to be proved.◀

Theorem 7-1 was utilized by Elmaghraby and Wig (1970) to aggregate several equations into a single equation simply by applying the theorem recursively to two equations at a time.

Although the theorem is applicable only when all the coefficients are strictly positive, Mathews observed that the w_1 and w_2 equations may be replaced, respectively, by $w_1 + w_2 = b_1 + b_2$ and $w_1 + 2w_2 = b_1 + 2b_2$, in which case the stipulated condition is satisfied. But Glover and Woolsey (1970) gave the following improved implementation in the sense that it leads to smaller coefficients. Suppose there are a total of m equations. The first step is to precondition these equations by replacing them with an equal number but such that the ith equation is the sum of the first i equations. Suppose these *preconditioned* equations are identified by w_1', w_2', ..., and w_m'. The aggregation is implemented by first applying the theorem to $w_1' + w_2'$ and w_2'. Suppose this yields w_1''. The theorem is again applied to $w_1'' + w_3'$ and w_3'. In general, if $v = c$ denotes the equation obtained by aggregating the first $k - 1$ preconditioned equations, and if $t = d$ denotes the kth preconditioned equation, then the theorem is applied to $v + t = c + d$ and $t = d$.

▶**EXAMPLE 7.2-1**   Consider the system

$$w_1 = x_1 + 2x_2 + 3x_3 = 6$$
$$w_2 = 2x_1 + 4x_3 + 4x_4 = 8$$
$$w_3 = x_1 + x_2 + 3x_3 + x_4 = 6$$
$$x_1, x_2, x_3, x_4 \geqslant 0 \quad \text{and integer}$$

The preconditioned system is given by (for simplicity, the variables are deleted from the equations):

$$
\begin{array}{llllll}
w_1' = 1 & 2 & 3 & 0 & : & 6 \\
w_2' = 3 & 2 & 7 & 4 & : & 14 \\
w_3' = 4 & 3 & 10 & 5 & : & 20
\end{array}
$$

Thus, aggregating,

$$
\begin{array}{llllll}
w_1'' = 4 & 4 & 10 & 4 & : & 20 \\
w_2' = 3 & 2 & 7 & 4 & : & 14
\end{array}
$$

one gets

$$\lambda > 14 \max\left\{\frac{4}{3}, \frac{4}{2}, \frac{10}{7}, \frac{4}{4}\right\} = 28$$

or $\lambda = 29$. This yields

$$w_1''' = 91x_1 + 62x_2 + 213x_3 + 120x_4 = 406$$

Now aggregating,

$$
\begin{array}{llllll}
w_2'' = w_1''' + w_3' = 95 & 65 & 223 & 125 & : & 426 \\
w_3' = 4 & 3 & 10 & 5 & : & 20
\end{array}
$$

one gets

$$\lambda > 20 \max\left\{\frac{95}{4}, \frac{65}{3}, \frac{223}{10}, \frac{125}{5}\right\} = 500$$

or $\lambda = 501$. This yields the final equation

$$2099x_1 + 1568x_2 + 5233x_3 + 2630x_4 = 10446$$

Notice that w_2'' could be taken directly equal to w_1''' since both w_1''' and w_3' have all positive coefficients. ◀

The disadvantage of this process is that the coefficients of the aggregated equation become significantly large even for a small number of equations. This limits the utilization of the results on the digital computer.

A new development by Glover (1972) shows how the coefficients of the aggregated equation can be reduced considerably, based on the following theorem:

▶THEOREM 7-2   Under the same conditions as in Theorem 7-1, the w_1 and w_2 equations are equivalent to $w_1 + \lambda w_2$ provided λ satisfies

$$\lambda > \max_j \{(b_2 - 1)(a_{1j}/a_{2j}) - b_1\}$$

and

$$\lambda > \max_j \{b_1 - (b_2 + 1)(a_{1j}/a_{2j})\}$$

The proof of the theorem is very similar to that of Theorem 7-1.◀

It is clear, by comparison with the bounds in Theorem 7-1, that Theorem 7-2 yields smaller lower bounds on λ and hence should generally lead to smaller coefficients of the aggregated equation. This does not say, however, that Theorem 7-2 will result in manageable coefficients. Indeed, the resulting coefficients may still be too large for the computer.

The preceding procedure can be extended to equations with negative coefficients provided that the associated variables have upper bounds. Thus, if a_{ij} is negative and $0 \leqslant x_j \leqslant u_j$, then the substitution $x_j' = u_j - x_j$ is used, which will make the coefficient of x_j' positive. Although it is true that the number of variables associated with negative coefficients will be doubled, the same number of variables in the original system can be recovered by resubstituting x_j' in terms of x_j in the final aggregated equation.

There are other direct methods that can handle equations with unrestricted coefficients, notable among which are those of Glover and Woolsey (1970), Bradley (1971c), Anthonisse (1970), and Padberg (1970). However, each of these methods requires that an explicit upper bound be set on *each* variable. These bounds are used directly in computing λ. Thus, if loose bounds are utilized, the resulting λ may be conservatively too large. On the other hand, computing tight bounds on the variables may require laborious and sophisticated computations. Consequently, there does not seem to be a real computational advantage in utilizing these methods. A version of Bradley's method is presented in Problem 7-3, however.

7.2.3   Algorithms for the Knapsack Problem

This section presents four different algorithms for solving the knapsack model, including the shortest route representation, dynamic programming, enumeration, and extreme point ranking. The shortest route and dynamic programming algorithms are very similar to those used in solving the asymptotic problem (Section 6.4) and hence will be discussed here only briefly. The extreme point ranking method is especially designed for the case where all the variables are binary. Other procedures for handling the binary knapsack problem were developed by Kolesar (1967) and Greenberg and Hegerich (1970). The Greenberg–Hegerich procedure is a branch-and-bound method that utilizes the Land–Doig algorithm (with the LIFO branching rule) in an almost straightforward manner (see Section 4.3.1-B). The primary difference, however, is that by using the special properties of the knapsack problem an easy method can be used to find the continuous solution of the linear programs at different nodes. This method of solving the linear program is summarized in Section 7.2.3-D. The Kolesar method, which is a branch-and-bound algorithm, is not as efficient as the Greenberg–Hegerich. Other methods for the general case are given by Pandit (1962), Greenberg (1969a), and Eilon and Cristofides (1971). Gilmore and Gomory (1966) give an interesting exposition of the theory and computations of knapsack functions.

A. Shortest Route Method†

For the purpose of the presentation, the knapsack problem is presented in the form:

minimize

$$z = \sum_{j=1}^{n} (-c_j) x_j$$

subject to

$$\sum_{j=1}^{n} a_j x_j + x_{n+1} = b$$

$$x_j \geqslant 0 \quad \text{and integer}$$

where x_{n+1} is a slack variable such that $c_{n+1} = 0$ and $a_{n+1} = 1$. It is assumed that b, $a_j > 0$ for all j and $c_j \geqslant 0$ for all j. All the coefficients are integers.

The representation of the problem as a *directed* network (since all $a_j > 0$) follows the same ideas applied in Section 6.4.1. The main difference is that regular arithmetic is used. There are $b + 1$ nodes in the network, which are identified as 0, 1, 2, ..., and b. These nodes represent the possible values of the left-hand side of the constraint resulting from assigning numerical values to the variables x_1, x_2, ..., x_{n+1}. At node i, $i = 0, 1, ..., b - 1$, $n + 1$ *directed* arcs must originate; each of these arcs is uniquely associated with a variable x_j. Since each arc represents a unit increment in a variable x_j, the terminating node k of arc (i, k) is given by $k = i + a_j$. Since $a_j > 0$, then $i < k$ and the network is acyclic. The "length" of the arc associated with x_j is equal to $-c_j$. There are no arcs leaving from node b or coming into node 0. The knapsack problem thus reduces to finding the shortest route from node 0 to node b.

† Dijkstra (1959).

▶Example **7.2-2**   Consider the knapsack problem:

minimize

$$z = -2x_1 - 3x_2 - 4x_3$$

subject to

$$x_1 + 2x_2 + 3x_3 + x_4 = 4$$
$$x_1, x_2, x_3, x_4 \geqslant 0 \quad \text{and integer}$$

The corresponding network is given in Fig. 7-1. Notice that x_4 need not be represented by an arc at any node, since x_1 creates the same arcs but at a lower cost ($c_1 = -2, c_4 = 0$).

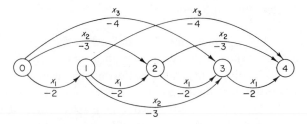

Figure 7-1

Because the network is acyclic, the method of finding the shortest route is quite simple. This is best illustrated directly in terms of the example.

Let d_i represent the shortest distance from node 0 to node i, $i = 1, 2, 3, 4$. The procedure starts at node 1 and terminates at (the last) node 4. Define $d_0 = 0$; then

$$d_1 = -2 \quad \text{(from node 0)}$$
$$d_2 = \min\{d_1 + (-c_1), d_0 + (-c_2)\}$$
$$\quad = \min\{-2 - 2, 0 - 3\} = -4 \quad \text{(from node 1)}$$
$$d_3 = \min\{d_2 + (-c_1), d_1 + (-c_2), d_0 + (-c_3)\}$$
$$\quad = \min\{-4 - 2, -2 - 3, 0 - 4\} = -6 \quad \text{(from node 2)}$$
$$d_4 = \min\{d_3 + (-c_1), d_2 + (-c_2), d_1 + (-c_3)\}$$
$$\quad = \min\{-6 - 2, -4 - 3, -2 - 4\} = -8 \quad \text{(from node 3)}$$

Thus the optimal solution is $x_1 = 1 + 1 + 1 + 1 = 4$ and $x_2 = x_3 = x_4 = 0$, with the optimum objective value equal to` -8.◀

The reader should notice the similarity between this procedure and the one devised for the group (knapsack) problem in Section 6.4.1.

B.   *Dynamic Programming Method* †

Let y_k be defined such that

$$\sum_{j=1}^{k} a_j x_j \leq y_k, k \leq n$$

that is, y_k is the constraint limitation on the first k variables. Define

$$f_k(y_k) = \min_{x_j}\left\{ \sum_{j=1}^{k} -c_j x_j \,\middle|\, \sum_{j=1}^{k} a_j x_j \leq y_k, x_j \text{ nonnegative integer}\right\}$$

The dynamic recursive equation is given by

$$f_1(y_1) = \min_{x_1}\{-c_1 x_1\}$$

$$f_k(y_k) = \min_{x_k}\{-c_k x_k + f_{k-1}(y_k - a_k x_k)\}, \qquad k = 2, \ldots, n$$

where $x_k = 0, 1, 2, \ldots, [y_k/a_k]$, and $[R]$ is defined as the largest integer not exceeding R. At any stage k of the calculation, $y_k = 0, 1, 2, \ldots, b$, for all k.

The knapsack problem may also be formulated using the same idea of the dynamic programming algorithm presented for the group minimization problem (Section 6.4.2). Another dynamic programming algorithm for the knapsack model is given by Gilmore and Gomory (1966) (see Problem 7-6).

C.   *Enumeration Algorithm* ‡

Cabot's algorithm is based on a (simple) enumeration procedure that excludes infeasible points by establishing adjustable upper and lower bounds on each variable. These bounds are adjusted based on the values assigned to some of the variables. The search for the optimal is made systematically in the range of values specified by the (adjustable) bounds.

To be more concrete, the enumeration procedure is developed as follows. Express the knapsack problem as:

$$\sum_{j=1}^{n}(-c_j)x_j \leq z$$

$$\sum_{j=1}^{n}a_j x_j \leq b$$

$$x_j \geq 0 \quad \text{and integer,} \qquad j = 1, 2, \ldots, n$$

† Bellman (1957), Dantzig (1957).
‡ Cabot (1970).

where it is required to determine minimum z. It is assumed that all a_j are positive, while c_j may be unrestricted in sign.

As will be shown, the bounds on x_j can be expressed as

$$L_j(x_{j+1}, \ldots, x_n, z) \leqslant x_j \leqslant U_j(x_{j+1}, \ldots, x_n, z), \qquad j = 1, 2, \ldots . n$$

which means that the bounds on x_j are functions of $x_{j+1}, \ldots, x_n$, and z only and that L_n and U_n, the bounds on x_n, are functions of z only. The development of these bounds is based on eliminating one variable at a time, starting with x_1. When x_n is reached, we have

$$L_n(z) \leqslant x_n \leqslant U_n(z)$$

This yields the valid inequality

$$L_n(z) \leqslant U_n(z)$$

which can be reduced to the form $z \geqslant z^0$, where z^0 is a *numerical* lower bound on z. Since z is minimized, then $z = z^0$ is a logical initial value for the enumeration procedure. Utilizing this value to determine numerical values for the bounds on x_n, a specific value of x_n $(= x_n{}^0)$ is then judiciously selected in this range. For example, if $c_n > 0$, then x_n is selected as large as possible, that is, $x_n{}^0 = U_n(z)$. Otherwise $x_n{}^0 = L_n(z)$ is used. Naturally, $x_n = x_n{}^0$ must satisfy $a_n x_n \leqslant b$. Next, utilizing $z = z^0$ and $x_n = x_n{}^0$, then $L_{n-1}(x_n{}^0, z^0)$ and $U_{n-1}(x_n{}^0, z^0)$ are determined, and a specific value $x_{n-1} = x_{n-1}{}^0$ is chosen in this range such that $a_{n-1}x_{n-1}{}^0 + a_n x_n{}^0 \leqslant b$. The process is continued so that given $z^0, x_n{}^0, \ldots, x_{j+1}{}^0$, then

$$L_j(x_{j+1}{}^0, \ldots, x_n{}^0, z^0)$$

and $U_j(x_{j+1}{}^0, \ldots, x_n{}^0, z^0)$ are determined, and $x_j{}^0$ is chosen in this range such that

$$a_j x_j{}^0 + \cdots + a_n x_n{}^0 \leqslant b, \qquad j = 1, 2, \ldots, n$$

If a feasible value can be determined this way for x_1, then $(x_1{}^0, x_2{}^0, \ldots, x_n{}^0, z^0)$ gives the optimal solution to the problem. However, if for any x_j, $L_j \not\leqslant U_j$, this would mean that the previously fixed values for $x_{j+1}, \ldots, x_n$, z, cannot yield a feasible solution. Thus one "backtracks" to select a new value of x_{j+1} $(= x_{j+1}{}^1 = x_{j+1}{}^0 + e$, where $e = -1$ if $x_j{}^0 = U_j$ and $e = 1$ if $x_j = L_j$) and new values for L_j and U_j are computed. If the condition $L_j \not\leqslant U_j$ persists, $x_{j+1}{}^2$ is selected and the process repeated. If none of the values of x_{j+1} in the range (L_{j+1}, U_{j+1}) breaks this condition, then the value of x_{j+2} must be changed. Eventually, it may be necessary to increase z above its lower bound z^0. In this case, the next value of z is $z^1 = z^0 + \Delta$, where Δ is the smallest amount by which z can be increased. This is given by $\Delta = 1$ if c_j are unrestricted in sign and $\Delta = $ least common multiple of c_j if all $c_j \geqslant 0$.

The procedure for determining L_j and U_j is now presented, based on what is known as the Fourier–Motzkin elimination method [see Dantzig (1963, pp. 84–85)].

From the objective constraint, if $c_1 > 0$, then

$$x_1 \geqslant \frac{1}{c_1}\left\{-z - \sum_{j=2}^{n} c_j x_j\right\}$$

If $c_1 < 0$, then

$$x_1 \leqslant \frac{1}{c_1}\left\{-z - \sum_{j=2}^{n} c_j x_j\right\}$$

Now from the constraint of the knapsack problem,

$$x_1 \leqslant \frac{1}{a_1}\left\{b - \sum_{j=2}^{n} a_j x_j\right\}$$

(notice that $a_1 > 0$). It then follows that

$$\max\left\{0, \frac{1}{c_1}\left(-z - \sum_{j=2}^{n} c_j x_j\right)\right\} \leqslant x_1 \leqslant \frac{1}{a_1}\left\{b - \sum_{j=2}^{n} a_j x_j\right\}, \qquad \text{if} \quad c_1 < 0$$

$$0 \leqslant x_1 \leqslant \min\left\{\frac{1}{c_1}\left(-z - \sum_{j=2}^{n} c_j x_j\right), \frac{1}{a_1}\left(b - \sum_{j=2}^{n} a_j x_j\right)\right\}, \qquad \text{if} \quad c_1 > 0.$$

$$0 \leqslant x_1 \leqslant \frac{1}{a_1}\left(b - \sum_{j=2}^{n} a_j x_j\right), \qquad \text{if} \quad c_1 = 0$$

These inequalities yield L_1 and U_1 as functions of $(x_2, x_3, \ldots, x_n, z)$.

To determine L_2 and U_2, consider the inequality

$$L_1(x_2, \ldots, x_n, z) \leqslant U_1(x_2, \ldots, x_n, z)$$

(Notice that this inequality implies $a_2 x_2 + \cdots + a_n x_n \leqslant b$.) By applying the same procedure followed in determining L_1 and U_1, L_2 and U_2 are evaluated. Notice that x_1 is not present in the above two inequalities, so that L_2 and U_2 are functions of $x_3, x_4, \ldots, x_n$, and z only. Notice also that if $c_1 = 0$, then L_1 and U_1 are not functions of z. In this case, the inequality $\sum_{j=2}^{n} c_j x_j \geqslant -z$ is used in conjunction with $L_1 \leqslant U_1$ to determine L_2 and U_2.

In general, L_j and U_j, $j = 1, 2, \ldots, n$, are determined from

$$L_{j-1}(x_j, x_{j+1}, \ldots, x_n, z) \leqslant U_{j-1}(x_j, x_{j+1}, \ldots, x_n, z)$$

which again implies

$$a_j x_j + a_{j+1} x_{j+1} + \cdots + a_n x_n \leqslant b$$

If $c_1 = c_2 = \cdots = c_{j-1} = 0$, then the following additional inequality is used:

$$-c_j x_j - c_{j+1} x_{j+1} - \cdots - c_n x_n \leqslant z$$

▶EXAMPLE 7.2-3 Consider Example 7.2-2, which may be written as

$$-2x_1 - 3x_2 - 4x_3 \leqslant z \tag{1}$$
$$x_1 + 2x_2 + 3x_3 \leqslant 4 \tag{2}$$

From (1)

$$x_1 \geqslant \tfrac{1}{2}(-z - 3x_2 - 4x_3) \tag{3}$$

and from (2)

$$x_1 \leqslant 4 - 2x_2 - 3x_3 \tag{4}$$

Thus

$$L_1 = \max\{0, -\tfrac{1}{2}z - \tfrac{3}{2}x_2 - 2x_3\} \tag{5}$$
$$U_1 = 4 - 2x_2 - 3x_3 \tag{6}$$

Now $L_1 \leqslant U_1$ is equivalent to

$$0 \leqslant 4 - 2x_2 - 3x_3 \tag{7}$$

and

$$-\tfrac{1}{2}z - \tfrac{3}{2}x_2 - 2x_3 \leqslant 4 - 2x_2 - 3x_3 \tag{8}$$

From (7)

$$x_2 \leqslant 2 - \tfrac{3}{2}x_3 \tag{9}$$

and from (8)

$$x_2 \leqslant 8 + z - 2x_3 \tag{10}$$

Hence

$$L_2 = 0 \tag{11}$$
$$U_2 = \min\{2 - \tfrac{3}{2}x_3, 8 + z - 2x_3\} \tag{12}$$

Next consider $L_2 \leqslant U_2$. This is equivalent to

$$0 \leqslant 2 - \tfrac{3}{2}x_3 \tag{13}$$
$$0 \leqslant 8 + z - 2x_3 \tag{14}$$

From (13)

$$x_3 \leqslant \tfrac{4}{3} \tag{15}$$

and from (14)

$$x_3 \leqslant 4 + \tfrac{1}{2}z \tag{16}$$

Thus

$$L_3 = 0 \tag{17}$$
$$U_3 = \min\{\tfrac{4}{3}, 4 + \tfrac{1}{2}z\} \tag{18}$$

Finally, $L_3 \leqslant U_3$ yields

$$0 < \tfrac{4}{3} \tag{19}$$
$$0 \leqslant 4 + \tfrac{1}{2}z \tag{20}$$

Inequality (19) is trivial, while (20) implies $z \geqslant -8$.
Let $z^0 = -8$; then from (17) and (18)

$$0 \leqslant x_3 \leqslant 0, \qquad \text{or} \quad x_3{}^0 = 0$$

Next, given $z^0 = -8$ and $x_3{}^0 = 0$, (11) and (12) yield $L_2 = 0$, $U_2 = 0$, and consequently $x_2{}^0 = 0$. Finally, using $z^0 = -8$, $x_3{}^0 = 0$, $x_2{}^0 = 0$, (5) and (6) yield $L_1 = 4$ and $U_1 = 4$, or $x_1{}^0 = 4$. It is obvious from the construction of the bounds that $x_1 = 4$, $x_2 = x_3 = 0$, is a feasible solution. Hence $z = -8$ is optimal.

Notice that it is unnecessary to backtrack in the above example since (1) all the computed bounds are consistent, and (2) all the computed bounds provide a single value for each variable. This obviously is not the case in general.◄

Cabot (1970) reports favorable computational experience with the algorithm. He uses three ordering schemes for the variables, namely, rank x_j based on: (i) ascending order of c_j/a_j, (ii) descending order of c_j/a_j, and (iii) descending order of $U_j - L_j$, where U_j and L_j are numerical values obtained by linear programming. The first two schemes prove much less efficient than the third. The third scheme is then tested with 50 randomly generated problems for $n = 10, 20, \ldots$, and 80 with b taken as twice the $\sum_{j=1}^{n} a_j$. The average computation time seems to grow as $n^{1.75}$. Perhaps one of the important advantages of Cabot's algorithm is that, unlike dynamic programming algorithms, the computation time tends to be unaffected by the increase in the right-hand side constant b.

D. *Extreme Point Ranking for Binary Problems*†

In Theorem 3-1 (Section 3.2), it was proved that the zero–one (linear) problem has the unique property that the optimal solution must occur at an extreme point of the convex polyhedron representing the continuous solution space. This theorem is the basis for the following algorithm.

---

† Taha (1971b).

Assume that the knapsack model is of the minimization type as defined above except that all the variables are binary. Let $E = \{e^0, e^1, \ldots, e^k\}$ be the set of extreme points of the continuous solution space and assume that e^i and e^{i+1} are defined such that $z^i \leqslant z^{i+1}$, $i = 0, 1, \ldots, k$. This means that the extreme points are ranked in ascending order according to the value of z. An extreme point e^i is said to be *binary feasible* if its associated x_j-values satisfy the binary restriction. Let e^* be a given feasible extreme point and define $\bar{z}$ as its associated objective value. Then $\bar{z}$ is an upper bound on the optimum binary solution. If no initial feasible solution is available, then $\bar{z} = \infty$ is taken as the initial upper bound. It is assumed, without loss of generality, that all c_j, a_j, and b are positive. (See Theorem 7-3 for a justification.)

The general outline of the algorithm is as follows:

Step 0 Solve the equivalent linear program with $x_j = (0, 1)$ replaced by $0 \leqslant x_j \leqslant 1, j = 1, 2, \ldots, n$, which yields e^0 and z^0. If e^0 is binary feasible, stop; e^0 is optimum. Otherwise set $i = 1$ and go to step 1.

Step 1 Determine e^i. Then z^i is a lower bound on the optimum objective value. If $z^i \geqslant \bar{z} - \Delta$, where Δ is the least common multiple of all c_j, go to step 2. If e^i is binary feasible, set $e^* = e^i$ and $\bar{z} = z^i$, then go to step 2. If $i = k$, terminate; no feasible solution exists. Otherwise, set $i = i + 1$ and repeat step 1.

Step 2 e^* and $\bar{z}$ yield the optimum solution. Stop.

The efficiency of the above algorithm depends on two basic points: (i) the determination of a good (tight) initial upper bound, and (ii) the determination of the next ranked extreme point e^i given e^{i-1}. Naturally, the smaller the initial upper bound, the faster the algorithm will terminate since this will involve ranking a smaller number of extreme points. On the other hand, it does not seem practical to rank all the extreme points in advance. Consequently, a (efficient) procedure is needed whereby the next ranked extreme point is generated only when it is needed. These two points are considered in detail next.

The preceding algorithm is basically an adaptation of Murty's algorithm (1968) for solving the fixed-charge problem. This problem is considered in further detail in Section 7.3.

1. *Determination of initial upper bound* $\bar{z}$†

For the sake of convenience, a "generalized knapsack" problem is considered in which all the coefficients c_j, a_j, and b are unrestricted in sign. Theorem 7-3 shows how such a problem may be reduced equivalently

† Glover (1965c).

to a knapsack problem in which all c_j, a_j, and b are positive. Theorem 7-4 then provides the basis for finding a (good) feasible solution and hence a (good) initial upper bound z^*.

▶**Theorem 7-3**   If the generalized knapsack problem has a feasible fractional solution when $x_j = (0, 1)$ is replaced by $0 \leqslant x_j \leqslant 1$, then it has an optimal fractional and an optimal binary solution in which the following binary assignment occurs:

$$x_j = \begin{cases} 1, & \text{if} \quad c_j \geqslant 0 \quad \text{and} \quad a_j \leqslant 0 \\ 0, & \text{if} \quad c_j \leqslant 0 \quad \text{and} \quad a_j \geqslant 0 \end{cases}$$

Proof   Let the fractional feasible solution to the knapsack problem be given by $\bar{x} = (\bar{x}_1, \ldots, \bar{x}_n)$. Then the *integer* solution $\bar{x}$ is also feasible, where fractional $\bar{x}_j$ is replaced by

$$\bar{x}_j = \begin{cases} 0, & \text{if} \quad a_j > 0 \\ 1, & \text{if} \quad a_j \leqslant 0 \end{cases}$$

Assume now that the binary assignment specified by the theorem is false. The following proof applies both to the fractional and integer cases. If $\bar{x}$ is an optimal solution and if there is any j such that $c_j \geqslant 0$ and $a_j \leqslant 0$ or such that $c_j \leqslant 0$ and $a_j \geqslant 0$, then, by assumption, for at least one such $j = r$, $\bar{x}_j$ does not agree with the value specified by the theorem. Define

$$\delta = \begin{cases} 1 - \bar{x}_r, & \text{if} \quad a_r \leqslant 0 \quad \text{and} \quad c_r \geqslant 0 \\ -\bar{x}_r, & \text{if} \quad a_r \geqslant 0 \quad \text{and} \quad c_r \leqslant 0 \end{cases}$$

and

$$x_j^* = \begin{cases} \bar{x}_j, & \text{if} \quad j \neq r \\ \bar{x}_j + \delta, & \text{if} \quad j = r \end{cases}$$

Then x_j^* has the value assigned to it by the theorem, and

$$\sum_{j=1}^n a_j x_j^* = \sum_{j=1}^n a_j \bar{x}_j + a_r \delta, \qquad \sum_{j=1}^n c_j x_j^* = \sum_{j=1}^n c_j \bar{x}_j + c_r \delta$$

If $a_r \leqslant 0$ and $c_r \geqslant 0$, then $\delta > 0$, $\sum_{j=1}^n a_j x_j^* \leqslant \sum_{j=1}^n a_j \bar{x}_j$, and

$$\sum_{j=1}^n (-c_j) x_j^* \leqslant \sum_{j=1}^n (-c_j) \bar{x}_j.$$

This shows that x^* is feasible and optimal. In the same manner, if $a_r \leqslant 0$ and $c_r \geqslant 0$, then $\delta < 0$ and a similar conclusion is reached. The procedure can be repeated if any of the elements of x^* does not comply with the theorem by simply ascribing to x^* the role of $\bar{x}$.◀

The implication of Theorem 7-3 is that any "generalized 0–1 knapsack" problem in which the coefficients c_j and a_j are unrestricted in sign can be reduced to an equivalent regular knapsack problem in which all c_j and a_j are positive. The substitution $x_j = 1 - x_j'$ is applied if necessary. This result is the basis for the next theorem.

▶**Theorem 7-4** (Dantzig, 1957)   Given that all a_j and c_j are positive, and assuming that the indices of x_j are arranged so that $c_1/a_1 \geqslant c_2/a_2 \geqslant \cdots \geqslant c_n/a_n$, define p as the least integer $(0 \leqslant p \leqslant n)$ such that $\sum_{j \leqslant p} a_j \geqslant b$. Then the optimal *fractional* solution to the knapsack problem is given by

$$x_j = \begin{cases} 1, & \text{if } j < p \\ \left(b - \sum_{k < p} a_k \right) \Big/ a_p, & \text{if } j = p \\ 0, & \text{if } j > p \end{cases}$$

If no p exists, then all $x_j = 1$, and if $x_p = 0$, then the resulting integer solution is optimal for the knapsack problem.

Proof   The problem is viewed as a linear program with bounded variables, $0 \leqslant x_j \leqslant 1, j = 1, 2, \ldots, n$. From the theory of linear programming, only one variable appears in the basis while all the remaining variables are nonbasic at level zero or level one. Let x_r be the basic variable. Then the reduced costs $z_j - c_j$ for $x_j, j \neq r$, must satisfy the following conditions for achieving optimality:

$$z_j - c_j = \begin{cases} (-c_r)(1/a_r)a_j - (-c_j) \leqslant 0, & \text{if } x_j = 0 \\ -(-c_r)(1/a_r)a_j + (-c_j) \leqslant 0, & \text{if } x_j = 1 \end{cases}$$

This means that $c_r/a_r \geqslant c_j/a_j$ for all j such that $x_j = 0$ and nonbasic, and $c_r/a_r \leqslant c_j/a_j$ for all j such that $x_j = 1$ and nonbasic. This agrees with the statement of the theorem. The selection of $r = p$ follows from the feasibility of the constraint.◀

Using the results of Theorem 7-4, the following approximation yields an (good) initial integer feasible solution and hence an initial upper bound $\bar{z}$.

Begin with $r = p$ and then increment r by unit steps to n, adding each a_r to $\sum_{k=1}^{p-1} a_k$, provided that the resulting sum—after including those already added—is less than or equal to b. Then the approximate solution is given by

$$x_j^* = \begin{cases} 1, & \text{if } j < p \quad \text{or} \quad \text{if } a_j \text{ is added to the summation} \\ 0, & \text{otherwise} \end{cases}$$

Thus, $\bar{z} = \sum_{j=1}^n (-c_j)x_j^*$ gives the initial upper bound.

The idea of the approximation is to add as many new variables as possible at level one provided feasibility is maintained since this leads to a smaller upper bound (notice that all $c_j > 0$). The scanning of the variables according to the auxiliary index (based on the ratios c_j/a_j) gives priority to the more attractive variables to be elevated to level one before infeasibility occurs.

►**EXAMPLE 7.2-4**   Minimize

$$v = -(-3y_1 + y_2 + 4y_3 - 6y_4 + 6y_5 - 4y_6 + 2y_7)$$

subject to

$$-2y_1 + 3y_2 + y_3 - 2y_4 + 3y_5 + 4y_6 - 5y_7 \leqslant -4$$
$$y_j = (0, 1), \qquad j = 1, 2, \ldots, 7$$

From Theorem 7-3, $y_6 = 0$ and $y_7 = 1$. Letting

$$x_j = \begin{cases} 1 - y_j, & j = 1, 4 \\ y_j, & j = 2, 3, 5 \end{cases}$$

the problem reduces to the following knapsack form. Given $x_j = (0, 1)$, $j = 1, 2, \ldots, 5$,

minimize

$$z = -(3x_1 + x_2 + 4x_3 + 6x_4 + 6x_5)$$

subject to

$$2x_1 + 3x_2 + x_3 + 2x_4 + 3x_5 \leqslant 5$$

auxiliary index   4   5   1   2   3

This yields $p = 3$, and the optimal continuous solution is $x_3 = x_4 = 1$, $x_5 = \frac{2}{3}, x_1 = x_2 = 0$, with $z = -14$. Thus, applying the approximation, $r = 3$ corresponds to x_5, which is set equal to zero since $(a_3 + a_4) + a_5 > b$. Next $r = 4$ corresponds to x_1, which is set equal to 1. Finally, $r = 5$ corresponds to x_2, which shows it must be zero. Thus

$$\bar{z} = -(c_1 + c_3 + c_4) = -13.$$

The approximation is usually very good. In fact, the above approximation coincides with the optimum of this example.◄

2.   *Ranking procedure*†

The ranking procedure starts with the continuous optimum solution of the knapsack problem, which may be determined based on Theorem 7-4. Assuming that such a solution is not all integer, let e^0 be the extreme

---

† Murty (1968).

point corresponding to the continuous optimum. From the theory of linear programming, e^0 has at most n distinct *adjacent* extreme points associated with $n - 1$ nonbasic variables and a slack variable (notice that the solution space is bounded). The next ranked extreme point e^1 is thus determined as the one yielding the smallest objective value among these adjacent extreme points.

To formalize the above idea, let z^0 be the optimum objective value at e^0 and $\pi_j{}^0$ the corresponding reduced cost $(= z_j - c_j)$ associated with non-basic x_j, with $\alpha_j{}^0$ its constraint coefficient. In other words, $\pi_j{}^0$ and $\alpha_j{}^0$ are defined by the associated simplex tableau. The current basic solution is $x_{k_0} = \beta_{k_0}{}^0$. Then the adjacent extreme point to e^0 is obtained as follows. For each $j, j \neq k_0$, determine

$$\theta_j{}^0 = \min \begin{cases} \theta_{1j}{}^0 = \beta_{k_0}{}^0/\alpha_j{}^0, & \text{if} \quad \alpha_j{}^0 > 0 \\ \theta_{2j}{}^0 = (U - \beta_{k_0}{}^0)/-\alpha_j{}^0, & \text{if} \quad \alpha_j{}^0 < 0 \\ \theta_{3j}{}^0 = U, & \text{if} \quad \alpha_j = 0 \end{cases}$$

where $U = 1$ if the nonbasic variable is binary and $U = \infty$ if it is a slack variable. The associated adjacent extreme point is given as:

(i)   If $\theta_j{}^0 = \theta_{1j}{}^0$, make x_j basic, with x_{k_0} nonbasic at level zero;
(ii) If $\theta_j{}^0 = \theta_{2j}{}^0$, make x_j basic, with x_{k_0} nonbasic at level one;
(iii) If $\theta_j{}^0 = \theta_{3j}{}^0$, x_{k_0} remains basic, but x_j becomes nonbasic at level one. (See Section 2.6.1 for the details.)

The next ranked extreme point e^1 is thus associated with the nonbasic variable x_r, yielding the *next largest* objective value z^1. This means that z^1 is the *smallest* $z \geqslant z^0$, that is,

$$z^1 = z^0 - \pi_r{}^0\theta_r{}^0 = \min_{j \neq k_0}\{z^0 - \pi_j{}^0\theta_j{}^0\}$$

(Notice that $\pi_j{}^0 \leqslant 0$ for all nonbasic x_j associated with e^0.)

Associated with the selection of x_r is a new simplex tableau that is generated by following the above rules for obtaining a new basic solution. This will yield $x_{k_1} = \beta_{k_1}{}^1$ as the new basic variable, with $\pi_j{}^1$ and $\alpha_j{}^1$, $j \neq k_1$, as the new nonbasic coefficients in the objective and constraints.

The determination of e^2 follows as the extreme point selected from adjacent extreme points to e^0 *and* e^1 as the one yielding the *smallest* objective value $z = z^2 \geqslant z^1$.

In general, suppose e^t is known. Then to determine e^{t+1}, consider the nonredundant adjacent extreme points to $e^0, e^1, \ldots,$ and e^t. Then e^{t+1} is the one having the *smallest* $z = z^{t+1} \geqslant z^t$ among all such adjacent extreme points. Nonredundancy here implies that none of the points $e^0, e^1, \ldots,$ and e^t is considered again.

For the purpose of computer programming, as the adjacent extreme points are generated they are stored in ascending order of the objective value so that at e^t all the adjacent extreme points to e^0, e^1, ..., and e^t are already in storage, with the first in the list being e^{t+1}. However, in order to economize the use of the computer memory, the following points are considered:

    (i)   At e^t, store only those adjacent points having $z \geqslant z^t$. When $z = z^t$ for any adjacent extreme point, make sure that such a point does not duplicate e^0, e^1, ..., or e^t. This is achieved by comparing the bases (together with the associated nonbasic variables at zero and one level).

    (ii)   Since the same extreme point may be adjacent to more than one of the *ranked* extreme points, redundancy must be avoided. Again this is done by comparing the bases with the already stored points having the same objective value as the newly generated one.

    (iii)   As the adjacent extreme points are generated, they are tested for integer feasibility. If any such point yields an objective value smaller than $\bar{z}$, then $\bar{z}$ should be updated accordingly.

    (iv)   Given a finite upper bound $\bar{z}$, any adjacent extreme point not satisfying the binary condition and yielding $z \geqslant \bar{z} + (-\Delta)$ should not be stored, where Δ is the least common multiple of all c_j in the original objective function. (Notice that all $c_j > 0$.) The reason here is that such an extreme point cannot improve the best available solution.

▶**EXAMPLE 7.2-5**   Minimize

$$z = -(6x_1 + 2x_2 + 10x_3 + 14x_4 + 2x_5 + 13x_6 + 2x_7 + 5x_8 + x_9)$$

subject to

$$2x_1 + x_2 + 8x_3 + 12x_4 + 2x_5 + 14x_6 + 3x_7 + 10x_8 + 11x_9 \leqslant 17$$
$$x_j = (0, 1), \qquad j = 1, 2, ..., 9$$

The variables are arranged such that $c_j/a_j \geqslant (c_{j+1})/(a_{j+1})$, $j = 1, ..., 9$.

The optimum continuous solution (using Theorem 7-4) is given in Table 7-1. The prime implies that the associated variable is at its upper bound.

The first extreme point e^0 is $x_1 = x_2 = x_3 = 1$, $x_4 = \frac{1}{2}$, $x_5 = x_6 = x_7 = x_8 = x_9 = 0$, with $z^0 = -25$.

Applying the approximation given after Theorem 7-4, one gets $x_1 = x_2 = x_3 = x_5 = x_7 = 1$, with the initial upper bound $\bar{z} = -22$.

TABLE 7-1

	x_1'	x_2'	x_3'	x_5	x_6	x_7	x_8	x_9	S
z	$-\frac{11}{3}$	$-\frac{5}{6}$	$-\frac{2}{3}$	$-\frac{1}{3}$	$-\frac{10}{3}$	$-\frac{3}{2}$	$-\frac{20}{3}$	$-\frac{71}{6}$	$-\frac{7}{6}$	-25
x_4	$-\frac{1}{6}$	$-\frac{1}{12}$	$-\frac{2}{3}$	$\frac{1}{6}$	$\frac{7}{6}$	$\frac{1}{4}$	$\frac{5}{6}$	$\frac{11}{12}$	$\frac{1}{12}$	$\frac{1}{2}$

The extreme points adjacent to e^0 are given in Table 7-2, which shows that only the adjacent extreme points associated with x_2', x_3', x_5, x_6, and x_7 need be stored for future scanning. These points are ordered according to the associated values of z.

TABLE 7-2

Nonbasic variable	θ^0	z	Remarks
x_1'	$\theta_{31}{}^0 = 1$	$-25 - (1) \times \left(-\frac{11}{3}\right) = -21\frac{1}{3}$	$z > \bar{z} + (-1)$, do not store
x_2'	$\theta_{32}{}^0 = 1$	$-25 - (1) \times \left(-\frac{5}{6}\right) = -24\frac{1}{6}$	
x_3'	$\theta_{23}{}^0 = \frac{3}{4}$	$-25 - \left(\frac{3}{4}\right) \times \left(-\frac{2}{3}\right) = -24\frac{1}{2}$	
x_5	$\theta_{35}{}^0 = 1$	$-25 - (1) \times \left(-\frac{1}{3}\right) = -24\frac{2}{3}$	$\leftarrow e^1$
x_6	$\theta_{16}{}^0 = \frac{3}{7}$	$-25 - \left(\frac{3}{7}\right) \times \left(-\frac{10}{3}\right) = -23\frac{4}{7}$	
x_7	$\theta_{37}{}^0 = 1$	$-25 - (1) \times \left(-\frac{3}{2}\right) = -23\frac{1}{2}$	
x_8	$\theta_{18}{}^0 = \frac{3}{5}$	$-25 - \left(\frac{3}{5}\right) \times \left(-\frac{20}{3}\right) = -21$	Do not store
x_9	$\theta_{19}{}^0 = \frac{6}{11}$	$-25 - \left(\frac{6}{11}\right) \times \left(-\frac{71}{6}\right) = -18\frac{6}{11}$	Do not store
S	$\theta_{1S}{}^0 = 6$	$-25 - (6) \times \left(-\frac{7}{6}\right) = 18$	Do not store

The next ranked extreme point e^1 corresponds to x_5. Since $\theta_5{}^0 = \theta_{35}{}^0 = 1$, the extreme point is generated by making x_5 nonbasic at level one, thus using the substitution $x_5 = 1 - x_5'$. This yields the associated simplex solution shown in Table 7-3, which gives e^1 as $x_1 = x_2 = x_3 = x_5 = 1$, $x_4 = \frac{1}{3}$, $x_6 = x_7 = x_8 = x_9 = 0$, with $z^1 = -24\frac{2}{3}$.

TABLE 7-3

	x_1'	x_2'	x_3'	x_5'	x_6	x_7	x_8	x_9	S
z	$-\frac{11}{3}$	$-\frac{5}{6}$	$-\frac{2}{3}$	$\frac{1}{3}$	$-\frac{10}{3}$	$-\frac{3}{2}$	$-\frac{20}{3}$	$-\frac{71}{6}$	$-\frac{7}{6}$	$-24\frac{2}{3}$
x_4	$-\frac{1}{6}$	$-\frac{1}{12}$	$-\frac{2}{3}$	$-\frac{1}{6}$	$\frac{7}{6}$	$\frac{1}{4}$	$\frac{5}{6}$	$\frac{11}{12}$	$\frac{1}{2}$	$\frac{1}{3}$

Next the extreme points adjacent to e^1 are generated (nonredundantly) and intermingled with those in Table 7-2 so that e^2 is the extreme point having the smallest $z \geqslant z^1$ from among all the extreme points adjacent to *both* e^0 and e^1.

The adjacent extreme point associated with x_3' yields the feasible solution $x_1 = x_2 = x_4 = x_5 = 1$ and $x_3 = x_6 = x_7 = x_8 = x_9 = 0$, with $z = -24$. This yields the new $\bar{z} = -24$. But since $z^1 = -24\frac{2}{3}$, this means that no better feasible solution can be encountered and thus it is not necessary to generate the remainder of the adjacent extreme points in Table 7-4. Moreover, none of the remaining extreme points in Table 7-2 need be considered. The optimal solution is associated with $\bar{z} = -24$, and the algorithm terminates.

TABLE 7-4

Nonbasic variable	θ^1	z	Remarks
x_1'	$\theta_{31}{}^1 = 1$	-21	$z > \bar{z} + (-1)$, do not store
x_2'	$\theta_{32}{}^1 = 1$	$-23\frac{5}{6}$	
x_3'	$\theta_{33}{}^1 = 1$	-24	feasible, set $\bar{z} = -24$

◀

The computation experience by Taha (1971b) reveals the interesting result that the average computation time based on 40 problems for each n using the ranking procedure varies linearly with the number of variables. The linear variation of the computation time with n is particularly interesting since all the other algorithms presented in this section experience some degree of polynomial or exponential variation. The result follows from the possibility that the ranking procedure may not be as sensitive to the increase in the number of variables n as the other procedures. Rather, the number of extreme points between the continuous optimum and the upper bound $\bar{z}$ appears to be the more prominent factor.

An important remark should be made here. In order to utilize the extreme point ranking procedure, an integer variable must be substituted in terms of equivalent 0–1 variables. Although this will increase the number of variables, this increase is only a multiple of approximately $\ln \prod_{j=1}^{n} (\mu_j - 1)/\ln 2$, where μ_j is the upper bound on the jth variable. This shows that the growth rate is not exponential.

The reader should observe the similarity between the ranking procedure and the Land–Doig algorithm. Actually, for the zero–one problem, the Land–Doig algorithm essentially ranks the extreme points but in a selective manner. Mainly, by optimizing the linear program at each subproblem, the Land–Doig algorithm seeks the next best extreme point satisfying a binary restriction. In other words, those extreme points violating the binary condition and having an objective value between those of the current node and the new node are implicitly ranked and need not be considered explicitly. But in the knapsack problem *with one constraint*, every extreme point must necessarily satisfy at least one binary restriction. Consequently there is no chance for implicit ranking, and the number of explicit extreme points ranked in both procedures is the same. However, notice that the determination of extreme points in the ranking procedure involves a single change of pivot as compared with the solution of a linear program in the Land–Doig procedure. Also, in the Land–Doig algorithm extreme points are determined by imposing a binary restriction on one variable at a time. In the ranking procedure, all the adjacent extreme points are enumerated, and this involves a wider range of search among all the extreme points. The major advantage here is the greater possibility of locating an improved upper bound and hence, in most cases, the termination of the algorithm. This point was illustrated in Example 7.2-4, but the extensive computations carried out by Taha (1971b) show that this result is generally true.

7.3   Fixed-Charge and Plant Location Problems

In Example 1.3-7 (Section 1.3), the fixed-charge problem is presented in the form of a mixed zero–one linear model. Although any of the appropriate algorithms in the preceding chapters may be used to solve the model, its special properties should generally yield more efficient specialized algorithms. In this section, a number of such algorithms is presented, categorized as *approximate* and *exact* methods. The first type does not yield the true optimum but generally provides good solutions. The advantage here is that it requires much less computation time than those algorithms seeking the exact optimum.

Plant location problems are actually a variation of the fixed-charge model. There is a number of receiving stations n and the demands at these destinations are satisfied from m potential plants or warehouses. Usually, n is considerably larger than m. It is required to decide on the location and capacity of each plant in order to satisfy the demand. There is installation (fixed) cost associated with the construction of each plant, and the objective is to minimize the total cost including the fixed costs and transportation costs between origins and destinations. The problem reduces to an ordinary transportation model with the exception that a fixed-charge term k_i appears in the objective function if for any j, $x_{ij} > 0$ and vanishes otherwise, where x_{ij} is the amount transported from origin i to destination j. Clearly, for prespecified fixed charges the problem is an ordinary transportation model.

There is a wide variety of algorithms, and a rich literature, for the plant location problems. They vary primarily in the details for constructing the objective and constraint functions, but the basic idea remains unchanged.

An excellent exposition of the different algorithms may be found in the work of Elshafei (1972).

As in the fixed-charge problem, there are approximate and exact algorithms for the plant location problems. Some of these algorithms will now be presented.

7.3.1 General Properties of the Fixed-Charge Problem

The fixed-charge problem defined as

minimize

$$z(x_1, \ldots, x_n) = \sum_{j=1}^{n} f_j(x_j) = \sum_{j=1}^{n} (k_j y_j + c_j x_j)$$

subject to

$$\sum_{j=1}^{n} a_{ij} x_j \leq b_i, \qquad\qquad i = 1, 2, \ldots, m$$

$$x_j \geq 0$$

$$y_j = \begin{cases} 0, & \text{if } x_j = 0 \\ 1, & \text{if } x_j > 0 \end{cases} \quad j = 1, 2, \ldots, n$$

actually involves the minimization of a concave objective function $z(x_1, \ldots, x_n)$ over a convex polyhedron $Q = \{x \mid \sum a_{ij} x_j \leq b_i, i = 1, \ldots, m, x_j \geq 0, j = 1, \ldots, n\}$. It is seen that $z(x_1, \ldots, x_n)$ is concave since it is the sum of n single variable functions $f_j(x_j)$, and $f_j(x_j)$ are obviously (at least from graphical representation) concave. It is clear that the same observation applies to plant location problems as defined above.

An important property of this type of problem is that the optimum solution must occur at an extreme point of the feasible space, that is, it must be associated with a feasible basic solution of Q. This means that the search for the global optimum can be restricted to considering the extreme points of Q only. This result is similar to that used with linear programming. However, because the objective function is concave, the associated algorithm is more complex.

A logical question that follows from this property is: When does the linear program obtained by ignoring the terms $k_j y_j$ yield the same optimum extreme point associated with the fixed-charge problem? The answer was observed by Hirsch and Dantzig (1954), who showed that a *sufficient* condition for this to occur is that all k_j are equal for all j and that all the extreme points of the convex polyhedron Q are nondegenerate. The nondegeneracy here is required since a degenerate extreme point has fewer *positive* basic variables and may thus be optimum for the fixed-charge problem but not for the linear program.

7.3.2 Algorithms for the Fixed-Charge Problem

A. *Approximate Algorithm†*

The general idea of this algorithm is to start with the optimum extreme points of the linear program ignoring the fixed charges. Such an extreme point sets a lower bound on the optimum objective value of the fixed-charge problem, since

$$\sum_{j=1}^{n} c_j x_j \leqslant \sum_{j=1}^{n} (c_j x_j + k_j y_j)$$

Because the optimum of the fixed charge problem must occur at an extreme point, intelligent criteria are developed to select another (hopefully improved) extreme point solution.

To formalize the above idea, let

$$w_1 = \sum_{j=1}^{n} c_j x_j, \qquad w_2 = \sum_{j=1}^{n} k_j y_j, \qquad w_3 = \sum_{j=1}^{n} (c_j x_j + k_j y_j)$$

Define $w_1{}^*$, $w_2{}^*$, and $w_3{}^*$ as the smallest values of w_1, w_2, and w_3 so far attained for any extreme point solution. In general, $w_1{}^*$, $w_2{}^*$, and $w_3{}^*$ need not occur at the same extreme point, and hence it is possible to have $w_3{}^* \neq w_1{}^* + w_2{}^*$. For convenience, let $w_1{}^\#$ and $w_2{}^\#$ be the values of w_1 and w_2 computed at the same extreme point associated with $w_3{}^*$; thus $w_3{}^* = w_1{}^\# + w_2{}^\#$. Also define $E_1{}^*$, $E_2{}^*$, and $E_3{}^*$ as the extreme points associated with $w_1{}^*$, $w_2{}^*$, and $w_3{}^*$, respectively.

The values of $w_1{}^*$, $w_2{}^*$, and $w_3{}^*$ are initialized as follows. Solve the linear program:

minimize

$$w_1 = \left\{ \sum_{j=1}^{n} c_j x_j \,\middle|\, x_j \in Q \right\}$$

where Q is the set of all feasible points as defined in Section 7.3.1. Let $e^0 = \{x_1{}^0, \ldots, x_n{}^0\}$ be the associated extreme point solution. Then

$$w_1{}^* = w_1{}^\# = w_1{}^0 \equiv \sum_{j=1}^{n} c_j x_j{}^0, \qquad w_2{}^* = w_2{}^\# = w_2{}^0 \equiv \sum_{j=1}^{n} k_j y_j{}^0$$

$$w_3{}^* = w_1{}^\# + w_2{}^\#$$

where $y_j{}^0$ is defined according to e^0. Initially, $E_1{}^* = E_2{}^* = E_3{}^* = e^0$.

The heuristic generates a sequence of extreme points starting with e^0. The best solution is given by $E_3{}^*$ after the algorithm terminates.

---

† Cooper and Derbes (1967).

The determination of e^{t+1}, given e^t, is achieved as follows. For all x_j, determine

$$c_j' = \begin{cases} c_j + (k_j/x_j'), & x_j' > 0 \\ c_j, & x_j' = 0 \end{cases}$$

Compute the familiar reduced costs $z_j - c_j$ for every nonbasic variable associated with e^t as

$$z_j^t - c_j^t = c_B^t B_t^{-1} P_j - c_j^t$$

where $c_B^t = (c_1^t, \ldots, c_m^t)$, computed c_j^t associated with basic x_j, B_t^{-1} is the inverse of the basis, and $P_j = (a_{1j}, \ldots, a_{mj})^{\mathrm{T}}$. Then

(a) If $z_j^t - c_j^t \leqslant 0$ for all nonbasic j, go to (c); otherwise,

(b) Change the basis by introducing the nonbasic variable having the largest $z_j^t - c_j^t$ and store the resulting extreme point as e^{t+1}. Compute w_k^{t+1}, $k = 1, 2, 3$.

   (i)   If $w_3^{t+1} < w_3^*$, set $w_3^* = w_3^{t+1}$, $E_3^* = e^{t+1}$, $w_1^\# = w_1^{t+1}$, $w_2^\# = w_2^{t+1}$, and go to (iii); otherwise,

   (ii) If $w_k^{t+1} < w_k^*$, $k = 1$ or 2, record $w_k^* = w_k^{t+1}$ and $E_k^* = e^{t+1}$, $k = 1$ or 2. Go to (iii).

   (iii) Perform (a) on e^{t+1}.

(c) Compute

$$\Delta_1 = w_1^* - w_1^\#, \qquad \Delta_2 = w_2^* - w_2^\#$$

   (i)   If $\Delta_1 = \Delta_2 = 0$, terminate.

   (ii) If $\Delta_1 < \Delta_2$, set $e^{t+1} = E_1^*$. Otherwise $e^{t+1} = E_2^*$. If $\Delta_1 = \Delta_2$, break the tie arbitrarily by selecting $e^{t+1} = E_1^*$ or E_2^*.

   (iii) Perform (a) on e^{t+1}.

The logic behind the determination of e^{t+1} should be obvious except perhaps for the details of (c). The idea here is that if $\Delta_1 < \Delta_2$, then it is better to choose e^{t+1} equal to E_1^* rather than E_2^*, since its w_1^* is closer to $w_1^\#$ associated with the best extreme point E_3^*.

It must be remarked that Cooper and Derbes developed more criteria for locating potentially better extreme points. These criteria are invoked in step (c) if the test in part (i) calls for terminating the search. For example, e^{t+1} may be generated by selecting the leaving basic variable in e^t as the one having the largest fixed charge k_j. Other details can be found in the work of Cooper and Derbes (1967).

The computational results reported by Cooper and Derbes appear promising. In a large sample of randomly generated problems, about 95%

of the tested problems yield the true optimal solution (for the purpose of comparison, the true optimum is determined by enumerating all the extreme points of Q). The computation time appears to increase rather exponentially with the size of the problem. For example, the average times for problems of sizes $(m \times n) = (5 \times 10)$, (7×15), and (15×30) are 0.33, 1.6, and 15.0 minutes, respectively.

The Cooper–Derbes heuristic was later "improved" by Cooper and Olson (date unknown). Other heuristics include Denzler (1969) and Steinberg (1970).

B.   *Exact Algorithm I*†

This algorithm is based on recognizing that the optimum solution occurs at an extreme point of the convex set representing the solution space. Moreover, all extreme points can be exhaustively identified by considering all the *basic* feasible solutions to the problem. However, rather than enumerating *all* the basic feasible solutions, a branch-and-bound algorithm is developed so that only a portion of the basic feasible solution is enumerated explicitly, while automatically accounting for the remaining ones. The basis for eliminating basic solutions is the development of bounds on the objective value.

The data of the problem are first arranged such that $k_j \geq k_{j+1}$ for $j = 1, 2, \ldots, n - 1$. Assuming there are m (independent) constraints, then every basis must include m (out of n) independent vectors. Naturally, if the optimal basic solution of the linear program, ignoring the fixed charges, is given by the variables $(x_{n-m+1}, x_{n-m+2}, \ldots, x_n)$, then it is also optimal for the fixed-charge problem, since no other combination can yield a smaller value for the fixed charges. The validity of this statement requires making the assumption that *all the extreme points of the solution space are nondegenerate*. This restriction is assumed to hold throughout the algorithm.

Initially, $(n - m + 1)$ branches emanate from the first node. The jth branch, $j = 1, 2, \ldots, n - m$, is associated with the vector $P_j = (a_{1j}, a_{2j}, \ldots, a_{mj})^{\mathrm{T}}$. The remaining m vectors $\{P_{n-m+1}, \ldots, P_n\}$ define the $(n - m + 1)$th branch. The vector P_j associated with each of the first $n - m$ branches indicates that all bases that may be generated from such a branch must each include P_j. On the other hand, the $(n - m + 1)$th branch defines a single "basis" so that no further branches can emanate from its end node. Such a node will be referred to as *terminal*.

The idea now is to add branches at each nonterminal node that will fix two vectors in a basis, then three, and so on. Eventually paths, each consisting of a string of at most m connected branches, will define bases.

† Bod (1970).

To avoid redundancy in creating such paths, a typical tree is shown in Fig. 7-2. The nodes marked terminal are those defining m vectors and hence a possible basis.

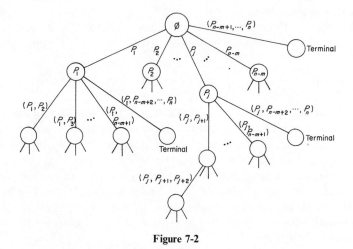

Figure 7-2

As in all branch-and-bound methods, the basic objective is to develop efficient criteria that will require testing some, but not all, of the branches in the tree. This is achieved based on two tests: (1) If the vectors (branches) defining any node are dependent, this node must be discarded since it cannot lead to a basis. (2) If the lower bound on the objective value (to be developed next) at a given node exceeds the best available objective value associated with a basic *feasible* solution, then the node must be discarded.

The development of a lower bound z at any node is achieved as follows: Let R define the set of indices of r $(< m)$ *independent* vectors associated with the node. Because all extreme points are assumed nondegenerate and by virtue of the fact that $k_j \geqslant k_{j+1}$ for all j, then a lower bound on the fixed charges associated with the node is

$$\underline{K} = \sum_{j \in R} k_j + \sum_{j = p}^{n} k_j$$

where $p = (n - m + 1) + r$. On the other hand,

$$z_0 = \min\left\{\sum_{j = 1}^{n} c_j x_j \,\middle|\, x_j \in Q, j = 1, 2, \ldots, n\right\}$$

provides a lower bound on the variable costs, $\sum_{j=1}^{n} c_j x_j$, at any extreme point of Q. Thus the overall lower bound is

$$\underline{z} = z_0 + \underline{K}.$$

There are two special cases for $\underline{z}$:

(i) If $R = \varnothing$, which can occur at the initial node only, then

$$\underline{K} = \sum_{j=n-m+1}^{n} k_j$$

(ii) If $r = m$ and the resulting basis yields a feasible extreme point $\bar{x} = (\bar{x}, \ldots, \bar{x}_n)$, then it is possible to determine the *exact* objective value as

$$z = \sum_{j=1}^{n} (c_j \bar{x}_j + k_j \bar{y}_j)$$

where $\bar{y}_j$ is determined according to $\bar{x}_j$.

The algorithm terminates when it becomes evident that all the remaining nodes cannot yield a better objective value, that is, when the lower bound associated with each unexplored node exceeds the best available objective value associated with a feasible solution.

▶**EXAMPLE 7.3-1** Minimize

$$z = -4x_1 - 3x_2 + 0(x_3 + x_4 + x_5) + y_1 + \tfrac{1}{2}y_2 + 0(y_3 + y_4 + y_5)$$

subject to

$$2x_1 + x_2 + x_3 = 4$$
$$x_1 + x_2 + x_4 = 3$$
$$2x_2 + x_5 = 5$$
$$x_j \geqslant 0, \qquad j = 1, 2, \ldots, 5$$
$$y_j = \begin{cases} 0 & \text{if } x_j = 0 \\ 1 & \text{if } x_j > 0, \end{cases} \qquad \text{for all } j$$

The solution tree is shown in Fig. 7-3. The optimal solution is associated with the basis $\{P_1, P_2, P_5\}$. This yields $x_1 = 1$, $x_2 = 2$, and $z = -8\tfrac{1}{2}$.◀

The example was especially designed to illustrate the inadequacy of the proposed algorithm, particularly when the fixed charges are small relative to the variable costs. In Fig. 7-3, it is seen that all the possible basic solutions of the problem $(=C_3{}^6 = 10$ bases) have been enumerated before

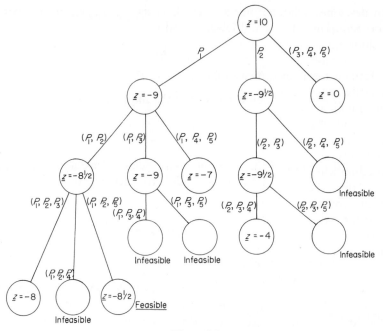

Figure 7-3

the optimum is determined. The fact that none of these bases is *implicitly* enumerated stems from the weakness of the lower bound $\underline{z}$ used to evaluate each node. Bod indicates that $\underline{z}$ will be more effective if the fixed charges are large relative to the variable costs, but he gives no computational evidence to support this statement.

One of the main disadvantages is that the algorithm requires all the extreme points of Q to be nondegenerate. The difficulty here is that there is no simple way to predict if degeneracy exists. In this case, and if degeneracy passes unchecked, the procedure may terminate with a nonoptimal solution. The suggestion of Bod to assign a very small $\varepsilon > 0$ to zero basic variables cannot rectify the problem since it may allow assigning a positive fixed charge to a zero (basic) variable.

C. *Exact Algorithm II†*

This algorithm is basically a branch-and-bound method that, starting from the extreme point associated with $\min\{\sum_{j=1}^{n} c_j x_j \,|\, x_j \in Q\}$, ranks the extreme points of Q according to their value of $\sum_{j=1}^{n} c_j x_j$. Proper tests are

† Taha (1973).

then developed so that, when a given ranked extreme point is reached, the algorithm terminates. The ranking of the extreme points is based on the development of a cutting plane procedure that employs Glover's convexity cut (see Section 5.2.5).

Actually the algorithm was originally proposed by Murty (1968) but was primarily restricted to fixed-charge transportation problems. His extreme point ranking procedure is based on generating all the adjacent extreme points and storing them in ascending order of the objective value. This exact method was adapted by Taha for solving the zero–one knapsack problem (see Section 7.2.3-D). But as mentioned previously, such a procedure will burden the computer memory, especially for problems with several (linear) constraints. Later Taha (1973) generalized Murty's algorithm for any concave minimization problem over a convex polyhedron. The algorithm reduces to solving a series of linear programs.

The idea is to define a *linear underestimator* $L(x)$ of $z(x)$ over the convex polyhedron Q, that is,

$$L(x) \leqslant z(x), \qquad x \in Q$$

where $z(x)$ is the objective function of the fixed-charge problem (as defined in Section 7.3.1). It then follows that

$$\min_x \{L(x) \mid x \in Q\} \leqslant \min_x \{z(x) \mid x \in Q\}$$

A typical linear underestimator is $L(x) = \sum_{j=1}^{n} c_j x_j$.

Starting with the extreme point x^0 satisfying $\min\{L(x) \mid x \in Q\}$, then $\underline{z} = L(x^0)$ is a lower bound on the optimum objective value. Also an associated upper bound is $\bar{z} = z(x^0)$. This means that the optimum extreme point x^* must satisfy $\underline{z} \leqslant z(x^*) \leqslant \bar{z}$. Initially $x^* = x^0$. At the ith *ranked* extreme point, $\underline{z} = L(x^i)$, while $\bar{z}$ is changed to $z(x^i)$ only if $z(x^i) < z(x^*)$. In this case, x^* is set equal to x^i. In other words, x^* keeps track of the best available solution. The algorithm terminates at the kth *ranked* extreme point x^k if $\underline{z} = L(x^k) \geqslant \bar{z}$, with the point x^* associated with $\bar{z}$ being the optimum. This follows since, for all the *remaining* extreme points,

$$z(x) \geqslant L(x^k) \geqslant \bar{z}$$

The method for generating the sequence of ranked extreme points $x^0, x^1, \ldots, x^k$ by using Glover's convexity cut is now presented. Referring to Section 5.2.5, which presents Glover's theory, the set S represents the extreme points of Q, while the set R is taken as the set Q itself. Clearly, it satisfies Glover's condition since it has no feasible points in its interior. Thus, starting with the extreme point x^0 as defined above, the convexity cut is defined so it will pass through its adjacent extreme points such that the point x^0 becomes infeasible. Then, by reoptimizing using the dual simplex method and $L(x)$ as the objective function, the new optimum feasible solution should yield x^1, the next ranked extreme point. The process is repeated at x^1 and successively until x^k is reached.

To formalize this discussion, let the current basic solution be defined by the set of equations

$$y_i = b_i - \sum_{j \in NB} \alpha_{ij} t_j, \qquad i = 1, 2, \ldots, m, \quad y_i, t_j \geq 0$$

where y_i are the basic and t_j the nonbasic variables. Then Glover's cut is given as

$$\sum_{j \in NB} t_j / t_j^* \geq 1$$

where t_j^* are given by

$$t_j^* = \begin{cases} \min_{i \in M}(b_i / \alpha_{ij}), & \text{if} \quad \alpha_{ij} > 0 \\ \infty, & \text{if} \quad \alpha_{ij} \leq 0 \end{cases}$$

Clearly t_j^* is strictly positive if $y_i = b_i > 0$ whenever $\alpha_{ij} > 0$. If $b_i = 0$ when $\alpha_{ij} > 0$, then $t_j^* = 0$ and the cut is not defined. This difficulty is avoided by using Balas' idea (Section 5.2.5), which calls for deleting those rows in the simplex tableau that are associated with *zero* basic variables. In this case the cut is always defined, but it may be weaker than the one that is constructed to pass through adjacent extreme points. This point is discussed further with the following numerical example.

▶**EXAMPLE 7.3-2**   Consider the same problem of Example 7.3-1. For convenience, the constraint $2x_2 \leq 5$ will not be considered explicitly in the simplex tableau. Rather, this constraint is implemented by using upper bounding techniques. Also the regular, rather than the lexicographic, simplex and dual simplex methods are used to solve the example. The different iterations are illustrated graphically in Fig. 7-4.

The problem may be written as:

minimize

$$z(x) = -4x_1 - 3x_2 + y_1 + \tfrac{1}{2}y_2$$

subject to

$$2x_1 + x_2 + S_1 = 4$$
$$x_1 + x_2 + S_2 = 3$$
$$x_2 \leq \tfrac{5}{2}$$
$$x_1, x_2, S_1, S_2 \geq 0$$

where $y_j = 1$ if $x_j > 0$ and zero otherwise, $j = 1, 2$.

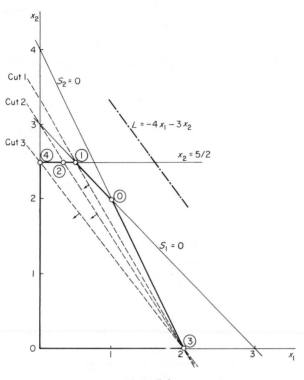

Figure 7-4

The linear underestimator is then given as

$$L(x) = -4x_1 - 3x_2$$

The first extreme point x^0 is then determined as shown in Table 7-5.

TABLE 7-5

	S_1	S_2	
L	-10	-1	-2
x_1	1	1	-1
x_2	2	-1	2

$x^0 = (1, 2)$; point ⓪
$\underline{z}(x^0) = -10$
$\bar{z}(x^0) = -10 + (1 + \frac{1}{2}) = -8\frac{1}{2}$
$\bar{z}(x^*) = \bar{z}(x^0) = -8\frac{1}{2}$

Cut 1 is developed now. The constants t_j^* of the cut are determined from the simplex tableau with upper bounded variables (Section 2.6.1). Thus

$$S_1^* = \min\{1/1, (2 - \tfrac{5}{2})/-1\} = \tfrac{1}{2}, \qquad S_2^* = \min\{\infty, 2/2\} = 1$$

and the cut is given by

$$(S_1/\tfrac{1}{2}) + (S_2/1) \geqslant 1$$

Or, given $S_3 \geqslant 0$, the cut is given in terms of x_1 and x_2 by

$$5x_1 + 3x_2 + S_3 = 10 \qquad \text{(Cut 1)}$$

Table 7-6 yields x^1 as a result of augmenting Table 7-5 by cut 1 and reoptimizing using the dual simplex method for upper bounded variables (Section 2.6.2).

TABLE 7-6

		S_2	S_3
L	$-9\tfrac{1}{2}$	$-\tfrac{3}{2}$	$-\tfrac{1}{2}$
x_1	$\tfrac{1}{2}$	$-\tfrac{3}{2}$	$\tfrac{1}{2}$
x_2	$\tfrac{5}{2}$	$\tfrac{3}{2}$	$-\tfrac{1}{2}$
S_1	$\tfrac{1}{2}$	$\tfrac{1}{2}$	$-\tfrac{1}{2}$

$x^1 = (\tfrac{1}{2}, \tfrac{5}{2})$; point ①
$z(x^1) = -9\tfrac{1}{2}$
$\bar{z}(x^1) = -9\tfrac{1}{2} + (1\tfrac{1}{2}) = -8 > \bar{z}(x^*)$
$\bar{z}(x^*) = \bar{z}(x^0)$

Notice that in Table 7-6 x_2 is basic at its upper bound. This means that the current solution is degenerate. Using Balas' condition, which in this case calls for ignoring the equations involving basic variables at upper bound or zero level, it is clear that the x_2-equation must be disregarded in developing cut 2. Thus

$$S_2{}^* = \min\{\infty, \infty, \tfrac{1}{2}/\tfrac{1}{1}\} = 1, \qquad S_3{}^* = \min\{\tfrac{1}{2}/\tfrac{1}{2}, \infty, \infty\} = 1$$

This yields a new cut, which when expressed in terms of x_1 and x_2 is given by

$$6x_1 + 4x_2 + S_4 = 12, \qquad S_4 \geqslant 0 \qquad \text{(Cut 2)}$$

Table 7-7 gives the new solution after cut 2 is effected. Notice that

TABLE 7-7

		x_2'	S_4
L	$-8\tfrac{5}{6}$	$-\tfrac{1}{3}$	$-\tfrac{2}{3}$
x_1	$\tfrac{1}{3}$	$-\tfrac{2}{3}$	$\tfrac{1}{6}$
S_2	$\tfrac{1}{6}$	$-\tfrac{1}{3}$	$-\tfrac{1}{6}$
S_1	$\tfrac{5}{6}$	$\tfrac{1}{3}$	$-\tfrac{1}{3}$
S_3	$\tfrac{5}{6}$	$\tfrac{1}{3}$	$-\tfrac{5}{6}$

$x^2 = (\tfrac{1}{3}, \tfrac{5}{3})$; point ②
$z(x^2) = -8\tfrac{5}{6}$
$\bar{z}(x^2) = -8\tfrac{5}{6} + 1\tfrac{1}{2} = -7\tfrac{1}{3} > \bar{z}(x^*)$
$\bar{z}(x^*) = \bar{z}(x^0)$

$x_2 = \frac{5}{2} - x_2'$. Notice also that since S_3 is associated with a previous cut and since it is basic, its corresponding equation can be dropped in future tableaus.

Cut 3 is now generated from Table 7-7, giving

$$30x_1 + 24x_2 \leqslant 60 \qquad \text{(Cut 3)}$$

The application of this cut will yield point ③ with $x^3 = (2, 0)$ and $\underline{z}(x^3) = -8$. Since $\underline{z}(x^3) > \bar{z}(x^*)$, the process terminates. Thus $x^* = x^0$ is the optimum solution.

Notice the effect of degeneracy at ①. Point ① is (over-) determined by the three lines $x_2 = \frac{5}{2}$, $x_1 + x_2 = 3$, and $6x_1 + 4x_2 = 12$. Balas' condition drops $x_2 = \frac{5}{2}$. Thus the half-lines $x_1 + x_2 \leqslant 3$ and $6x_1 + 4x_2 \leqslant 12$ produce cut 2. The optimum point ② is a new extreme point which does not belong to the original solution space. However, points such as ② cannot be optimum since this contradicts the condition that the global optimum must occur at an extreme point of the original convex polyhedron Q.◀

The tightness of the linear underestimator $L(x)$ has a direct effect on the efficiency of the algorithm. Thus $L_1(x)$ is a better underestimator than $L_2(x)$ if

$$L_2(x) \leqslant L_1(x) \leqslant z(x)$$

This means that $L_1(x)$ may produce the condition for terminating the algorithm [namely, $L_1(x^k) \geqslant \bar{z}(x^*)$] faster than $L_2(x)$. A better underestimator for the fixed-charge problem can be defined as follows. Let $x_j \leqslant u_j$ represent an upper bound on x_j such that none of the feasible points in Q is eliminated. Then a legitimate underestimator is

$$L(x) = \sum_{j=1}^{n} [c_j + (k_j/u_j)]x_j$$

which is better than the previous underestimator since

$$\sum_{j=1}^{n} c_j x_j \leqslant \sum_{j=1}^{n} [c_j + (k_j/u_j)]x_j$$

for all $(x_1, x_2, \ldots, x_n) \in Q$.

The computer results indicate that the algorithm can be used to solve relatively large problems. There is no difficulty resulting from exceeding the computer storage since the storage requirement is primarily controlled by the dimensions of the original problem. But the major difficulty stems from the degeneracy problem, particularly with the machine rounding error. Degeneracy has the effect of creating new extreme points and hence slowing down the termination of the algorithm. It is evident from the computational experience that the smaller the fixed charges the faster the algorithm converges. Also the use of the tighter lower bound generally results in improved computation time.

Other exact algorithms are due to Gray (1971), Steinberg (1970), and Cabot (1972). Gray's algorithm is based on the mixed integer programming formulation presented in Example 1.3-7. His idea is to generate a sequence of values for the zero–one variables and then solve the resulting linear program. The idea is somewhat similar to those of Driebeek and Benders (Section 3.7). Steinberg's algorithm is a branch-and-bound method where at each node two branches associated with $x_j = 0$ and $x_j > 0$ for some j must emanate. Finally, Cabot's algorithm utilizes cuts similar to those given with exact algorithm II, except that the cuts are generated from the objective function. The algorithm is based on a general method by Tuy (1964). A later work (Zwart, 1973), however, shows that Tuy's algorithm may not converge in general.

7.3.3   Algorithms for Plant Location Problems

A simple form of the plant location model may be given as follows:

minimize

$$z(x) = \sum_{i=1}^{m} \sum_{j=1}^{n} c_{ij} x_{ij} + \sum_{i=1}^{m} k_i y_i$$

subject to

$$\sum_{j=1}^{n} x_{ij} = a_i, \qquad i = 1, 2, \ldots, m$$

$$\sum_{i=1}^{m} x_{ij} = b_j, \qquad j = 1, 2, \ldots, n$$

$$x_{ij} \geq 0 \text{ for all } i \text{ and } j$$

and $y_i = 1$ if $\sum_{j=1}^{n} x_{ij} > 0$ and zero otherwise.

The coefficients c_{ij} represent the cost of transporting one unit of a common product from source i, $i = 1, \ldots, m$, to destination $j, j = 1, \ldots, n$. A fixed charge k_i is incurred if source i supplies a positive quantity to any destination j. The constants a_i and b_j represent the supply limitation at source i and the demand requirement at destination j. The objective is decide on the location and size of plant i so that the total costs are minimized. (If $y_i = 0$, or $\sum_{j=1}^{n} x_{ij} = 0$, no plant is assigned to location i.) Clearly, in the absence of the fixed charges, the problem reduces to a regular transportation model.

This model can be further complicated by assuming that the term $\sum_{i=1}^{m} \sum_{j=1}^{n} c_{ij} x_{ij}$ in the objective function is replaced by some nonlinear function (mostly concave). It can also be further simplified by relaxing the capacity restriction on the sources, that is, by eliminating the restrictions $\sum_{j=1}^{n} x_{ij} = a_i$, $i = 1, 2, \ldots, m$. One approximate and one exact algorithms involving these situations are now presented.

A. *Approximate Algorithm*†

Balinski's algorithm is developed for the problem:

minimize

$$z(x) = \sum_{i=1}^{m} \sum_{j=1}^{n} c_{ij} x_{ij} + \sum_{i=1}^{m} \sum_{j=1}^{n} k_{ij} y_{ij}$$

subject to

$$\sum_{j=1}^{n} x_{ij} = a_i, \qquad i = 1, \ldots, m$$

$$\sum_{i=1}^{m} x_{ij} = b_j, \qquad j = 1, \ldots, n$$

$$x_{ij} \geqslant 0$$

and $y_{ij} = 1$ if $x_{ij} > 0$ and $y_{ij} = 0$ if $x_{ij} = 0$. The assumption here is that each delivery from a source i to a destination j will incur a fixed cost k_{ij}. This is contrary to the previous formulation, where a source (plant) i requires a fixed installation cost k_i if it produces any positive quantity, regardless of the destination j.

Balinski's approximate method is based on the development of a linear estimate of the (concave) objective function. He stipulates that the number of sources m is usually much smaller than the number of destinations n, so that the optimal solution will have the property that the demand at a given destination can only be supplied by one source, except possibly for a very few cases. As a result, it can be assumed that the amount shipped from source i to destination j will frequently be $x_{ij} = \min\{a_i, b_j\} = u_{ij}$. He then replaces the objective function by the linear estimate

$$\sum_{i=1}^{m} \sum_{j=1}^{n} c_{ij} x_{ij} + \sum_{i=1}^{m} \sum_{j=1}^{n} (k_{ij}/u_{ij}) x_{ij} = \sum_{i=1}^{m} \sum_{j=1}^{n} [c_{ij} + (k_{ij}/u_{ij})] x_{ij}$$

The optimal solution to the transportation problem using the linear estimate provides the approximate solution to the original problem.

† Balinski (1961).

A more straightforward, and perhaps enlightening, way of reaching Balinski's result is as follows. Regardless of the assumption of whether or not m is much smaller than n, it is true that

$$0 \leqslant x_{ij} \leqslant u_{ij} \equiv \min\{a_i, b_j\}$$

Thus, using the idea employed with exact algorithm II (Section 7.3.2), a linear underestimator of $z(x)$ is given by

$$L(x) = \sum_{i=1}^{m} \sum_{j=1}^{n} [c_{ij} + (k_{ij}/u_{ij})]x_{ij} \leqslant z(x)$$

Let $\bar{x}$ be the optimum extreme point of the transportation problem using the underestimator $L(x)$. If z^* is the optimum objective value of the fixed-charge problem, then

$$L(\bar{x}) \leqslant z^* \leqslant z(\bar{x})$$

Consequently, the determination of $\bar{x}$ automatically provides bounds on the optimum solution to the fixed-charge problem. In this case, the aproximation is good if the lower and upper bounds, $L(\bar{x})$ and $z(\bar{x})$, respectively, are close to each other. This same result was provided by Balinski but in a more laborious manner.

This algorithm can easily be extended to problems in which only one fixed charge k_i is associated with source i regardless of the number of destinations it feeds.

Other heuristic algorithms for the problem are developed by Baumal and Wolfe (1958), Kuehn and Hamburger (1963), Sà (1969), and Shannon and Ignizio (1970).

B. *Exact Algorithm†*

The exact algorithm is concerned with the uncapacitated plant location problem, which is formulated as follows: Let P_j be the set of indices of those plants that can supply destination j, D_i the set of indices of those destinations that can be supplied from plant i, $n_i = |D_i|$ the number of elements in D_i, and x_{ij} the fraction of the total demand at j supplied by plant i. If k_i is the fixed charge for plant i, and assuming that there is no limit on the amount supplied from each plant, the model becomes:

minimize

$$z = \sum_{i,j} c_{ij} x_{ij} + \sum_i k_i y_i$$

† Efroymson and Ray (1966).

subject to

$$\sum_{i \in P_j} x_{ij} = 1, \qquad j = 1, 2, \ldots, n$$

$$\sum_{j \in D_i} x_{ij} \leqslant n_i y_i, \qquad i = 1, 2, \ldots, m$$

$$y_i = (0, 1), \qquad i = 1, 2, \ldots, m$$

$$x_{ij} \geqslant 0, \qquad \text{for all } i \text{ and } j$$

The solution procedure proposed by Efroymson and Ray is based on the Land–Doig algorithm as improved by Dakin (Section 4.3.2-A). The branching occurs on the zero–one variables y_i. The main contribution of Efroymson and Ray is that they developed an efficient method for solving the continuous linear program at each node. Simple tests are also provided for terminating the search along different branches.

To show how the continuous optimum solution at any node is obtained, let S_1 be the associated set of fixed variables at level one, and S_2 the set of free variables. If $y_i^* > 0$ is the optimum continuous value of y_i, then it must be true that optimal x_{ij}^* satisfies

$$\sum_{j \in D_i} x_{ij}^* = n_i y_i^*, \qquad i \in S_2$$

For example, suppose that $\sum_{j \in D_i} x_{ij}^* < n_i y_i^*$; then y_i^* cannot be optimal since it can be reduced to y_i^{**}, so that $\sum_{j \in D_i} x_{ij} = n_i y_i^{**}$ and such that the objective value is reduced. It follows that

$$y_i^* = \sum_{j \in D_i} x_{ij}^*/n_i, \qquad i \in S_2$$

To obtain the values of x_{ij}^*, substitute for y_i^* in the objective function. The problem reduces to:

minimize

$$z = \sum_{i \in S_1} k_i + \sum_{\substack{i \in S_1 \\ j \in D_i}} c_{ij} x_{ij} + \sum_{\substack{i \in S_2 \\ j \in D_i}} [c_{ij} + (k_i/n_i)] x_{ij}$$

subject to

$$\sum_{i \in P_j} x_{ij} = 1, \qquad j = 1, \ldots, n$$

$$x_{ij} \geqslant 0$$

Since all the coefficients of the objective function are nonnegative, optimal x_{ij} are given by

$$
x_{ij}{}^* = \begin{cases} 1, & \text{if } x_{ij} \text{ has the smallest coefficient in the objective} \\ & \text{function, where } i \in (S_1 \cup S_2) \cap P_j \\ 0, & \text{otherwise} \end{cases}
$$

The ties for the smallest objective coefficient can be broken arbitrarily.

It is also possible to develop simple tests that will simplify the search procedure:

Test 1   At a given node, a variable y_i is fixed at level 1 for all the emanating branches if the reduction in variable cost as a result of shipping from plant i exceeds the fixed cost k_i charged for "opening" the ith plant. This can be expressed mathematically as follows:

Since the exact reduction is not known at this point, a lower bound may be used. Define

$$
\Delta_{ij} = \min_{\substack{k \in (S_1 \cup S_2) \cap P_j \\ k \neq i}} \{\max(c_{kj} - c_{ij}, 0)\}, \qquad i \in S_2, \quad j \in D_i
$$

This gives a lower bound on the reduction in variable costs as a result of supplying destination j from plant i rather than from plants k, $k \in (S_1 \cup S_2) \cap P_j$, $k \neq i$. It then follows that $y_i = 1$ if $\sum_{j \in D_i} \Delta_{ij} > k_i$.

Test 2   This test provides a means for reducing the number of elements in D_i $(=n_i)$ as follows. If

$$
\min_{k \in S_1}\{c_{kj} - c_{ij}\} < 0, \qquad i \in S_2, \quad j \in D_i
$$

then it is not promising for plant i to supply destination j, and the element j is dropped from the set D_i. Naturally, if this condition holds for all $j \in D_i$, then $y_i = 0$ for all the emanating branches.

Test 3   This test is related to test 1. Let Δ_{ij} be defined as in test 1 but with $k \in S_1 \cap P_j$ only. Then an upper bound on the cost reduction for opening a plant i is $\sum_{j \in D_i} \Delta_{ij}$. Then $y_i = 0$ if $\sum_{j \in D_i} \Delta_{ij} < k_i$.

Naturally, these three tests are used in conjunction with the regular features of the (improved) Land–Doig algorithm in order to economize on both storage requirements and computation time.

▶**EXAMPLE 7.3-3**   The data of the problem are summarized in Fig. 7-5. The associated elements of the sets P_j and D_i are defined as follows:

$$
D_1 = \{1, 2, 5\}, \qquad D_2 = \{1, 3\}, \qquad D_3 = \{2, 3, 4, 5\}
$$

$$
P_1 = \{1, 2\}, \qquad P_2 = \{1, 3\}, \qquad P_3 = \{2, 3\}, \qquad P_4 = \{3\}, \qquad P_5 = \{1, 3\}
$$

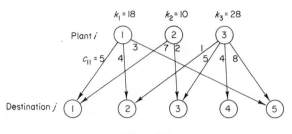

Figure 7-5

The complete model of the problem is given by:

minimize

$$z = 5x_{11} + 4x_{12} + 3x_{15} + 7x_{21} + 2x_{23} + 1x_{32} + 5x_{33}$$
$$+ 4x_{34} + 8x_{35} + 18y_1 + 10y_2 + 28y_3$$

subject to

$$x_{11} + x_{21} = 1$$
$$x_{12} + x_{32} = 1$$
$$x_{23} + x_{33} = 1$$
$$x_{34} \qquad = 1$$
$$x_{15} + x_{35} = 1$$
$$x_{11} + x_{12} + x_{15} \leqslant 3y_1$$
$$x_{21} + x_{23} \qquad \leqslant 2y_2$$
$$x_{32} + x_{33} + x_{34} + x_{35} \leqslant 4y_3$$
$$x_{ij} \geqslant 0 \qquad \text{and} \qquad y_i = (0, 1) \qquad \text{for all } i \text{ and } j$$

The solution tree for the problem is given in Fig. 7-6. The optimal solution occurs at the node yielding $z = 60$. Because of the small number of y-variables, there is no chance for using the simplifying three tests given previously.◀

Apparently Efroymson and Ray have experienced good computational results with their algorithm. They report that a number of 50-plant, 200-destination problems are solved (on the IBM 7094) with an average computation time of 10 minutes each. This success stems from the efficiency by which the linear program at each node is solved.

The preceding model is an improved version of another suggested by Balinski (1964). Balinski's solution method employs Benders' cuts (Section 3.7.3), which leads to solving a sequence of increasingly more complex zero–one linear programs. The method is not feasible computationally.

There is a host of other exact algorithms for the generalized (capacitated) plant location problem in which constraints are specified to limit the capacity of each plant and, moreover, to limit the numer of plants that can be constructed. Space limitations do not allow the presentation of these methods, but the general idea is to establish an efficient enumeration scheme for searching among all the binary (y_i) variables associated with the fixed charges with the objective of discarding most of

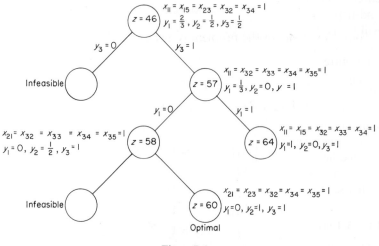

Figure 7-6

these assignments implicitly. The procedure also relies on the fact that a complete specification of the y_i-variables reduces the problem to an ordinary transportation model. Among these are the algorithms of Manne (1964), Feldman *et al.* (1966), Spielberg (1969a, b), P. S. Davis and Ray (1969), Sà (1969), Curry and Skeith (1969), Ellwein and Gray (1971), and Mallette and Francis (1972). There are also other related algorithms, but these are concerned primarily with the selection of the optimal size of the plant [see, for example, J. A. White and Francis (1971)]. These models are not directly related to the integer problem and hence are not treated here.

7.4   Traveling Salesman Problem

Example 1.3-5 introduces the traveling salesman problem as determining the shortest round-trip for visiting n cities provided each city is visited exactly once. A basic assumption for this formulation is that the distance matrix $C = (c_{ij})$ satisfies the so-called *triangular inequality*. This inequality specifies that if c_{ij} is greater than $c_{ij}' = c_{ii_1} + c_{i_1i_2} + \cdots + c_{i_kj}$, then the element c_{ij} should be replaced by c_{ij}' in C. In other words, the element of the matrix C should give the shortest distance between two cities. This can always be achieved by utilizing a shortest-route algorithm (see, for example, Section 7.2.3).

The triangular inequality assumption allows one to specify that each city is visited exactly once. Otherwise, if this inequality is not satisfied, then each city may be visited more than once in order to achieve the shortest round-trip. Since C can always be made to satisfy the triangular inequality, it will be assumed throughout this section that each city is visited exactly once.

The pure zero–one integer model given in Example 1.3-5 is too large to be computationally feasible, even for a moderate value of n. The objective of this section is to introduce different formulations of the problem that are computationally more tractable. These formulations can be divided into two main types: cutting-plane models and branch-and-bound models. Other formulations, which are based on dynamic programming, will not be presented here. An outline of a formulation due to Bellman (1962) is given in Problem 7-16.

This section is based on a survey by Bellmore and Nemhauser (1968).

7.4.1   A Formulation Based on the Assignment Model

The well-known assignment model can be adpated to yield a particularly appealing formulation of the traveling salesman problem. This formulation is the basis for both the cutting-plane and branch-and-bound methods.

Before introducing the formulation, a definition of a tour and a subtour is given. A solution to the traveling salesman problem is said to constitute a *tour* if, starting from an arbitrary city i_0, every city is visited exactly once before returning to i_0. (It is assumed that the distance matrix C satisfies the triangular inequality.) A *subtour* may be defined as a tour comprising k cities, where k is strictly less than n. Thus a subtour is not a feasible solution to the problem.

Let $x_{ij} = 1$ if the salesman travels directly from city i to city j, and $x_{ij} = 0$ otherwise. A necessary condition for a tour is that city i connects to one city only and that city j is reached from exactly one city. This yields the formulation:

minimize

$$z = \sum_{i=1}^{n} \sum_{j=1}^{n} c_{ij} x_{ij}, \qquad \text{where} \quad c_{ij} = \infty \quad \text{for} \quad i = j$$

subject to

$$\sum_{j=1}^{n} x_{ij} = 1, \qquad i = 1, 2, \ldots, n$$

$$\sum_{i=1}^{n} x_{ij} = 1, \qquad j = 1, 2, \ldots, n$$

$$x_{ij} = (0, 1)$$

solution is a tour

The condition $c_{ij} = \infty$ for $i = j$ guarantees that $x_{ii} = 0$, as should be desired in a tour solution.

Except for the requirement that the solution be a tour, the formulation is an assignment model. Unfortunately, there is no guarantee that the optimal solution of the assignment model will be a tour. Most likely such a solution will consist of subtours. The purpose of the algorithms presented next is to show how the tour restriction may be imposed explicitly so that the optimal solution is a tour. One must note that for an n-city problem, there are $(n - 1)!$ possible tours, so that explicit enumeration of all such tours is not practical in general (e.g., $10! = 3,628,800$).

7.4.2   "Cutting-Plane" Methods

The two methods presented here are not truly of the cutting-plane type of Chapter 5. Rather, explicit constraints are imposed that guarantee a tour solution. In this respect, the additional constraints "cut off" the infeasible (subtour) solutions. Actually, the two "methods" are formulations rather than algorithms, since the solution employs a (modified) Gomory cut that takes advantage of the special formulation of the problem.

A.   *Dantzig–Fulkerson–Johnson Method*†

A tour solution is guaranteed if all subtours are eliminated. Define $N = \{1, 2, \ldots, n\}$ and let S be the set of indices representing the cities associated with a subtour. Then, by definition, S is disconnected from $\overline{S} = N - S$, that is, no path exists between any of the cities in S and those in $\overline{S}$. Thus, for $i \in S$ and $j \in \overline{S}$, $x_{ij} = 0$. This means that S is a subtour if and only if

$$\sum_{i \in S} \sum_{j \in \overline{S}} x_{ij} = 0$$

† Dantzig *et al.* (1954, 1959).

Consequently, a subtour S is eliminated if and only if

$$\sum_{i \in S} \sum_{j \in \bar{S}} x_{ij} \geqslant 1$$

There are $2^n - 2$ possible subtours and hence constraints. This creates a very large number of constraints if all such constraints are imposed simultaneously. (For $n = 10$, $2^n - 2 = 1022$.) Also, the optimal linear programming solution of the assignment problem with its subtour restrictions may not be all integer.

What Dantzig *et al.* did in 1954 was to start with a small number of subtour (or loop) restrictions, then add new ones only when a subtour is to be blocked. Certain combinatorial arguments were used to eliminate fractional solutions. This experience led them to develop special cuts that would eliminate fractional solutions, an idea that was later generalized by the developments of Gomory's cuts. Although they succeeded in solving a 42-city problem, apparently the overall behavior of the method was not encouraging.

B.   *Miller–Tucker–Zemlin Method*†

This formulation requires a slightly different definition of the traveling salesman problem. City i_0 (i_0 arbitrary) from which the trip starts is artificially split into two cities, city 0 and city n. The salesman's trip starts at 0 and ends at n. This means that the *closed* tour that starts and ends at i_0 is now replaced by the *open* tour that starts at 0 and ends at n.

Associate with every city i, $i = 0, 1, 2, \ldots, n$, a real number v_i such that $0 \leqslant v_i \leqslant n$. The following constraint eliminates all subtours:

$$v_i - v_j + nx_{ij} \leqslant n - 1, \qquad i = 0, 1, 2, \ldots, n - 1, \quad j = 1, 2, \ldots, n$$

The index i varies from 0 to $n - 1$ because of the definition of open tour. To show that these constraints eliminate (closed) subtours, let cities $1, 2, \ldots,$ and k form a subtour. (Notice that cities 0 and n cannot be in a subtour.) Thus $x_{12} = x_{23} = \cdots = x_{k1} = 1$, and the constraints become

$$v_1 - v_2 \qquad\qquad\qquad \leqslant -1$$
$$v_2 - v_3 \qquad\qquad \leqslant -1$$
$$\ddots$$
$$v_k - v_1 \leqslant -1$$

These constraints are inconsistent because adding the left- and right-hand sides yields $0 \leqslant -k$, which is impossible.

† C. E. Miller *et al.* (1960).

Now we show that an open tour (starting at 0 and ending at n) always satisfies the given constraints. Let $v_i = r$ if city i is the rth city of the tour $(v_0 = 0, v_n = n)$. This condition can be implemented by utilizing the constraints

$$(v_i + 1) - M(1 - x_{ij}) \leqslant v_j \leqslant (v_i + 1) + M(1 - x_{ij}), \qquad \text{all } i \text{ and } j$$

where $M > 0$ is sufficiently large, $v_0 = 0$, and $v_n = n$. If $x_{ij} = 0$, this constraint is redundant but the subtour constraint becomes $v_i - v_j \leqslant n - 1$ $(i \neq n, j \neq 0)$, which will always be satisfied, since by definition $v_i - v_j$ cannot exceed $n - 1$. If $x_{ij} = 1$, that is, city j is reached directly from city i, then the last constraint yields $v_i + 1 \leqslant v_j \leqslant v_i + 1$, or $v_j = v_i + 1$. This automatically satisfies the subtour constraints.

The number of constraints in this formulation is considerably smaller than those in Section A. C. E. Miller *et al.* experimented with the formulation using a Gomory cutting-plane algorithm to eliminate fractional solutions. The computational results were rather discouraging. However, Martin (1966) was able to solve Dantzig's 42-city problem on the IBM 7094 in less than five minutes. His success is probably due to his judicious use of a combination of Dantzig's and Miller's constraints together with an accelerated version of Gomory's cut (see Section 5.2.1-C). Martin discovered that although Dantzig's formulation results in a large number of constraints, Dantzig's constraints are stronger than Miller's. To illustrate this point, suppose a subtour includes cities 1, 2, and 3. The corresponding Miller's constraints when added together give

$$x_{12} + x_{23} + x_{31} \leqslant 3 - 3/n$$

while Dantzig's constraints effectively reduce to

$$x_{12} + x_{23} + x_{31} \leqslant 2$$

which not only is stronger, but also its strength is independent of n. Unfortunately, no general conclusions can be drawn from Martin's experience since he only solved one 42-city problem. But it is also possible that his accelerated cut may be effective with the traveling salesman special formulation.

7.4.3 Branch-and-Bound Algorithms

The three methods to be presented here are based on the assignment model formulation of the traveling salesman problem. Indeed, the problem at each node (of the branch-and-bound algorithm) is essentially an assignment model that modifies the original $n \times n$ assignment problem so that subtours are eliminated.

A.   *Subtour Elimination Algorithm†*

This algorithm is based on the general branch-and-bound principle of Section 4.2. The only details required to complete the algorithm are:

(1)   determination of upper and lower bounds on the optimum objective of the traveling salesman problem; that is, given that z^* is the optimum objective value associated with a tour, then $\underline{z}$ and $\bar{z}$ are determined such that $\underline{z} \leqslant z^* \leqslant \bar{z}$, and

(2)   specifying the exact procedure for branching at each node.

Initially, set $\bar{z} = \infty$. However, if the (feasible) tour $(1, 2, \ldots, n, 1)$ has $c_{12} + c_{23} + \cdots + c_{n1} < \infty$, then $\bar{z}$ may be set equal to this amount. The initial value of $\underline{z}$ is determined by solving the assignment model associated with the (original) traveling salesman problem. If z^0 is the optimal objective value of the assignment problem, then $\underline{z} = z^0$ is a legitimate initial lower bound.

Naturally, if the solution associated with z^0 is a tour, the computations terminate. Otherwise, the given solution consists of at least two subtours. Select the subtour with the least number of cities. Let such a subtour include cities $i_1, i_2, \ldots,$ and i_k. Then $x_{i_1 i_2} = x_{i_2 i_3} = \cdots = x_{i_k i_1} = 1$. Branching is designed so that the assignment problems associated with subsequent nodes emanating from the current node will eliminate this subtour. This can be effected by setting one of the variables $x_{i_1 i_2}, \ldots, x_{i_k i_1}$ equal to zero, which gives rise to k branches or subproblems. In order for each subproblem to remain an assignment model, the condition $x_{ij} = 0$ can be effected by setting $c_{ij} = \infty$ in the C-matrix at the immediately preceding node.

The exact algorithm is summarized as follows: At the tth iteration (node) let z^t be the optimum objective value of the associated assignment problem. Because the lower bound on z^* changes with the node, z^t will automatically define $\underline{z}$ at the tth node. (Notice that $\bar{z}$ is the same for all nodes.)

Step 0   Determine z^0. If the associated solution is a tour, stop. Otherwise, record $\bar{z} = c_{12} + c_{23} + \cdots + c_{n1}$ and $(1, 2, \ldots, n, 1)$ as its associated tour. Set $t = 0$, then go to step 1.

Step 1   Select a subtour solution associated with z^t that has the smallest number of cities and initiate as many branches as the number of x_{ij}-variables at level one that define the subtour. For the (i, j)-branch define a new cost matrix that differs from the one from which it is generated in that $c_{ij} = \infty$. Set $t = t + 1$ and go to step 2.

† Eastman (1958).

Step 2   Select one of the "unbranched" nodes. If none is left, stop; the tour associated with $\bar{z}$ is optimum. Otherwise, go to step 3.

Step 3   Solve the assignment problem associated with the selected node. Three cases will result:

(i)   If $z^t \geq \bar{z}$, then the current node is fathomed since it cannot yield a better tour than that associated with $\bar{z}$. Set $t = t + 1$, then go to step 2.

(ii)   If $z^t < \bar{z}$ and the associated solution is a tour, then set $\bar{z} = z^t$ and record the associated tour as the best solution so far available. Set $t = t + 1$ and go to step 2.

(iii)   If $z^t < \bar{z}$ but the associated solution is not a tour, then set $t = t + 1$ and go to step 1.

▶EXAMPLE 7.4-1   Consider the traveling salesman problem whose C-matrix is given as:

$$C = \begin{pmatrix} \infty & 2 & 0 & 6 & 1 \\ 1 & \infty & 4 & 4 & 2 \\ 5 & 3 & \infty & 1 & 5 \\ 4 & 7 & 2 & \infty & 1 \\ 2 & 6 & 3 & 6 & \infty \end{pmatrix}$$

The solution tree is given in Fig. 7-7. The initial value of $\bar{z} = c_{12} + c_{23} + c_{34} + c_{45} + c_{51} = 10$ and its associated tour is given by $(1, 2, 3, 4, 5, 1)$. C^0 is the same as C above. This yields $z^0 = 8$ and the solution $x_{12} = x_{21} = 1$, $x_{34} = x_{45} = x_{53} = 1$, which consists of two subtours. Since subtour $(1, 2, 1)$ has a fewer number of cities, it is used for branching. This creates the two nodes corresponding to $x_{12} = 0$ (or, equivalently, $c_{12} = \infty$ in the C^0-matrix) and $x_{21} = 0$ (or, equivalently, $c_{21} = \infty$ in the C^0-matrix).

Selecting the node associated with $x_{12} = 0$, the resulting assignment problem yields $z^1 = 9$ with a tour solution $(2, 1, 3, 4, 5, 2)$. Thus $\bar{z} = 9$ and the tour is recorded. Now, the only remaining branch $x_{21} = 0$, where

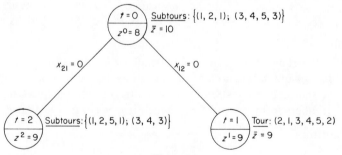

Figure 7-7

C^2 is obtained from C^0 by setting $c_{21} = \infty$, gives $z^2 = 9$. Since $z^2 = \bar{z}$, node 2 cannot produce an improved solution and the search terminates. The optimum solution is $(2, 1, 3, 4, 5, 2)$ with $z^* = 9.$◀

In this example, the selection of the next node to be branched is arbitrary. But as is the case with any branch-and-bound algorithm, it is crucial. Unfortunately, there are no specifics for selecting the next node since the lower bound z^t is determined only by solving an assignment model. This point is taken care of in the next algorithm by developing a simpler estimate of a lower bound.

The reason for selecting the subtour with the smallest number of cities is purely intuitive, as this may lead to generating a smaller number of branches before the algorithm terminates. Generally, however, this may not be true.

Computational experience with this algorithm and its extension was reported by D. Shapiro (1966), who reports considerable difficulty, in general, presumably because an assignment problem is optimized at each node. He further states that a symmetric C-matrix poses a particular difficulty in computations mainly because the number of subtours including two cities is excessive. Shapiro suggests a number of ramifications for this problem.

Another algorithm based on solving an assignment model at each node was developed by Bellmore and Malone (1971). The branching at each node is based on eliminating subtours that will create as many new branches as the number of elements in a subtour, but the conditions defining each branch are different. The computational experience reported shows that the algorithm is successful in solving relatively small-sized problems ($n \leqslant 30$). For larger problems, the ratio of the number of successfully solved problems to those attempted decreases with the increase in n. In addition, the computation time increases at an exponential rate with the increase in n.

Other algorithms were developed by Held and Karp (1970, 1971) for symmetric problems ($c_{ij} = c_{ji}$). Problems with n varying from 20 to 64 were attempted. But it appears that the computation time, as usual, is data dependent, since one 64-city problem was solved in less than 3 minutes, while a 46-city problem took about 15 minutes.

B. *Tour Building Algorithm*†

In the subtour elimination algorithm an assignment model is solved at each node. This may be too costly from the computational viewpoint. In this algorithm, a new branching rule is developed that does not depend

† Little *et al.* (1963).

on solving an assignment model at each node. A lower bound on the optimum objective value z^* of the traveling salesman problem is estimated by using the following simple observation:

Let $p_i = \min_j\{c_{ij}\}$ and $q_j = \min_i\{c_{ij} - p_i\}$, that is, p_i is the minimum entry of the ith row of the C-matrix, while q_j is the minimum entry of column j of the C-matrix after the p_i's are subtracted. Define

$$c_{ij}' = c_{ij} - p_i - q_j$$

It follows by construction that c_{ij}' must be nonnegative for all i and j. Thus

$$z = \sum_i \sum_j c_{ij} x_{ij} = \sum_i c_{ij}' x_{ij} + \sum_i p_i + \sum_j q_j \geqslant \sum_i p_i + \sum_j q_j$$

This means that $\underline{z} = \sum_i p_i + \sum_j q_j$ is a proper lower bound on z.

The branching process is based on the idea of building tours, that is, successively assigning value 1 to proper x_{ij} variables. This must be done while ensuring that no subtours are created by such assignment. The following procedure accomplishes this result. A variable is said to be free if it has no binary assignment in the tree. At any node t, exactly two branches will emanate. A first branch is associated with $x_{kr} = 1$, where x_{kr} is a free variable with $c_{kr} < \infty$. A second branch should naturally correspond to $x_{kr} = 0$. This condition is effected by updating the C^t-matrix so that $c_{kr} = \infty$.

Two conditions must be imposed on the selection of x_{kr} to ensure that no subtours are formed:

(1) By following the branches starting from node t back to the first node of the tree, the variables set to one along these branches should not form a subtour when x_{kr} is set equal to 1; for example, if $n > 3$ and if $x_{12} = 1$ and $x_{23} = 1$ lead to the current node, then $x_{31} = 1$ is not appropriate for creating a new branch.

(2) In order to commemorate that $x_{rk} = 1$ is unacceptable for any branch reached by $x_{kr} = 1$ [otherwise, a subtour (k, r, k) is created], the C^t-matrix at node t is updated by setting $c_{rk} = \infty$.

A note about the lower bound $\underline{z}$ is now in order. Since the matrix C is updated to C^t at node t and since the lower bound is computed from C^t, it is appropriate to designate it by $\underline{z}^t$. Because the branches leading to node t assign value one to some of the variables, this is equivalent to eliminating some of the rows and columns of the original matrix C. For example, if $x_{ij} = 1$, then the size of the C-matrix is reduced by eliminating row i and column j. In general, let $(i_1, i_2), \ldots, (i_{k-1}, i_k)$ be the deleted routes that lead to C^t. Then p_i^t and q_j^t are defined relative to C^t. But since $x_{i_1 i_2} = \cdots = x_{i_{k-1} i_k} = 1$, then a lower bound on z^* at t is given by

$$\underline{z}^t = \left(c_{i_1 i_2} + \cdots + c_{i_{k-1} i_k}\right) + \sum_i p_i^t + \sum_j q_j^t$$

where i and j are defined over the (undeleted) rows and columns of C^t.

The steps of the algorithm are as follows:

Step 0   Determine $\bar{z} = c_{12} + c_{23} + \cdots + c_{n1}$ and record its associated tour $(1, 2, \ldots, n, 1)$. Set $t = 0$ and go to step 1.

Step 1   Select a free variable x_{kr} with $c_{kr} < \infty$ and create two branches at node t, one associated with $x_{kr} = 1$ and $c_{rk} = \infty$, and the other associated with $x_{kr} = 0$, or equivalently $c_{kr} = \infty$. The variable x_{kr} must be proper in the sense that its branch does not lead to the formation of a subtour. Set $t = t + 1$ and go to step 2.

Step 2   Select one of the "unbranched" nodes. If none exists, stop; the tour associated with $\bar{z}$ is optimum. Otherwise go to step 3.

Step 3   Determine $\underline{z}^t$. (i) If $\underline{z}^t \geq \bar{z}$, then the current node is fathomed; set $t = t + 1$ and go to step 2. (ii) If $\underline{z}^t < \bar{z}$ and the branches leading to node t yield a tour, set $\bar{z} = \underline{z}^t$ and record its tour. Set $t = t + 1$ and go to step 2. (iii) If $\underline{z}^t < \bar{z}$ but node t does not yield a tour, then set $t = t + 1$ and go to step 2.

It is important to notice that this algorithm (as compared with the one above) greatly simplifies the computations at each node. However, this simplification is at the expense of producing an enlarged search tree. In particular, a tour is formed only after adding at least $n - 1$ *successive* branches to the tree.

The rule used to select the next node to be branched (step 2) is as follows. When the two branches at a current node are formed, select the node associated with the branch yielding the smaller lower bound. The remaining node is stored in a list in the order it is generated. If the selected node is "fathomed," the stored list is scanned on a last-in–first-out (LIFO) basis. The LIFO rule was devised by Little *et al.* because it proved effective in retrieving stored problems from computer tape units quickly. However, this factor was important in 1963 and with the development of the magnetic disc storage units whose access time is virtually independent of the order in which problems are stored, this rule is by no means essential. In other words, there may be advantages in considering the node having the smallest lower bound in the *entire* tree since this could lead to an improved (tour) solution, thus enhancing the probability of fathoming nodes with large lower bounds.

In the following example, the LIFO rule is not utilized. Rather, the next node is selected as the one having the smallest $\underline{z}$ among all un-branched nodes.

▶**EXAMPLE 7.4-2**   Consider Example 7.4-1. The associated solution tree is given in Fig. 7-8.

Initially, $\bar{z} = 10$ corresponds to $x_{12} = x_{23} = x_{34} = x_{45} = x_{51} = 1$. The lower bound $\underline{z}^0 = 7$ is computed as follows. From the C-matrix, $p_1{}^0 = 0$, $p_2{}^0 = 1$, $p_3{}^0 = 1$, $p_4{}^0 = 1$, $p_5{}^0 = 2$. This leads to the "reduced" matrix:

$$
(c_{ij} - p_i{}^0) =
\begin{vmatrix}
\infty & 2 & 0 & 6 & 1 \\
0 & \infty & 3 & 3 & 1 \\
4 & 2 & \infty & 0 & 4 \\
3 & 6 & 1 & \infty & 0 \\
0 & 4 & 1 & 4 & \infty
\end{vmatrix}
$$

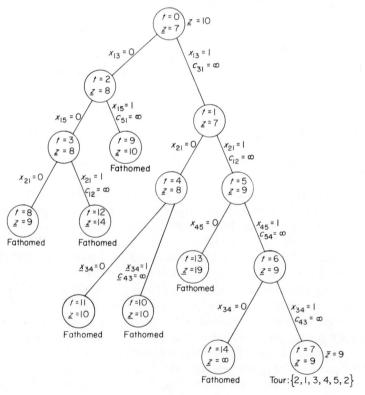

Figure 7-8

From this we get $q_1{}^0 = 0$, $q_2{}^0 = 2$, $q_3{}^0 = 0$, $q_4{}^0 = 0$, $q_5{}^0 = 0$. Hence

$$\underline{z}^0 = \sum_{i=1}^{5} p_i + \sum_{j=1}^{5} q_j = 7$$

This means that the optimal z^* of the traveling salesman problem must satisfy $7 \leqslant z^* \leqslant 10$.

The selection of x_{13} for branching at $t = 0$ is arbitrary. But a useful rule of thumb is to select a "proper" free variable having the smallest c_{ij} among all the free ones.

To illustrate how $\underline{z}$ is computed at some later node, consider $t = 4$. Since $x_{13} = 1$, this is equivalent to deleting row 1 and column 3 from the C^0-matrix. In addition, by setting $c_{31} = \infty$ (that is, blocking the route 3, 1), this yields the C^1-matrix as

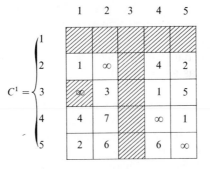

Next C^4 is obtained from C^1 by setting $c_{21} = \infty$. Thus

where $p_1{}^4$ and $q_3{}^4$ are undefined, $p_2{}^4 = 2$, $p_3{}^4 = 1$, $p_4{}^4 = 1$, $p_5{}^4 = 2$, $q_1{}^4 = 0$, $q_2{}^4 = 2$, $q_4{}^4 = 0$, $q_5{}^4 = 0$. This gives

$$\underline{z}^4 = c_{13} + (p_2{}^4 + p_3{}^4 + p_4{}^4 + p_5{}^4) + (q_1{}^4 + q_2{}^4 + q_4{}^4 + q_5{}^4)$$
$$= 0 + (2 + 1 + 1 + 2) + (0 + 2 + 0 + 0) = 8$$

A feasible solution is determined at $t = 7$ when it becomes evident that, by setting $x_{52} = 1$, this node will yield the tour (2, 1, 3, 4, 5, 2) with an improved objective value $z = 9$. This changes the upper bound to $\bar{z} = 9$. It is interesting to note that 7 more nodes were considered before verifying that $t = 7$ yields the optimal.◄

A comparison of Figs. 7-7 and 7-8 reveals that the simplicity of computing lower bounds $\underline{z}$ is achieved at the expense of enlarging the tree. Thus, while the computation time consumed in optimizing the assignment model at each node is reduced drastically, this may be offset by the increase in the number of nodes. In addition, it would appear that this algorithm might pose a problem due to excessive requirement of computer storage.

Little *et al.* report solving a 42-city problem by their algorithm. However, this is not sufficient information to draw general conclusions.

In both the subtour elimination and tour building algorithms, the determination of a good initial upper bound $\bar{z}$ is very crucial in effecting early fathoming of nodes. Yet neither one of the two algorithms provides an effective method for determining a *good* initial upper bound. In other words, a good upper bound is only produced as a byproduct of the basic steps of the algorithm.

The best way so far available for producing a good initial upper bound is to use a heuristic. Lin (1965) gives an approximate method that appears promising. Others include Karg and Thompson (1964), Krolak *et al.* (1971), and Webb (1971). No computational experience is reported about the impact of utilizing such heuristics with the above branch-and-bound algorithms, but one would expect that combining of the heuristic and exact methods would lead to an improved algorithm.

7.5   Set Covering Problem

The set covering model is introduced in Example 1.3-6 as an application to the delivery problem, the airline crew scheduling problem, and graphs. Problems 1-8 and 1-16 cite applications of the model to political districting and information retrieval. Other applications include line balancing (Salveson, 1955), switching theory (Hohn, 1955; J. P. Roth, 1958), and capital investment (Valenta, 1969), among others.

The mathematical definition of the set covering (SC) problem is given as:

minimize

$$z = \{cy \,|\, Ey \geqslant e, \, y = (0, 1)\} \tag{SC}$$

where $c = (c_j)$ is an n-vector, $E = (e_{ij})$ an $m \times n$ matrix whose elements are zeros and ones, and e an m-column vector of ones. It is assumed that $c_j > 0$ for all j, since if $c_j \leqslant 0$, then y_j can be set equal to one, which makes the satisfied constraints redundant.

If all the inequalities in the SC problem are replaced by equations, the resulting problem is called a *set partitioning* (SP) problem. For example, in the airline crew scheduling problem (Example 1.3-6), all constraints are equations if each flight leg must be assigned to exactly one crew.

The discussion in the following sections concentrates primarily on the properties and algorithms of the SC problem. This will be followed by a brief discussion on the properties and algorithms of the SP problem.

The material in this section is based partly on a survey article by Garfinkel and Nemhauser (1972a).

7.5.1 Useful Properties of the SC Problem†

Consider the following problem:

minimize

$$x_0 = \{cx \,|\, Ex \geqslant e, x \geqslant 0\} \tag{LPC}$$

In this problem all the variables are continuous. However, the main constraints $Ex \geqslant e$ are the same as in the set covering problem. The problem is thus a linear program associated with the SC problem (henceforth referred to as LPC) and possesses important properties that relate directly to the SC problem.

PROPERTY I An optimal feasible solution of LPC automatically satisfies $x_j \leqslant 1$ for all j.

This property is proved as follows. If $x_j > 1$ is feasible, then so is $x_j = 1$. Because $c_j > 0$ for all j, it follows that $x_j > 1$ cannot be optimal.

PROPERTY II Every extreme point solution of LPC must satisfy $0 \leqslant x_j \leqslant 1$.

This property follows directly from Property I, which states that an optimum solution to LPC satisfies $0 \leqslant x_j \leqslant 1$. From linear programming theory such a solution must be an extreme point. But since the optimum solution to LPC can occur at any of its extreme points depending on the gradient of the objective function, Property II must be true.

In preparation for the remaining properties, define $S = Ex - e$, and for a given solution $x = \bar{x}$ let $J = \{j \,|\, x_j = 1\}$ and $K = \{k \,|\, S_k > 0\}$. The solution $\bar{x}$ is an extreme point if and only if the columns $E_j, j \in J$, and the columns $I_k, k \in K$, of the m-identity matrix I are linearly independent. Further, define $K_j = \{k \,|\, e_{kj} = 1\}$ and $H = \{j \in J \,|\, K_j \subset K\}$. Thus the next property is given as follows:

† Lemke *et al.* (1971), Glover (1971b), Bellmore and Ratliff (1971a).

PROPERTY III   If x' is feasible for LPC and $H \neq \emptyset$, then for each $j \in H$ there is a feasible solution $x'' \leq x'$ such that $x_j'' < x_j'$.

Let $\theta_j = \min\{S_k' \mid k \in K_j\}$ where S_k' is the kth slack given x'. (If $K_j = \emptyset$, $\theta_j = \infty$.) It follows that feasibility is maintained if for $j \in H$, x_j' can be decreased by at most $\min\{\theta_j, x_j'\}$.

PROPERTY IV   If $\bar{x}$ is an integer feasible solution for the SC problem, then $\bar{x}$ is an extreme point of LPC if and only if $H = \emptyset$.

The "only if" part is proved as follows. Multiples of the I_k, $k \in K$, can be used to transform E to $\bar{E}$ containing all zeros in the rows $k \in K$. In particular, the columns $\bar{E}_j$ of $\bar{E}$ for all $j \in H$ become null vectors. Hence the matrix associated with $\bar{x}$ cannot form a basis.

The "if" part follows by noting that because $\sum_{j \in J} e_{kj} \bar{x}_j = 1$ for all $k \notin K$, there is exactly one $j \in J - H$ such that $e_{kj} = 1$ for each $k \notin K$. Hence the columns E_j, $j \in J - H$, are always linearly independent. This means that if $H = \emptyset$, all positive $\bar{x}_j$ are associated with independent E_j, $j \in J$.

Property IV reveals the following powerful result: If the SC problem has a (integer) feasible solution that is not an extreme point of LPC, then an *improved* SC solution can be obtained by successively applying Property III until $H = \emptyset$, in which case the resulting SC solution becomes an extreme point of LPC. This extreme point solution is superior (in terms of the objective value) to any other nonextreme point solution of SC since it is effected by reducing the level of some variables from one to zero. This emphasizes the importance of LPC in the development of an enumeration algorithm for the SC problem (see next section) since the optimal SC solution must also be an extreme point of LPC. Indeed, the following property elaborates on how LPC can be used to find an initial feasible solution for the SC problem.

PROPERTY V   Let x' be a feasible solution of LPC, and let $\bar{x}$ be the "round-up" solution obtained by setting x_j equal to the smallest *integer* greater than or equal to x_j', for all j. Then, for each j such that $0 < x_j' < 1$, the solution obtained by decreasing the jth component of $\bar{x}$ to zero level is integer feasible.

This property follows because $x_j' < \bar{x}_j$ for noninteger x_j'. Thus $\bar{S}_k > 0$ for all $k \in K_j$, where $j \in H$. Property III then shows that $\bar{x}_j$ can be reduced to zero level.

By applying Property V until $H = \emptyset$, the variables at level one in the resulting $\bar{x}$ vector define the basic variables associated with an extreme point. Generally, in order to complete the basis some of the slack variables must be basic. In order to determine such variables, the following procedure is used. Let $J^* = \{j|\bar{x}_j = 1\}$, where $\bar{x}_j$ is the jth component of $\bar{x}$. Furthermore, define $I_r = \{i|e_{i_r} = 1$ and $e_{ij} = 0$ for all $j \in J^* - r\}$, $r \in J^*$.

    (a)   Select S_k as a basic variable if, given $\bar{x}_j$, $j \in J^*$, S_k is positive; and,

    (b)   For every $r \in J^*$ such that $|I_r| \geqslant 2$, select S_k, $k \in \bar{I}_r$ as basic variables, where $\bar{I}_r$ is the same as I_r less any one element of I_r; that is, $\bar{I}_r = I_r - \{p\}$, where $p \in I_r$.

This procedure adds the proper number of basic slack variables. It remains now to show that the resulting matrix forms a basis. The following property proves this and also provides additional characteristics of the basic matrix that will be useful in developing a cutting plane algorithm for the SC problem (see next section).

PROPERTY VI Given that B is the matrix associated with the solution obtained by the preceding properties, if $\hat{B}$ is obtained from B through suitable permutations of rows, then $\hat{B}^{-1}$ exists and, moreover, $\hat{B}^{-1} = \hat{B}$.

It is always possible to rearrange the rows of B so it will appear as

$$\hat{B} = \left(\begin{array}{c|c} I_{p \times p} & 0 \\ \hline D & -I_{q \times q} \end{array}\right)$$

where $I_{p \times p}$ and $I_{q \times q}$ are identity matrices, and p is the number of unit *row* vectors appearing in B, while q is the number of slacks in the basic solution. Notice that the "x" basic variables are associated with the first p columns while the slack basic variables are associated with the remaining q columns. From the structure of $\hat{B}$, it is clear that $\hat{B}^{-1} = \hat{B}$. The matrix B is thus said to be *involutory*.

▶**EXAMPLE 7.5-1** Consider the SC problem whose information is given as follows:

$$C = (5, 7, 8, 6, 10, 9, 12, 6, 11)$$

$$E = \begin{pmatrix} 1 & 0 & 1 & 0 & 1 & 1 & 0 & 0 & 1 \\ 1 & 0 & 0 & 1 & 1 & 0 & 0 & 1 & 1 \\ 0 & 1 & 1 & 0 & 0 & 0 & 1 & 0 & 1 \\ 1 & 1 & 1 & 0 & 1 & 0 & 1 & 1 & 0 \\ 0 & 0 & 0 & 1 & 0 & 0 & 0 & 0 & 0 \\ 0 & 1 & 1 & 0 & 0 & 0 & 0 & 1 & 0 \\ 0 & 1 & 0 & 0 & 1 & 1 & 0 & 1 & 1 \end{pmatrix}$$

The optimal solution to the LPC problem is given by

$$x' = (0, \tfrac{1}{2}, \tfrac{1}{2}, 1, 0, \tfrac{1}{2}, 0, 0, 0)$$

The "round-up" solution is then given as

$$\bar{x} = (0, 1, 1, 1, 0, 1, 0, 0, 0)$$

where $J = (2, 3, 4, 6)$.

Now, given $\bar{x}$, the slack vector is

$$S = (1, 0, 1, 1, 0, 1, 1)$$

Hence

$$K = \{1, 3, 4, 6, 7\}, \qquad K_2 = \{3, 4, 6, 7\}$$
$$K_3 = \{1, 3, 4, 5\}, \qquad K_4 = \{2, 5\}, \qquad K_6 = \{1, 7\}$$

From the definition of K and K_j, $j \in J$, it follows that $H = \{2, 6\}$. This means that either x_2 or x_6 can be decreased to zero level.

Set $x_6 = 0$. This will yield $H = \varnothing$. Thus,

$$x = (0, 1, 1, 1, 0, 0, 0, 0)$$

is an extreme point solution with x_2, x_3, x_4 basic variables, which yields $J^* = \{2, 3, 4\}$.

To determine the associated basic slacks, notice that S_3, S_4, and S_6 are positive and hence must be basic. Next, compute I_r, $r \in J^*$. Thus

$$I_2 = \{7\}, \qquad I_3 = \{1\}, \qquad I_4 = \{2, 5\}$$

Since $|I_4| = 2$, either S_2 or S_5 can be made basic. Selecting S_2 arbitrarily, it follows that the basic variables identifying the above extreme points are

$$(x_2, x_3, x_4, S_2, S_3, S_4, S_6) = (1, 1, 1, 0, 1, 1, 1)$$

The basic matrix B associated with the above basic solution is now given as

$$B = \begin{pmatrix}
0 & 1 & 0 & 0 & 0 & 0 & 0 \\
0 & 0 & 1 & -1 & 0 & 0 & 0 \\
1 & 1 & 0 & 0 & -1 & 0 & 0 \\
1 & 1 & 0 & 0 & 0 & -1 & 0 \\
0 & 0 & 1 & 0 & 0 & 0 & 0 \\
1 & 1 & 0 & 0 & 0 & 0 & -1 \\
1 & 0 & 0 & 0 & 0 & 0 & 0
\end{pmatrix}$$

The involutory matrix $\hat{B}$ is then obtained from B as

$$\hat{B} = \left(\begin{array}{ccc|cccc}
1 & 0 & 0 & 0 & 0 & 0 & 0 \\
0 & 1 & 0 & 0 & 0 & 0 & 0 \\
0 & 0 & 1 & 0 & 0 & 0 & 0 \\
\hline
0 & 0 & 1 & -1 & 0 & 0 & 0 \\
1 & 1 & 0 & 0 & -1 & 0 & 0 \\
1 & 1 & 0 & 0 & 0 & -1 & 0 \\
1 & 1 & 0 & 0 & 0 & 0 & -1
\end{array} \right)$$

The order of the basic variables in relationship to the columns of $\hat{B}$ remains the same as in B since the rows only are permutated.◀

7.5.2 Algorithms for the SC Problem

This section primarily presents two types of algorithms for the SC problem: a cutting plane method and an implicit enumeration method. Before presenting these methods, however, procedures for eliminating some of the rows and columns in the E-matrix of the SC problem are developed. This should prove useful since it ultimately reduces the number of both variables and constraints. Next a heuristic is introduced for finding a good solution for the SC problem. This is particularly important since it improves the performance of the implicit enumeration algorithm.

A. *Elimination Tests*†

E1 The SC problem is feasible if and only if every row of the matrix E is not null. The usefulness of this elimination may not be apparent when applied to the SC problem since it implies a logical inconsistency in the construction of the model. But as will be shown later, the implicit enumeration algorithms deal with subproblems whose E-matrix is a subset of the E-matrix for the SC problem. Under such conditions, this elimination may be applicable.

E2 If for rows p and q $(p \neq q)$, $e_{pj} \geqslant e_{qj}$, for all j, then row p may be deleted since constraint q is satisfied by a fewer number of variables at level one.

E3 If in the pth row, $e_{pj} = 1$ for $j = k$ and $e_{pj} = 0$ for $j \neq k$, then $x_k = 1$ in every feasible solution and the pth row can be dropped. Also, if $e_{ik} = 1$ for any i, then row i can be deleted. (This also follows from E2.)

E4 Given a set of columns R, if for column k, $\sum_{j \in R} e_{ij} \geqslant e_{ik}$, for all i, and $\sum_{j \in R} c_j \leqslant c_k$, then column k may be deleted.

▶**EXAMPLE 7.5-2** Consider the E-matrix of Example 7.5-1. Assume that the C-vector is changed to

$$C = (1, 3, 10, 4, 4, 3, 10, 8, 15)$$

The elimination rules apply as follows. Each new elimination test is applied to the *updated* E matrix resulting from implementing all the preceding eliminations, if any:

E2: $e_{4j} \geqslant e_{6j}$, for all j; delete row 4.
E4: $c_1 + c_2 < c_3$ and $e_{i1} + e_{i2} \geqslant e_{i3}$, for all i; delete column 3.

---

† Garfinkel and Nemhauser (1972a).

E3:   Row 5 is a unit vector with $e_{54} = 1$. Thus $x_4 = 1$; delete row 5. Since $e_{24} = 1$, delete row 2 also.

E4: $c_2 < c_7$ and $e_{i2} \geqslant e_{i7}$, for all i; delete column 7.

E4: $c_1 + c_2 = c_5$ but $e_{i1} + e_{i2} \geqslant e_{i5}$, for all i; delete column 5.

E4: $c_1 + c_2 < c_9$ and $e_{i1} + e_{i2} \geqslant e_{i9}$, for all i; delete column 9.

E2: $e_{7j} \geqslant e_{6j}$, for all j; delete row 7.

E2: $e_{6j} \geqslant e_{3j}$, for all j; delete row 6.

E4: $c_1 < c_6$ and $e_{i1} = e_{i6}$, for all i; delete column 6.

The original E-matrix thus reduces to

$$E = \begin{matrix} 1 \\ 3 \end{matrix} \begin{pmatrix} \overset{x_1}{1} & \overset{x_2}{0} \\ 0 & 1 \end{pmatrix}$$

The associated c-vector is $(1, 3)$. This shows that $x_1 = x_2 = 1$. In this case the elimination tests happen to produce the optimal solution $x_1 = x_2 = x_4 = 1$ with all the remaining variables equal to zero. ◀

B.   *Approximate Solution for the SC Problem*

Properties I–V (Section 7.5.1) are the basis for this heuristic. The idea is that given a nonextreme point $(H \neq \varnothing)$ integer feasible solution, an extreme point solution can be extracted by reducing some $x_j, j \in H$, to zero level and repeating the process until $H = \varnothing$.

Let the superscript t represent the tth iteration of the heuristic. At the tth iteration, a variable $x_j, j \in H^t$ (currently equal to 1) is reduced to zero level if it is associated with the largest *cost per oversatisfied constraint*. The kth constraint is said to be oversatisfied if its slack variable S_k is positive. An empirical measure of the cost per oversatisfied constraint is

$$d_j = c_j / \sum_{i \in K^t} e_{ij}, \qquad j \in H^t$$

where $K^t = \{i \,|\, S_i{}^t > 0\}$. Notice that $S_i{}^t$ is computed given $\bar{x}^t$, where $\bar{x}^t$ is the vector of nonzero $\bar{x}_j$ at the tth iteration.

The steps of the algorithm are thus given as follows:

Step 0   Solve the LPC and assuming that the optimum solution x' is noninteger obtain the round-up solution $\bar{x}^0$; that is, for every $x_j{}'$ such that $0 < x_j{}' < 1$, obtain $x_j = \bar{x}_j{}^0 = 1$. Compute S_i for all i, given $\bar{x}^0$. Set $t = 0$ and go to Step 1.

Step 1   Determine H^t. If $H^t = \varnothing$, stop; $x^* = \bar{x}^t$ is the required extreme point. Otherwise, go to Step 2.

Step 2 Compute d_j, $j \in H^t$ and determine j^* such that $d_{j^*} = \max\{d_j | j \in H^t\}$. (If there is more than one j corresponding to max d_j, break the tie by selecting j^* such that $c_{j^*} = \max\{c_j | j \in H^t\}$. If the tie persists, break it arbitrarily.) Obtain $\bar{x}^{t+1}$ by setting $x_{j^*} = 0$ in $\bar{x}^t$. Compute S_i^{t+1} given $\bar{x}^{t+1}$, and go to Step 1.

▶**EXAMPLE 7.5-3** The data of the SC problem are:

$$C = (4, 2, 6, 4, 3)$$

$$E = \begin{vmatrix} 1 & 0 & 1 & 1 & 0 \\ 0 & 1 & 0 & 1 & 1 \\ 0 & 0 & 1 & 0 & 1 \\ 1 & 1 & 1 & 1 & 1 \end{vmatrix}$$

The optimal solution to the LPC problem is

$$x' = (0, 0, \tfrac{1}{2}, \tfrac{1}{2}, \tfrac{1}{2})$$

so that the round-up solution is given by

$$\bar{x}^0 = (0, 0, 1, 1, 1)$$

The slack vector associated with $\bar{x}^0$ is $S^0 = (1, 1, 1, 2)$. Hence

$$K^0 = (1, 2, 3, 4)$$

and

$$K_3{}^0 = (1, 3, 4), \qquad K_4{}^0 = (1, 2, 4), \qquad K_5{}^0 = (2, 3, 4)$$

so that $H^0 = (3, 4, 5)$.

Since $d_3 = \tfrac{6}{3}$, $d_4 = \tfrac{4}{3}$, and $d_5 = \tfrac{3}{3}$, it follows that x_3 is reduced to zero level. This yields $\bar{x}^1 = (0, 0, 0, 1, 1)$. Thus $S^1 = (0, 1, 0, 1)$ and $K^1 = (2, 4)$. Since $K_4{}^1 = (1, 2, 4)$ and $K_5{}^1 = (2, 3, 4)$, it follows that $H^1 = \emptyset$ and $\bar{x}^1 = (0, 0, 0, 1, 1)$ is the solution obtained by the heuristic.◀

The preceding heuristic is based on Property IV, which is due to Glover (1971b). Lemke *et al.* (1971) develop a different procedure, which selects the variables to be set equal to one from among those variables that have positive values in the optimal LPC solution. The selection is based on a slightly different empirical measure, where a variable is set equal to one if it yields the *smallest* cost per *satisfied* constraint (compare with the definition of d_j given in the preceding heuristic).

C. *Implicit Enumeration Algorithm*†

This algorithm utilizes the same enumeration scheme presented in Section 3.4. The fathoming tests are based primarily on solving a linear program of the LPC type at each node.

†Lemke *et al.* (1971).

Let J_t be the current partial solution so that the variables x_j, $j \notin J_t$, are *free*. Define $\text{LPC}(J_t)$ as its associated linear program; that is, given the SC problem as

$$\min z = \left\{ \sum_{j=1}^{n} c_j y_j \,\middle|\, \sum_{j=1}^{n} e_{ij} y_j \geq 1, \, i = 1, 2, \ldots, m, \, y_j = (0, 1), \text{ for all } j \right\}$$

then $\text{LPC}(J_t)$ is

$$\min z_t = \left\{ \sum_{j \notin J_t} c_j x_j + \sum_{j \in J_t^+} c_j \,\middle|\, \sum_{j \notin J_t} e_{ij} x_j \geq 1 - \sum_{j \in J_t^+} e_{ij}, \right.$$

$$\left. i = 1, 2, \ldots, m, \, x_j \geq 0 \right\}$$

where J_t^+ is a subset of J_t such that for $j \in J_t^+$, $x_j = 1$. Notice that if for any i, $\sum_{j \in J_t^+} e_{ij} \geq 1$, then the ith constraint of $\text{LPC}(J_t)$ is redundant and should be deleted. If all the constraints are redundant, then the obvious solution to $\text{LPC}(J_t)$ is $x_j = 0$ for all $j \notin J_t$.

Initially, $J_0 = \varnothing$. However, a possible J_0 can be obtained by applying the above heuristic to $\text{LPC}(\varnothing)$. The fathoming of partial solution J_t is effected as follows. Let $z_{\min}$ be the current best objective value (initially, $z_{\min} = \infty$ if $J_0 = \varnothing$, or $z_{\min} = \sum_{j \in J_0^+} c_j$ if $J_0 \neq \varnothing$). Solve $\text{LPC}(J_t)$ and let z_t^* be its associated optimal objective value. Then J_t is fathomed if:

(i) $\text{LPC}(J_t)$ has no feasible solution,
(ii) $\text{LPC}(J_t)$ has a feasible *noninteger* solution and $z_t^* \geq z_{\min} - 1$, or
(iii) $\text{LPC}(J_t)$ has a feasible *integer* solution but $z_t^* \geq z_{\min}$.

Suppose J_t is not fathomed; then a new variable must be augmented at level one to the current partial solution. A reasonable choice is to select the variable having the *smallest* fractional value in the optimal solution to $\text{LPC}(J_t)$. The justification here is that this may cause the largest alteration in $\text{LPC}(J_{t+1})$ as compared with $\text{LPC}(J_t)$, thus leading to new information. Observe that before $\text{LPC}(J_{t+1})$ is considered, it may be worthwhile to apply the heuristic to the solution of $\text{LPC}(J_t)$ to see if it produces an improved $z_{\min}$ for the SC problem. Keep in mind, however, that although the heuristic finds an extreme point solution for $\text{LPC}(J_t)$, this, together with J_t, may not be an extreme point solution for $\text{LPC}(\varnothing)$. Thus it may be necessary to reduce the overall solution to an extreme point of $\text{LPC}(\varnothing)$, which in turn should yield a lower objective value.

The termination of the algorithm occurs in the same manner given in Section 3.4. In this case $z_{\min}$ and its associated partial solution give the optimal solution.

Lemke *et al.* give further ramifications of the algorithm that include, for example, reducing the size of the E matrix associated with $LPC(J_t)$ by using the eliminations in Section 7.5.2-A. They also recommend that if J_{t+1} is reached from J_t by setting $x_{j*} = 1$, and if J_{t+1} is fathomed, then J_{t+2} is obtained from J_{t+1} by setting $x_{j*} = 0$. The objective here is that $LPC(J_{t+2})$ may be drastically different from $LPC(J_{t+1})$. This is actually equivalent to using the LIFO rule, which is automatically employed in Glover's enumeration scheme (Section 3.4).

▶**EXAMPLE 7.5-4**

$$C = (5, 7, 8, 10, 9, 12, 6, 11)$$

$$E = \begin{pmatrix} 1 & 0 & 1 & 1 & 1 & 0 & 0 & 1 \\ 0 & 1 & 1 & 0 & 0 & 1 & 0 & 1 \\ 1 & 1 & 1 & 1 & 0 & 1 & 1 & 0 \\ 0 & 1 & 1 & 0 & 0 & 0 & 1 & 0 \\ 0 & 1 & 0 & 1 & 1 & 0 & 1 & 1 \end{pmatrix}$$

The optimum continuous solution [that is, that of $LPC(\varnothing)$] is given by $x' = (0, \frac{1}{2}, \frac{1}{2}, 0, \frac{1}{2}, 0, 0, 0)$, and the round-up solution is $\bar{x} = (0, 1, 1, 0, 1, 0, 0, 0)$. Using the heuristic, this is reduced to the extreme point solution $(0, 1, 1, 0, 0, 0, 0, 0)$ with $z = 15$.

Let $J_0 = \{2, 3\}$; thus $z_0^* = 15 = z_{min}$. Because J_0 is feasible for the SC problem, all the constraints of $LPC(J_0)$ are redundant. Thus the optimal solution to $LPC(J_0)$ gives all zero variables, which means that J_0 has no better completions and hence is fathomed; backtrack.

Consider $J_1 = \{2, -3\}$. Then $LPC(J_1)$ is given by:

minimize

$$z_1 = 5x_1 + 10x_4 + 9x_5 + 12x_6 + 6x_7 + 11x_8 + 7$$

subject to

$$x_1 + x_4 + x_5 + x_8 \geqslant 1$$
$$x_1, \ldots, x_8 \geqslant 0$$

The optimal solution is $x_1 = 1$ with all the remaining $x_j = 0$ and $z_1^* = 12$. The solution $x_1 = 1$, together with $x_2 = 1$ as given by $J_1 = (2, -3)$, happens to be an extreme point of $LPC(\varnothing)$. [This is checked as usual by showing that for $LPC(\varnothing)$, the solution $x_1 = x_2 = 1$ and $x_3 = \cdots = x_8 = 0$ produces $H = \varnothing$.] Since $z_1^* < z_{min}$, let $z_{min} = z_1^* = 12$. J_1 is fathomed; backtrack.

Consider $J_2 = \{-2\}$. Thus $LPC(J_2)$ is given by:

minimize

$$z_2 = 5x_1 + 8x_3 + 10x_4 + 9x_5 + 12x_6 + 6x_7 + 11x_8$$

subject to

$$
\begin{pmatrix}
1 & 1 & 1 & 1 & 0 & 0 & 1 \\
0 & 1 & 0 & 0 & 1 & 0 & 1 \\
1 & 1 & 1 & 0 & 1 & 1 & 0 \\
0 & 1 & 0 & 0 & 0 & 1 & 0 \\
0 & 0 & 1 & 1 & 0 & 1 & 1
\end{pmatrix}
\begin{pmatrix}
x_1 \\ x_3 \\ x_4 \\ x_5 \\ x_6 \\ x_7 \\ x_8
\end{pmatrix}
\geq
\begin{pmatrix}
1 \\ 1 \\ 1 \\ 1 \\ 1
\end{pmatrix}
\qquad x_1, x_3, \ldots, x_8 \geq 0
$$

The optimal solution is $x' = (0, \frac{1}{2}, 0, 0, 0, 0, \frac{1}{2}, \frac{1}{2})$ with $z_2{}^* = 12.5$. Since x' is noninteger and $z_2{}^* > z_{\min} - 1$, J_2 is fathomed and the enumeration is complete. The optimal solution is $z^* = 12$, $x_1 = x_2 = 1$, with all the remaining variables equal to zero.◀

The computational experience reported by Lemke *et al.* (1971) indicates that the algorithm is very efficient. Indeed, they indicate that, according to their experience, the first application of the heuristic usually produces a very tight bound that effectively terminates the enumeration rapidly. For example, a problem with 134 constraints and 1642 variables took only 25 minutes to solve on the IBM 360/50. This appears quite remarkable as compared with the regular integer programming codes. The authors also indicate that the effectiveness of the algorithm may be enhanced by employing a more sophisticated linear programming code since the fathoming is primarily dependent on the speed of solving $LPC(J_t)$. In this respect, the speed of solving $LPC(J_t)$ must be data dependent, since a lower density E-matrix is likely to yield a faster solution.

D. *"Cutting-plane" Algorithm*†

This algorithm is not a cutting-plane method in the sense given in Chapter 5. Rather, the cuts are constraints that are constructed so that those integer feasible solutions previously considered cannot be encountered again. These cuts are in the form of regular SC constraints, and hence their augmentation to the original problem will again result in an SC-type problem.

The development of the cut is based on the idea that a current integer feasible solution cannot be encountered again if it is stipulated that every new feasible solution must yield an *improved* objective value. This idea is further enhanced by Property IV, where it is shown that the optimal SC solution must be an extreme point of the LPC problem. This means that the theory of linear programming can be used advantageously to derive the required cut.

† Bellmore and Ratliff (1971a).

Let B be the basic matrix associated with an extreme point solution of LPC and assume that c_B is the basic cost vector associated with B. From the theory of linear programming, the extreme point solution is given by $x^* = B^{-1}e$, where e is a vector of all ones. The objective equation of LPC expressed in terms of the current extreme point is given by

$$z = c_B B^{-1}e - \sum_{j \in NB} (c_B B^{-1}E_j - c_j)x_j$$

where NB represents the nonbasic x_j only (no slacks are included) and E_j is the jth column of the matrix E. At the current extreme point all nonbasic $x_j, j \in NB$, are zero and the associated objective value $z^* = c_B B^{-1}e$. (The reason for excluding nonbasic slacks is justified below.)

Now any other *feasible* solution x^0 to LPC will yield an *improved* objective value z^0 (as compared with $z^* = c_B B^{-1}e$) if

$$z^0 - z^* = - \sum_{j \in NB} (c_B B^{-1}E_j - c_j)x_j^{0} < 0$$

(Note that this is a necessary but not sufficient condition.) This is equivalent to saying that *at least* one of the variables $x_j, j \in NB$, having $c_B B^{-1}E_j - c_j > 0$ must be positive. But since one is interested in integer solutions only, this can be translated to the condition

$$\sum_{j \in Q} x_j \geqslant 1$$

where

$$Q = \{j \in NB \mid c_B B^{-1}E_j - c_j > 0\}$$

The definition of Q ignores the fact that some slacks are nonbasic. The next lemmas utilize the involutory property of $\hat{B}$, which is obtained by permutating the rows of B (Property V, Section 7.5.1), to show that Q can never include slacks. They also show how Q can be determined by simple computations.

▶**LEMMA 1** Given c_B as defined above, then $c_B \hat{B}^{-1} = c_B$. The proof follows from the definition of $\hat{B}^{-1}$. Partition c_B to $(c_B{}^x, c_B{}^S)$, where $c_B{}^S = 0$ is associated with the basic slack variables. Thus

$$c_B \hat{B}^{-1} = (c_B{}^x, 0) \left(\begin{array}{c|c} I & 0 \\ \hline D & -I_q \end{array} \right) = (c_B{}^x, 0) = c_B$$

Notice that there are q slack variables, and hence the number of elements in $c_B{}^S$ is equal to q.◀

▶**LEMMA 2**   Let $\hat{E}$ be the matrix E with the same row permutation used to obtain $\hat{B}$ from B. Then $c_B B^{-1} E_j = c_B \hat{E}_j$.

Since the same rearrangements are made in the rows of E_j to obtain $\hat{E}_j$ as were made in the rows of B to obtain $\hat{B}$, and since this permutation in the rows of B results in a corresponding interchange in the columns of its inverse, it follows that $B^{-1} E_j = \hat{B}^{-1} \hat{E}_j$. Thus $c_B B^{-1} E_j = c_B \hat{B}^{-1} \hat{E}_j = c_B \hat{E}_j$ by Lemma 1. ◀

Lemma 2 reveals that no slack variables can be included in the set Q. This follows because $z_j - c_j$ for any nonbasic slack variable is $c_B(-e_j)$, which is always negative.

The algorithm is now presented. Let LPC^0 be the LPC problem defined in Section 7.5.1 and define LPC^t as LPC^0 after t cuts are augmented.

Step 0   Let $z^* = \infty$ and let x^* be the vector that keeps track of the best solution vector. Set $t = 0$ and go to Step 1.

Step 1   Find an integer extreme point solution for LPC^t (using the heuristic given in Section 7.5.2-B). If $z^t < z^*$, let $z^* = z^t$ and record the corresponding solution vector in x^*. Go to Step 3.

Step 2   Determine Q^t. If $Q^t = \varnothing$, terminate; (x^*, z^*) gives the optimal solution. Otherwise, go to Step 3.

Step 3   Form LPC^{t+1} by augmenting the constraint $\sum_{j \in Q^t} x_j \geq 1$ to LPC^t. Set $t = t + 1$ and go to Step 1.

The finiteness of the algorithm is guaranteed since the tth cut ensures that the *integer* extreme point solution of LPC^t is excluded in LPC^k, $k \geq t + 1$, and there are at most 2^n such solutions. Notice that it is conceivable to generate as many as 2^n cuts before the algorithm terminates.

▶**EXAMPLE 7.5-5**   Consider the problem of Example 7.5-4, which defines LPC^0. A first extreme point solution is shown to be $x^0 = (0, 1, 1, 0, 0, 0, 0, 0)$ with $z^0 = 15$. Set $x^* = x^0$ and $z^* = z^0$. The basis associated with x^0 is constructed according to Property V and is given by

$$
B^0 = \begin{pmatrix}
x_2 & x_3 & S_2 & S_3 & S_4 \\
0 & 1 & 0 & 0 & 0 \\
1 & 1 & -1 & 0 & 0 \\
1 & 1 & 0 & -1 & 0 \\
1 & 1 & 0 & 0 & -1 \\
1 & 0 & 0 & 0 & 0
\end{pmatrix}
$$

The involutory basis $\hat{B}^0$ is obtained by interchanging the rows of B^0 as follows: 5 to 1, 1 to 2, 2 to 3, 3 to 4, and 4 to 5 (see Property VI). The same permutation is applied to the rows of E to obtain $\hat{E}$. Thus $c_B \hat{E}_j - c_j$, $j \in NB$, is computed as follows:

$$(7, 8, 0, 0, 0) \begin{pmatrix} 0 & 1 & 1 & 0 & 1 & 1 \\ 1 & 1 & 1 & 0 & 0 & 1 \\ 0 & 0 & 0 & 1 & 0 & 1 \\ 1 & 1 & 0 & 1 & 1 & 0 \\ 0 & 0 & 0 & 0 & 1 & 0 \end{pmatrix} - (5, 10, 9, 12, 6, 11)$$

$$\begin{array}{cccccc} x_1 & x_4 & x_5 & x_6 & x_7 & x_8 \end{array}$$

$$= (3, 5, 6, -12, 1, 4)$$

so that

$$Q^0 = \{1, 4, 5, 7, 8\}$$

and the first cut is

$$x_1 + x_4 + x_5 + x_7 + x_8 \geqslant 1 \qquad (Cut\ 1)$$

Thus LPC^1 is formulated by augmenting this cut to LPC^0. The continuous solution of LPC^1 is

$$x^1 = (1, 1, 0, 0, 0, 0, 0, 0), \qquad z^1 = 12$$

which obviously must be an extreme point solution. Since $z^1 < z^*$, set $z^* = z^1$ and $x^* = x^1$.

The basis B^1 is now given as

$$\begin{array}{ccccccc} & x_1 & x_2 & S_1 & S_3 & S_4 & S_5 \end{array}$$

$$B^1 = \begin{pmatrix} 1 & 0 & -1 & 0 & 0 & 0 \\ 0 & 1 & 0 & 0 & 0 & 0 \\ 1 & 1 & 0 & -1 & 0 & 0 \\ 0 & 1 & 0 & 0 & -1 & 0 \\ 0 & 1 & 0 & 0 & 0 & -1 \\ 1 & 0 & 0 & 0 & 0 & 0 \end{pmatrix}$$

The number of rows in B^1 is increased by one over those of B^0 because of the additional cut. (Notice that there is more than one B^1 depending on the choice of the basic zero slacks. This in turn may have an effect on the subsequent calculations and hence the speed of convergence of the algorithm. There is no easy way to check which basis is better, but intuitively the basis that results in a *smaller* number of elements in Q may be preferred since it may result in a "stronger" cut. Recall that the cut is of the form $\sum_{j \in Q} x_j \geqslant 1$, so that the smaller the number of elements in Q, the more restrictive is the cut.)

$\hat{B}^1$ is obtained by interchanging the rows of B^1 as follows: 6 to 1, 1 to 3, 3 to 4, 4 to 5, and 5 to 6. For $j = 3, 4, \ldots, 8$, $c_B \hat{E}_j - c_j$ is given as

$$(7, 5, 5, 7, 5, 12) - (8, 10, 9, 12, 6, 11) = (-1, -5, -4, -5, -1, 1)$$

so that $Q^1 = \{8\}$ and the second cut is

$$x_8 \geqslant 1 \qquad (cut\ 2)$$

Augmenting cut 2 to LPC^1 to obtain LPC^2, the solution of LPC^2 yields the solution

$$x^2 = (0, 0, 0, 0, 0, 0, 1, 1), \qquad z^2 = 17$$

Since $z^2 > z^*$, x^* and z^* remain unchanged.

The basis B^2 associated with x^2 is given by:

$$B^2 = \begin{array}{c} \begin{array}{ccccccc} x_7 & x_8 & S_1 & S_2 & S_4 & S_5 & S_6 \end{array} \\ \begin{pmatrix} 0 & 1 & -1 & 0 & 0 & 0 & 0 \\ 0 & 1 & 0 & -1 & 0 & 0 & 0 \\ 1 & 0 & 0 & 0 & 0 & 0 & 0 \\ 1 & 0 & 0 & 0 & -1 & 0 & 0 \\ 1 & 1 & 0 & 0 & 0 & -1 & 0 \\ 1 & 1 & 0 & 0 & 0 & 0 & -1 \\ 0 & 1 & 0 & 0 & 0 & 0 & 0 \end{pmatrix} \end{array}$$

$\hat{B}^2$ is obtained from B^2 by interchanging the rows as follows: 3 to 1, 7 to 2, 1 to 3, 2 to 4, 4 to 5, 5 to 6, and 6 to 7. Computing $c_B \hat{E}_j - c_j$ for $j = 1, 2, \ldots, 6$, we have

$$(6, 6, 6, 6, 0, 6) - (5, 7, 8, 10, 9, 12) = (1, -1, -2, -4, -9, -6)$$

so that $Q^2 = \{1\}$, and the third cut is

$$x_1 \geqslant 1 \qquad (cut\ 3)$$

Augmenting cut 3 to LPC^2, the solution of LPC^3 yields

$$x^3 = (1, 0, 0, 0, 0, 0, 1, 1), \qquad z^3 = 22$$

which again leaves z^* and x^* unchanged. x^3 produces the fourth cut

$$x_2 + x_3 \geqslant 1 \qquad (cut\ 4)$$

The solution of LPC^4 together with the auxiliary computations of $c_B \hat{E}_j - c_j$, $j \in NB$, show that $Q = \emptyset$. Thus the algorithm terminates, and $x^* = (1, 1, 0, 0, 0, 0, 0, 0)$ and $z^* = 12$ is the optimal solution. ◀

It is interesting to notice that cut t is constructed based on comparing the objective value of future feasible solutions with z^t rather than with z^*, the best available objective value. This is an inherent weakness in the construction of the cut since it does not take advantage of the best information so far available. In this respect, the cut is "memoryless," a situation illustrated in this example where cut 3 is developed with respect to $z^2 = 17$ rather than with respect to $z^* = 12$. Perhaps the behavior of this cut as compared with those developed in Chapter 5 shows the important difference between the two methods, and consequently the reason for not classifying the present algorithm as a truly cutting method:

(i)   In Chapter 5, a cut guarantees that the (minimization) objective value is always monotone increasing or decreasing (depending on whether the cut is dual or primal, respectively).

(ii) In the cutting methods, the optimal solution is not available until the algorithm terminates, while in the present algorithm the optimal solution may be available (see the preceding example), but several additional cuts may be needed to effect termination. In this respect, the present algorithm behaves more as a branch-and-bound method.

This example shows that LPC^t is obtained by adding t successive cuts to LPC^0. This could result in a severe taxation of the computer memory due to the increase of the size of the problem. The computational experience reported by Bellmore and Ratliff indicates that this may actually be a serious problem. Some of the problems they experimented with added several hundred cuts without finding an improved feasible solution. (Recall that the cut represents only a necessary but not sufficient condition for improving the objective value.) They indicate that the storage problem may be improved if a method is devised by which "redundant" constraints are eliminated [cf. the cutting-plane methods, Chapter 5, where the number of additional constraints (or cuts) is limited]. But it is not apparent that this could be accomplished since redundancy has to be established relative to newly constructed cuts. It thus appears proper to conclude that this algorithm is not as effective as the enumeration algorithm.

E.   *Other Algorithms*

Other developments for the SC problem include the work of Lawler (1966). The algorithm is enumerative and relies on taking advantage of certain dominance properties of the columns of the E-matrix. There is no computational experience reported on this method.

House *et al.* (1966) developed an algorithm that may be classified as enumerative. It proposes the solution of successive problems so that the feasible solutions of a current problem constitute a subset of the immediately preceding one. They report that a problem with $m = 81$ and $n = 499$ was reduced to an (81×180) problem by precomputational analyses (see Section 7.5.2-A), which was then solved in about 75 seconds on the IBM 7094.

Martin (1963) has developed a special type of accelerated cut for the set covering problem (see Section 5.2.1-C) using the dual cutting-plane method (Chapter 5). Apparently the application was very successful since problems with thousands of variables were reported solved in a relatively short time.

Along the same line of cutting-plane algorithms, Salkin and Koncal (1973) developed a code for the SC problem that operates as follows: Perform the dual simplex on LPC so long as the pivot element is equal to 1 (observe that LPC automatically has a starting *dual feasible* solution). If the pivot element is other than 1, develop a Gomory all-integer cut (Section 5.2.2) and augment it to the problem. This will always guarantee that all the coefficients will remain integral. The experience reported by the authors indicates that the algorithm is efficient for relatively small problems. They suggest that it may then be imbedded in some enumerative algorithms where small subproblems may arise in the course of computations. This, however, does not seem to provide a real advantage unless the basic enumerative scheme is itself efficient.

7.5.3 Set Partitioning Problem

The set partitioning (SP) problem is given as:

minimize

$$z = \{cx \,|\, Ex = e, x = (0, 1)\} \qquad\qquad \text{(SP)}$$

where c, E, and e are as defined in the SC problem (Section 7.5). The following discussion presents the basic relationships between the SC and SP problems. A brief survey of the SP algorithms is also given.

A. *Elimination Tests*†

Some of the elimination tests of the SC problem are directly applicable to the SP problem.

A cross reference is thus used to avoid repetition.

P1 Same as E1 (Section 7.5.2-A).

P2 Same as E2.

P3 If P2 shows that the pth constraint can be dropped because the qth constraint has $e_{qj} \leqslant e_{pj}$ for all j, then any column k in which $e_{pk} = 1$ and $e_{qk} = 0$ can be deleted. This follows because there must exist another column

---

† Garfinkel and Nemhauser (1969).

$r \neq k$ such that $e_{qr} = e_{pr} = 1$. Thus, if $x_r = 1$, both rows p and q are satisfied simultaneously, while if $x_k = 1$ only row p is satisfied. The latter cannot be optimal.

P4 Same as E3.

P5 If $x_k = 1$ must be satisfied in every feasible solution (this is usually established by P4), then any other column $p \neq k$ for which $e_{ip} = e_{ik} = 1$ for any i can be eliminated. This follows from all the SP constraints being equations so that x_p must be zero.

P6 Same as E4 except that the inequalities $\sum_{j \in R} e_{ij} \geq e_{ik}$ for all i is replaced by strict equations.

▶**EXAMPLE 7.5-6** Consider Example 7.5-2 but as an SP problem:

P2 $e_{4j} \geq e_{6j}$ for all j; delete row 4.
P3 Delete columns 1, 5, and 7.
P4 $x_4 = 1$, delete row 5. Row 2 is deleted because $e_{24} = 1$.
P5 Delete columns 8 and 9.
P2 Rows 3 and 6 are identical, delete row 6.

The reduced E matrix is thus given as

$$E = \begin{matrix} & x_2 & x_3 & x_6 & \\ & \begin{pmatrix} 0 & 1 & 1 \\ 1 & 1 & 0 \\ 1 & 0 & 1 \end{pmatrix} & & \begin{matrix} 1 \\ 3 \\ 7 \end{matrix} \end{matrix}$$

and the associated c-vector is $(3, 10, 3)$. It is clear from the E-matrix that the SP problem cannot have a feasible solution.◀

B. *SC Equivalence of a Feasible SP†*

Every SP problem *having a feasible solution* can be converted to an equivalent SC problem. Equivalence here implies that both problems have the same optimum solution(s), even though SC may have a more relaxed solution space.

The procedure for conversion is as follows. Let $c = (c_j)$ and $E = (e_{ij})$ define the parameters of the SP problem. Define $h_j = \sum_{i=1}^{m} e_{ij}$ and choose any $d > \sum_{j=1}^{n} c_j$. Then an equivalent SC problem is given by:

minimize

$$z^* = \{c^*x \mid Ex \geq e, x = (0, 1)\}$$

where $c^* = (c_j^*)$ and $c_j^* = c_j + dh_j$.

† Lemke *et al.* (1971).

The proof (Garfinkel and Nemhauser, 1972a) shows that a solution that is feasible for SC but not for SP cannot yield an objective value smaller than the best feasible solution of SP. Let x^p be a feasible solution for SP and let x^c be feasible for SC but not for SP. By definition,

$$\sum_{j=1}^{n} e_{ij} x_j^c \geq 1$$
$$= 1 + \Delta_i$$

where $\Delta_i \geq 1$ for at least one i. Otherwise, if $\Delta_i = 0$ for all i, then the SC and SP feasible solutions are the same and the conversion is trivial.

Now

$$\sum_{j=1}^{n} c_j^* x_j^c = \sum_{j=1}^{n} c_j x_j^c + d \sum_{j=1}^{n} h_j x_j^c$$
$$\geq d \sum_{i=1}^{m} \left(\sum_{j=1}^{n} e_{ij} x_j^c \right) = d \sum_{i=1}^{m} (1 + \Delta_i) \geq d(m + 1)$$

On the other hand,

$$\sum_{j=1}^{n} c_j^* x_j^p = \sum_{j=1}^{n} c_j x_j^p + d \sum_{i=1}^{m} \left(\sum_{j=1}^{n} e_{ij} x_j^p \right)$$
$$= \sum_{j=1}^{n} c_j x_j^p + dm \leq \sum_{j=1}^{n} c_j + dm < d + dm = d(m + 1)$$

This shows that $c^* x^c > c^* x^p$, which means that a feasible solution to SC only cannot be optimal for SP.

Notice that while every SC problem is feasible, the same is not true for SP problems. But in applying the preceding conversion procedure, it may be difficult to determine in advance whether SP is infeasible. It is thus necessary after solving the equivalent SC to make sure that the resulting optimal solution is feasible for the SP problem.

C.   *Algorithms for SP Problem*

Because the SP problem has a more restricted feasible space than in the SC problem, it is possible to develop specialized SP algorithms that may be more efficient computationally. Several such algorithms were developed, for example, those of Pierce (1968), Garfinkel and Nemhauser (1969), Pierce and Laskey (1970), and Marsten (1974). All these algorithms are of the implicit enumeration type. Unfortunately, the reported computational results show that the solution time is very data dependent. In particular, the density of the matrix E is an important factor. But while Garfinkel and Nemhauser report more favorable results as the density increases, Marsten's algorithm seems to work successfully for small density only. However, this result is not general, since the reported computations indicate that problems with the same size and same density may still exhibit drastic differences in solution time. Apparently the specific distribution of the elements of E has a direct effect on the complexity of the problem.

7.6 Concluding Remarks

This chapter has presented a number of specialized algorithms for certain well-known integer models. Although the new algorithms do not resolve the computational difficulties completely, they do present the advantage that could mean the difference between being able to solve problems of certain sizes and the impossibility of solving such problems.

It is interesting to observe that almost all the specialized algorithms in this chapter that can be classified as promising are mainly of the branch-and-bound type. Those that are classified as cutting plane do not really utilize the cutting methods advanced in Chapters 5 and 6. Perhaps this reflects the general conclusion that cutting methods cannot be employed separately to produce effective computational results for integer programming problems. However, as will be shown in Section 8.3, ideas from cutting methods can be used to enhance the effectiveness of branch-and-bound algorithms.

Problems

7-1 A factory produces standard 50-in. wide paper reels. The following (nonstandard) order must be filled:

Width (in.)	Number of reels
11	100
12	200
35	150

The objective is to minimize the trim loss. Obtain a starting basis of three legitimate cutting patterns. Then by using the method in Section 7.2.1, define the knapsack problem that can be used to generate the most promising pattern associated with the given basis.

7-2 Consider the following equations:

$$
\begin{aligned}
2x_1 + 3x_2 \quad\quad + x_4 &= 10 \\
x_1 + x_2 + x_3 \quad\quad &= 5 \\
3x_2 + x_3 + 2x_4 &= 8 \\
x_j \geqslant 0 \text{ and integer for all } j
\end{aligned}
$$

(a)   Apply Mathews' Theorem 7-1 for aggregating the three equations.
(b)   Solve the same problem by using Glover's Theorem 7-2 and compare the results.

7-3   (Bradley, 1971c).   Given the integer program:

maximize

$$z = cx$$

subject to

$$Ax = b$$
$$0 \leqslant x \leqslant d, \qquad x \quad \text{integer}$$

where x is an n-column vector and A is an $m \times n$ matrix. The elements of c, A, b, and d are integer constants. Prove that the problem can be replaced by the following equivalent knapsack problem:

maximize

$$z' = hx$$

subject to

$$p \leqslant hx \leqslant q$$
$$0 \leqslant x \leqslant d, \qquad x \quad \text{integer}$$

where h is an integer vector and p and q are integer constants.

7-4   Consider the following knapsack problem:

maximize

$$z = 18x_1 + 14x_2 + 8x_3 + 4x_4$$

subject to

$$15x_1 + 12x_2 + 7x_3 + 4x_4 + x_5 \leqslant 45$$
$$x_1, \ldots, x_5 \geqslant 0 \quad \text{and integers}$$

Solve the problem by (a) the shortest route algorithm, (b) the dynamic programming algorithm, and (c) the enumeration algorithm. Compare the three methods.

7-5   Solve the following zero–one knapsack problem by the ranking procedure:

minimize

$$z = x_1 + 2x_2 + 3x_3 - 5x_4 + 6x_5$$

subject to

$$-x_1 + 5x_2 + 3x_3 + 3x_4 + x_5 \geqslant 8$$

$$x_1, \ldots, x_5 = 0 \quad \text{or} \quad 1.$$

7-6 (Gilmore and Gomory, 1966). Consider the knapsack problem:

maximize

$$z = \sum_{j=1}^{n} c_j x_j$$

subject to

$$\sum_{j=1}^{n} a_j x_j \leqslant b$$

$$x_j \geqslant 0 \quad \text{and integer} \quad \text{for all } j$$

where $c_j \geqslant 0$, $a_j > 0$, and $b > 0$ are all integer constants. Define

$$f_k(y) = \max\left\{ \sum_{j=1}^{k} c_j x_j \;\middle|\; \sum_{j=1}^{k} a_j x_j \leqslant y, \, 0 \leqslant k \leqslant n, \, 0 \leqslant y \leqslant b \right\}$$

Show that the knapsack problem can be solved by considering the following recursive equation:

$$f_k(y) = \max\{ f_{k-1}(y), f_k(y - a_k) + c_k \}, \qquad k = 1, 2, \ldots, n$$

Solve Problem 7-4 by using this algorithm. (*Hint:* Compare with the algorithm in Section 6.4.2.)

7-7 In the knapsack problem defined in Problem 7-6, let $r_j = c_j/a_j$, $j = 1, 2, \ldots, n$, and assume that $r_1 \geqslant r_2 \geqslant \cdots \geqslant r_n$. Prove that x_1 must be positive in the optimal solution if

$$b \geqslant r_1 a_1/(r_1 - r_2)$$

In this case, the dynamic programming formulation in Problem 7-6 implies that

$$f_n(b) = c_1 + f_n(b - a), \qquad \text{for} \quad b > r_1 a_1/(r_1 - r_2)$$

Next, for b large enough as given immediately above, define

$$g(b) = r_1 b - f_n(b)$$

which is the difference between the optimum objective values of the knapsack problem without and with the integer restriction on the variables. Show that

$$g(b - a_1) = g(b)$$

and hence $g(b)$ is periodic with period a_1.

7-8 (Faaland, 1973). The knapsack problem in Problem 7-3 can be converted so that all the elements of the vector h are positive by using a proper upper bound substitution. Then the problem can be written as

maximize

$$z = \{hx \mid hx \leqslant r, \ 0 \leqslant x \leqslant d, \ x \text{ integer}\}$$

Consider

$$\sum_{j=1}^{k} h_j x_j = i, \qquad i = 1, 2, \ldots, r, \quad k = 1, \ldots, n$$

$$0 \leqslant x_j \leqslant d_j, \quad x_j \text{ integer}, \quad j = 1, \ldots, k$$

Let $f_n(i)$ be the number of solutions to the above system given n and i and define

$$x_k'(i) = \min\left\{d_k, \left[\frac{i}{h_k}\right]\right\}, \qquad k = 1, \ldots, n, \quad i = 0, 1, \ldots, r$$

Thus a solution x_k to the above system assumes one of the values $0, 1, \ldots, x_k'(i)$. Show that

$$f_k(i) = f_{k-1}(i), \qquad\qquad\qquad\qquad i = 0, 1, \ldots, h_k - 1$$

$$f_k(i + h_k) = \begin{cases} f_{k-1}(i + h_k) + f_k(i), & 0 \leqslant i \leqslant h_k d_k \\ f_{k-1}(i + h_k) + f_k(i) - f_{k-1}(i - h_k d_k), & h_k d_k \leqslant i \leqslant r - h_k \end{cases}$$

Then using this information develop a procedure for solving the (value-independent) knapsack problem defined above.

7-9 The fixed-charge problem can be written as a mixed zero–one problem as shown in Example 1.3-7. Show how Benders' decomposition approach (Section 2.5.6) can be applied to this problem.

7-10 Consider the following fixed-charge problem:

minimize

$$z = 2x_1 + 5x_2 + 6x_3$$

subject to

$$2x_1 + 3x_2 + x_3 + x_4 \qquad = 180$$
$$x_1 + 2x_2 + 2x_3 \qquad + x_5 = 150$$
$$3x_1 + 2x_2 + x_3 \qquad = 120$$
$$x_1, \ldots, x_5 \geqslant 0$$

where $(k_1, k_2, \ldots, k_5) = (1000, 500, 400, 20, 10)$. Solve the problem by (a) the heuristic of Section 7.3.2-A, (b) the exact algorithm of Section 7.3.2-B, and (c) the exact algorithm of Section 7.3.2-C.

7-11 Resolve Example 7.3-2 by using the improved linear underestimator

$$L(x) = \sum_{j=1}^{n} [c_j + (k_j/u_j)]x_j$$

where u_j is determined from the feasible region of the problem.

7-12 (Murty, 1968). Solve the following fixed-charge transportation problem by using the extreme point ranking procedure (cf. Section 7.2.3-D):

	1	2	3	4	a_i
1	6 16	8 13	0 12	3 6	23
2	35 17	4 40	5 15	1 8	26
3	9 19	11 90	24 8	16 29	38
b_j	22	9	35	21	

Key:

k_{ij}	
	c_{ij}

7-13 Find an approximate solution to Problem 7-12 by using Balinski's method (Section 7.3.3-A).

7-14 (Balinski, 1964). A simplified (earlier) version of Efroymson and Ray's plant location model (Section 7.3.3-B) is given as follows:

minimize

$$z = \sum_i \sum_j c_{ij} x_{ij} + \sum_i k_i y_i$$

subject to

$$\sum_i x_{ij} \geq 1$$
$$0 \leq x_{ij} \leq y_i, \qquad y_i = 0 \quad \text{or} \quad 1 \quad \text{for all } i$$

The linear programming solution to this problem may not always be integer. Develop Benders' cut for this problem and give an economic interpretation of this cut in terms of the plant location problem.

7-15 Solve the following traveling salesman problem by the branch-and-bound methods of Section 7.4.3:

$$C = \begin{pmatrix} \infty & 10 & 10 & 25 & 25 \\ 1 & \infty & 2 & 10 & 15 \\ 8 & 9 & \infty & 10 & 20 \\ 10 & 8 & 24 & \infty & 15 \\ 14 & 10 & 25 & 27 & \infty \end{pmatrix}$$

7-16 (R. Bellman, 1962). In the traveling salesman problem, let $f_{k-1}(i_m|i_1, \ldots, i_{m-1}, i_{m+1}, \ldots, i_{k-1})$ be the length of the shortest path that starts at node 1, passes through $(i_1, \ldots, i_{m-1}, i_{m+1}, \ldots, i_{k-1})$ and terminates at node i_m. Show that a shortest path from node 1 to node j that passes through $i_1, \ldots, i_{k-1}$ can be determined from

$$f_k(j|i_1, \ldots, i_{k-1}) = \min_{m=1, \ldots, k-1} \{f_{k-1}(i_m|i_1, \ldots, i_{m-1}, i_{m+1}, \ldots, i_{k-1}) + c_{i_m j}\}$$

where the calculations start from

$$f_2(j|i_1) = c_{1i_1} + c_{i_1 j}, \qquad \text{for all } i_1, \quad j \neq 1, \quad i_1 \neq j$$

and terminates when $f_n(1|i_1, \ldots, i_{n-1})$ is computed. Discuss the computational difficulties associated with this formulation and, in particular, the computer storage requirements.

7-17 Prove that the procedure given with Property V, Section 7.5.1, adds the right number of basic slack variables.

7-18 In Example 7.5-1, suppose C is changed to

$$C = (1, 5, 14, 3, 1, 2, 18, 5, 16)$$

Find an integer feasible solution to the problem by using the round-up solution to find an extreme point solution. Determine the associated involutory matrix.

7-19 Apply the elimination tests of Section 7.5.2-A to Problem 7-18.

7-20 Solve Problem 7-18 by using the heuristic in Section 7.5.2-B. How does this solution compare with the exact solution?

7-21 Solve Problem 7-18 by the implicit enumeration algorithm of Section 7.5.2-C.

7-22 (Bellmore and Ratliff, 1971a). Solve the following SC problem by the cutting-plane method (Section 7.5.2-D):

$$C = (2, 1, 1, 3)$$

$$E = \begin{vmatrix} 1 & 1 & 0 & 1 \\ 0 & 1 & 1 & 1 \\ 1 & 0 & 1 & 1 \end{vmatrix}$$

7-23 Suppose Problem 7-22 is treated as an SP problem. Find its SC equivalence.

7-24 (Garfinkel and Nemhauser, 1969). Prove the following elimination test for the SP problem. Suppose the two *row* vectors R_i and R_k (of E) are not comparable, that is, it is not true that $R_i \geqslant R_k$ or $R_k \geqslant R_i$. Let

$K = \{j | e_{kj} > e_{ij}\}$ and $I = \{j | e_{kj} < e_{ij}\}$. If there exists an R_r with $e_{rj} = 1$ for all $j \in K$ and $e_{rj} = 1$ for at least one $j \in I$, say $j = t$, then $x_t = 0$.

7-25 (Guha, 1973). Prove that the test in Problem 7-24 can be replaced by the following uniformly stronger test. If $x_t = 1$ causes a number of columns of E to be eliminated such that at least one row vector R_i becomes null, then $x_t = 0$.

CHAPTER 8

Computational Considerations in Integer Programming

8.1 Introduction

In the preceding chapters, computational experiences with the different integer algorithms were cited frequently. These experiences may serve as some guide for comparing the relative performances of the algorithms, but one must observe that this information is of limited usefulness. The basic issue here is that these experiences are often based on sample problems that may not be totally representative of the common types encountered in applications. A major limitation is the size of the problem. Almost invariably, reported computations apply to relatively small-sized problems and although one method of solution may be superior to another when applied to these sample problems, this does not say how effective this relative improvement is when applied to problems of larger sizes. There may well be relative improvements, but the resulting computational times may still be excessively too long.

The basic disadvantage of comparing different computational experiences is that generally there are no common grounds for the comparison. For example, the computational times are usually dependent on the type of the computer used, the selection of the parameters for the problem, as well as the skill of the programmer coding the algorithm. What is even more frustrating is that an algorithm may perform successfully on certain problems, but would prove a complete failure when only some of the parameters are changed slightly or even when the variables or constraints are rearranged [see Trauth and Woolsey (1969)]. Such inconsistency in the efficacy of the integer programming algorithms shows that it is misleading to rely solely on the computational experiences in selecting a code for solving an integer problem.

In view of this, the author does not find it necessary to elaborate further on the computational experiences reported in the literature. The results cited throughout the book should suffice in this respect. The objective of the remainder of this chapter is to develop general strategies that will allow the users to select a code (or a combination of codes) that will most probably be effective in solving his problem. These strategies are developed based on the theoretical aspects of the different methods presented in the preceding chapters. The ultimate goal is the development of some kind of "composite" algorithm that includes all (or most of) the desired properties in the different algorithms presented in this book. These properties are incorporated in one code so that when a certain method ceases to perform "satisfactorily," the deadlock can be broken by invoking a different strategy. For example, the use of penalties in the branch-and-bound algorithms (Section 4.3.2-B) is known to lose its effectiveness when the current continuous optimum is dual degenerate (has alternative solutions). This difficulty can be eliminated by applying a proper cutting plane of the types presented in Chapter 5. Of course, criteria must be developed for "measuring" when a certain course of action would cease to be "satisfactory." These details will be included in Section 8.3.

While much concern has been expressed in the literature about the efficiency of an integer *code*, there seems to be a general unawareness of the fact that it may be the poor formulation of the model (rather than the efficiency of the code) that is responsible for inadequate performance. In Section 8.2, a number of guidelines are provided as to the particular properties that are desired in formulating an integer model. The objective, of course, is to make the model less complex from the computational standpoint.

The reader must not be deluded into believing that the development in this chapter will solve the computational problem of integer codes once and for all. The only guarantee here is an improved chance that the integer problem will be solved.

8.2   Model Formulation in Integer Programming

As can be seen from the developments in the preceding chapters, the computational complexity of an integer model is affected primarily by the increase in the number of integer variables. In cutting-plane algorithms, a cut is weakened as the number of integer variables increases, since it must satisfy more integer restrictions and relationships. In branch-and-bound (and implicit enumeration) methods, the increase in the number of integer variables generally signifies an increase in the number of branches in the tree, and hence the storage space and computation time.

This observation indicates that in formulating a mathematical model the number of integer variables should be reduced to a minimum. However, there are also other factors that, for all practical purposes, have the same effect on computation as the increase in the number of variables. A list of the important factors that should be taken into account in modeling integer problems is given next.

(i) *"Think noninteger" first.*   Operations researchers realize that there is usually more than one way to formulate a model. Some formulations may not even involve integer variables at all and should thus be given first priority. A typical example is illustrated by the knapsack problem (Section 7.2). The problem can be solved as a shortest-route model, a dynamic programming model, or as some form of a direct integer programming model. Of course, all these formulations are equivalent. But in general the shortest-route model (or the dynamic programming model) is known to produce more consistent results. This means that at least one can predict in advance whether a given problem size can be handled by such methods. There are also many transportation-type problems that can be formulated as direct integer models typically involving thousands of variables, but that can be reformulated as network models whose solution time takes no more than a few minutes (possibly seconds) on the computer.

(ii) *Approximate certain integer variables by continuous ones.*   In some models, the integer variables may be of the direct type (see the classification in Section 1.2.1), that is, they represent a number of men, machines, etc. In this case, if the optimal value of such variables is expected to be large, then by allowing these variables to be continuous, the error resulting from taking the nearest integer to the optimum continuous value will most likely not be appreciable.

(iii) *Restrict the feasible range for integer variables.*   The (needless) increase in the feasible range of an integer variable has the same general effect as the increase in the number of variables. This is seen especially in the branch-and-bound method, where the increase in the feasible ranges will generally allow the creation of more branches. Also, in converting a regular integer model to a binary model, the number of zero–one variables is directly dependent on the size of the feasible range.

One way of alleviating this situation is to impose proper restricting (upper and lower) bounds on the integer variables. Another possibility is to introduce more constraints relating *integer* variables, with the net effect of reducing the feasible space.

(iv) *Avoid the use of auxiliary binary variables.* In Section 1.3.2, examples of "transformed" models are introduced that originally do not include integer variables. However, for the sake of securing a manageable formulation, auxiliary variables (typically binary) are utilized as an integral part of the model. An interesting observation on some of these formulations is that during the course of computation, the optimum solution to the original problem may be available, but additional computation may still be needed to force the (extraneous) binary variables to assume integer values. But it is generally difficult to predict when computations should be stopped prior to natural termination. The main conclusion then is to avoid using such variables whenever possible. A typical example is illustrated by the fixed-charge problem (Example 1.3-7). In Section 7.3, algorithms are developed for tackling the problem directly without utilizing the auxiliary variables.

(v) *Avoid nonlinearity.* In Example 1.3-3, by redefining the zero–one variables used, it is possible to obtain a completely linear model. Non-linearity should be avoided whenever possible since it has the same general effect of increasing the number of variables. Indeed, in Section 3.5.3, it was shown that the polynomial zero–one algorithm can be converted to a linear model at the expense of adding new variables (and constraints).

(vi) *Reduce the magnitudes of the parameters.* In most cutting-plane algorithms, it is required that all the parameters (in the pure integer models) be all integer. This restriction can be satisfied at the expense of increasing the magnitudes of the different parameters, but the problem is that such an increase may have a direct effect on the machine round-off error. On the other hand, it was seen in the cutting-plane methods that the effectiveness of the algorithms is hampered by the increase in the magnitude of the absolute value of the determinant associated with the current basis. The increase in the magnitude of the parameters may then increase the absolute value of the determinant and hence decrease the efficiency of the algorithms. One possible way of overcoming this difficulty is to scale down these parameters as much as is practically possible without severely affecting the accuracy of the model.

(vii) *"Organize" the data.* In the implicit enumeration algorithms (Chapter 3), the variables are usually scanned from left to right and the constraints are examined from top to bottom. As evident from the nature of the fathoming tests, the specific arrangement of the variables and the constraints may have a direct effect on the efficacy of the tests. For example, under minimization, a desirable arrangement is that the variables be listed in ascending order of their objective coefficients. The constraints, on the other hand, may be arranged so that the most restrictive constraints (e.g., multiple choice constraints or constraints expressing complex logical relationships) are the first to be examined. This illustrates the importance of "organizing" the data in a way that will enhance the effectiveness of the fathoming tests.

8.3   A Composite Algorithm

The long computational experience with the cutting methods and their variants (Chapters 5 and 6) shows clearly that these methods yield erratic and unrealiable performance, even for small-sized problems. Only search methods (branch-and-bound and implicit enumeration) appear to give some hope of yielding consistent results even though these methods are still plagued by severe limitations, including primarily the taxation of the computer memory. It is no wonder then that most commercial codes (including **OPHELIE MIXED, MPSX-MIP,** and **UMPIRE**) are all of the branch-and-bound type. This does not mean that cutting methods are without value, for it was shown previously that many ideas are borrowed from these methods to improve the branch-and-bound methods. It does indicate, however, that the sole implementation of cutting methods cannot be very promising in solving practical problems. The final conclusion then is that any code must be centered around a branch-and-bound procedure with some of the useful properties of the cutting methods being implemented to alleviate its overall performance. This will be the basic guideline for the composite algorithm that will be presented here. It is not the intention, however, to present detailed steps, but rather a general outline of the desired properties of a composite algorithm will be given.

Experience with practical integer problems shows that if a solution is to be found, manual intervention during the course of the computations is a must. This means that depending on the progress of the calculations, the user may find it necessary to change the search strategy in order to take advantage of the available information. This emphasizes the importance of including as many feasible options as possible in the integer program-ming code. These options should be designed to exploit the different techniques available for solving integer programming problems, including heuristics. The collection of these options, together with the manual inter-vention by the user, produce the so-called "composite" algorithm. Naturally, the specific steps of the algorithm are not fixed in advance but will primarily depend on the experience of the user in selecting the most effective strategies for directing the search toward finding the optimal solution. These strategies are usually based on the information feedback from the computer and also on the type of problem under investigation.

To illustrate how some of these options are linked together, consider the determination of the bounds at a node in the branch-and-bound procedure. One may start by computing the upper bound by the method of penalties (Section 4.3.2-B). If, for example, the problem is dual degenerate, the penalties may be zero, in which case no information can be deduced from the penalties. A possible strategy for this situation is to use the option of generating legitimate cuts and applying them to the problem until dual degeneracy is eliminated. The use of cuts is also consistent with the desirable condition of producing smaller upper bounds at the node.

In order to decide whether the penalties approach is effective in designating the "correct" path to optimality, the user may wish to experiment with some of the nodes by solving the exact linear programs associated with them and then comparing with those produced by the penalties. If the use of penalties is frequently incorrect (in the sense that it does not identify the cheaper branch correctly), then the user may decide to abandon the penalties in favor of using the more effective heuristics of Section 4.3.3.

Of course, the options for manual intervention should not be limited to the choice of solution steps (by using different techniques) only. Rather, it may be necessary to simplify the formulation of the model or simply to change some of the data.

The implication of this discussion is that a knowledge of the different integer programming techniques may not be sufficient to solve an integer problem. Rather, a user must develop experience in how these different techniques can be combined in a *dynamic* fashion in order to effect the desired result. Also, one may be disappointed if he thinks that an integer model can be formulated and then immediately solved, as is usually the case in, for example, linear programming. Instead, interaction between the user and the "integer code" is necessary if progress toward solving the problem is to be achieved.

8.4 "General" Approximate Methods for Integer Programming

Throughout this book, different heuristics are presented but mainly within the context of some optimization procedure. For example, the heuristics in Section 4.3.3 were developed to enhance the effectiveness of the branch-and-bound algorithm.

There are approximate methods designed to find "good" solutions to the (linear) integer problem that may or may not be optimal. These methods are of the *direct search* type and are supposedly intended for use when all others fail to solve the problem.

In this section, general outlines of these methods are presented. Because a working algorithm involves a large number of fine details, it will be rather difficult to present a complete method here. Moreover, these details are based on intuitive rules rather than theory; hence their presentation may not be of too much value.

Two heuristics will be outlined: The first is due to Echoles and Cooper (1968) and the second to Hillier (1969a). A major drawback of the two methods (and indeed almost every heuristic along these lines) is the basic assumption that the solution space must have an interior. This rules out implicit and explicit inclusion of equality constraints in the model. Implicit inclusion means that an equation cannot be replaced by two inequalities since from the viewpoint of the direct search technique the solution space still has no interior. Since almost every nontrivial integer model includes equality constraints of some sort, it appears that the use of these heuristics is not as general as may be claimed.

One advantage of these heuristics, however, is that when they work, they produce "good" solutions quickly. Thus they may be used as an option in a branch-and-bound algorithm for the possibility of producing good bounds at a node.

8.4.1   Method I†

The general idea is to start with a feasible integer point. Then an integer variable having the best objective coefficient is selected and its value increased by as much as the feasibility of the solution allows. At this point, the next best integer variable (from the standpoint of its contribution to the objective function) is selected and its value again increased to the maximum limit allowed by feasibility. The process is repeated so long as the objective value can be improved. The first phase is terminated when the indicated moves do not produce better integer feasible points. Other phases are then initiated that attempt to find a better solution by "arbitrating" the values of the variables associated with the current feasible point.

The application of the method is illustrated by an example. Consider the problem:

maximize

$$z = 7x_1 + 9x_2$$

subject to

$$-x_1 + 3x_2 \leqslant 6$$
$$14x_1 - 8x_2 \leqslant 35$$
$$x_1, x_2 \geqslant 0 \quad \text{and integer}$$

† Echoles and Cooper (1968).

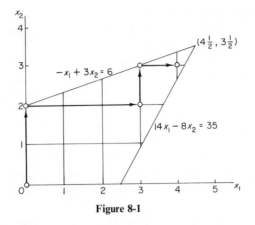

Figure 8-1

Figure 8-1 gives a graphical representation of the solution space. Let the initial integer feasible solution be $(x_1 = 0, x_2 = 0)$. Now increase x_2 to a maximum value $\frac{6}{3} = 2$ allowed by the first constraint. Thus the next solution point is $(0, 2)$. Fixing $x_2 = 2$, then x_1 can be increased to a maximum value of $(35 + 2 \times 8)/14 = 3.64$. This means that x_1 can be increased from zero to $[3.64] = 3$.

Now, fixing $x_1 = 3$, this shows that the value of x_2 can be increased from 2 to 3. Finally, x_1 is increased from 3 to 4. At this point, any improved integer solution will be infeasible and the first phase terminates. The resulting solution $x_1 = 4$ and $x_2 = 3$ happens to be optimal.

Of course, the heuristic will not always work as smoothly as the one in Fig. 8-1. The graphical example in Fig. 8-2 illustrates a possible difficulty. If x_1 is increased to its maximum integer value 5, it will not be possible to increase the value of x_2 in the next move, since the search is trapped at

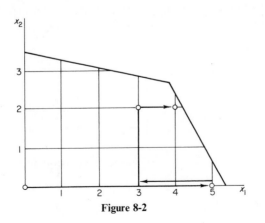

Figure 8-2

a *local* optimum point. What the heuristic does at this point is to backtrack by reducing the value of x_1 (usually by halving its current value and then rounding it to the upper integer value, if necessary). For example, x_1 in Fig. 8-2 can be reduced from 5 to 3. At this point the procedure used with the problem in Fig. 8-1 can be applied successfully as indicated by arrows in Fig. 8-2.

8.4.2 Method II†

This method is also based on the direct search technique. The main difference is that it seeks an initial feasible solution that is "close" to the continuous optimum. Let R be the set of the basic variables at the continuous optimum point x^* excluding the slacks. Suppose the continuous optimum point x^* has p equations associated with it. These equations are identified by the constraints for which the slack variables are equal to zero. Assume that these equations are given by the top p rows of the tableau and define

$$b_i' = b_i - \tfrac{1}{2}|R|^{1/2}, \qquad i = 1, 2, \ldots, p$$

where b_i is the right-hand side element of constraint i in the *original* problem and $|R|$ is the number of elements of R.

Let x' be the solution of the linear programming problem given that b_i is replaced by b_i' for $i = 1, 2, \ldots,$ and p. The new right-hand side has the property that its solution point x', when rounded, will always satisfy the constraints passing through the point x^*. However, the rounded solution may not satisfy the constrains that are not binding at x^*. But if it does, then an initial feasible solution is available. If the rounded solution is not feasible, then it is advisable to scan the points on the line segment joining x' and x^*. A rounded solution of a scanned point may be feasible with respect to the original problem. However, this result is not guaranteed.

To illustrate the application of this method, consider the problem of Fig. 8-1. The set $R = \{x_1, x_2\}$, and the binding equations are

$$-x_1 + 3x_2 = 6$$
$$14x_1 - 8x_2 = 35$$

which gives the solution $x^* = (4\tfrac{1}{2}, 3\tfrac{1}{2})$. Now

$$\begin{pmatrix} b_1' \\ b_2' \end{pmatrix} = \begin{pmatrix} 6 \\ 35 \end{pmatrix} - \begin{pmatrix} \tfrac{1}{2}\sqrt{2} \\ \tfrac{1}{2}\sqrt{2} \end{pmatrix} = \begin{pmatrix} 5.29 \\ 34.29 \end{pmatrix}$$

This yields $x' = (4.26, 3.18)$. The rounded solution $(4, 3)$ is feasible with respect to the entire problem.

† Hillier (1969a).

8.5   Concluding Remarks

In spite of the tremendous importance of integer programming, its methods of solution are still inadequate, particularly from the computational standpoint. The different methods presented throughout this book are incapable of alleviating this difficulty when each is considered separately, but the situation is improved when these techniques are combined together in a manner that exploits the advantages of each technique.

Perhaps an optimal algorithm capable of solving large practical problems efficiently will not be available in the foreseeable future; but because of the importance of the integer problem, research on both the theoretical and computational aspects of integer programming must continue. The continual advancement in the capabilities of the digital computer may perhaps produce the kind of computational speed and storage capacity that will mitigate the difficulties currently plagueing the (promising) branch-and-bound methods.

References†

Abadie, J. (1969). Une méthode arborscente pour les programmes partiellement discrets. *R.I.R.O.* **3**, 24–50.

Abadie, J., ed. (1970). *Integer and Nonlinear Programming.* Amer. Elsevier, New York.

Abadie, J. (1971). Une méthode de résolution de programmes nonlineare partiellement discrets san hypothèse de convexité. *R.I.R.O.* **1**, 23–38.

Adams, W., Gewirtz, A., and Quintas, L. (1969). *Elements of Linear Programming.* Van Nostrand-Reinhold, Princeton, New Jersey.

Again, J. (1966). Optimum seeking with branch-and-bound. *Management Sci.* **13**, B176–B185.

Alcaly, R. E., and Klevorick, A. V. (1966). A note on the dual prices of integer programs. *Econometrica* **34**, 206–214.

Allen, S. G. (1962). Computation for the redistribution model with setup charge. *Management Sci.* **8**, 482–489.

Anderson, I. (1971). Perfect matchings in a graph. *J. Combinatorial Theory* **10**, 183–186.

Anthonisse, J. M. (1970). A note on reducing a system to a single equation. Mathematisch Centrum report, BN 1/70, Amsterdam.

Aoki, M. (1971). *Introduction to Optimization Techniques.* Macmillan, New York.

Appelgren, L. H. (1971). Integer programming methods for a vessel scheduling problem. *Transportation Sci.* **5**, 64–78.

Arabeyre, J. P., Fearnley, J., Steiger, F. C., and Teather, W. (1969). The airline crew scheduling problem: A survey. *Transportation Sci.* **3**, 140–163.

Archer, J. (1965). *Linear Algebra and Linear Programming.* Dunod, Paris.

Aris, R. (1964). *Discrete Dynamic Programming.* Ginn (Blaisdell), Boston.

† The first comprehensive references list is due to Balinski and Spielberg (1969). A later list appears in the work of Garfinkel and Nemhauser (1972b).

Armour, G. C., and Buffa, E. S. (1965). A heuristic algorithm and simulation approach to relative location of facilities. *Management Sci.* **9**, 294–309.

Armstrong, R. D., and Sinha, P. (1973). Improved penalty calculations for a mixed integer branch-and-bound algorithm. Tech. Rep. No. 8. Univ. of Massachusetts, Amherst.

Arnold, L., and Bellmore, M. (1974a). Iteration skipping in primal integer programming. *Oper. Res.* **22**, 129–136.

Arnold, L., and Bellmore, L. (1974b). A generated cut for primal integer programming. *Oper. Res.* **22**, 137–143.

Arnold, L., and Bellmore, L. (1974c). A bounding minimization problem for primal integer programming. *Oper. Res.* **22**, 34–349.

Aronofsky, J., ed. (1969). *Progress in Operations Research—Volume III: O. R. and the Computer.* Wiley, New York.

Arthanari, T. S., and Ramamurthy, K. G. (1970). A branch-and-bound algorithm for sequencing n-Jobs on m-parallel processors. *Opsearch* **7**, 147–156.

Ashour, S. (1970a). A branch-and-bound algorithm for flow shop scheduling problems. *A.I.I.E. Trans.* **2**, 172–176.

Ashour, S. (1970b). An experimental investigation and comparative evaluation of flow shop scheduling techniques. *Oper. Res.* **18**, 541–548.

Ashour, S. (1972). *Sequencing Theory.* Springer-Verlag, Berlin and New York.

Ashour, S., and Char, A. R. (1971). Computational experience on 0–1 programming approaches to various combinatorial problems. *J. Oper. Res. Soc. Jap.* **13**, 78–107.

Ashour, S., and Parker, R. (1971). A precedence graph algorithm for the shop scheduling problem. *Operational Res. Quart.* **22**, 165–175.

Baker, K. E. (1975a). A comparative study of flow-shop algorithms. *Oper. Res.* **23**, 62–74.

Baker, K. E. (1975b). An elimination method for flow-shop problem. *Oper. Res.* **23**, 159–161.

Balas, E. (1964). Extension de l'Algorithme Additif en Nombre Entiers et a la Programmation Nonlineare. *C. R. Acad. Sci.* **258**, 5136–5139.

Balas, E. (1965). An additive algorithm for solving linear programs with zero–one variables. *Oper. Res.* **13**, 517–546.

Balas, E. (1967a). Finding a minimaximal path in a disjunctive pert network. *Theorie de Graphes.* Dunod, Paris.

Balas, E. (1967b). Discrete programming by the filter method. *Oper. Res.* **15**, 915–957.

Balas, E. (1968). A note on the branch-and-bound principle. *Oper. Res.* **16**, 442–445; errata *Oper. Res.* **16**, 886.

Balas, E. (1969a). Duality in discrete programming II: The quadratic case. *Management Sci.* **16**, 14–32.

Balas, E. (1969b). Machine sequencing via disjunctive graphs: An implicit enumeration algorithm. *Oper. Res.* **17**, 941–957.

Balas, E. (1970a). Duality in discrete programming. *Proc. Symp. Math. Programming, Princeton, 1970* (H. W. Kuhn, ed.), pp. 179–198. Princeton Univ. Press, Princeton, New Jersey.

Balas, E. (1970b). Machine sequencing, disjunctive graphs and degree constrained subgraphs. *Nav. Res. Logist. Quart.* **17**, 1–10.

Balas, E. (1970c). Minimax and duality for linear and nonlinear mixed-integer programming. In *Integer and Nonlinear Programming* (J. Abadie, ed.), pp. 385–418. Amer. Elsevier, New York.

Balas, E. (1971a). Intersection cuts—A new type of cutting planes for integer programming. *Oper. Res.* **19**, 19–39.

Balas, E. (1971b). A duality theorem and an algorithm for (mixed) integer nonlinear programming. *Linear Algebra Appl.* **4**, 341–352.

Balas, E. (1972). Integer programming and convex analysis. *Math. Programming* **2**, 330–381.

Balas, E. (1973). A note on the group theoretic approach to integer programming and the 0–1 case. *Oper. Res.* **21**, 321–322.

Balas, E., and Jeroslow, R. (1969). Canonical cuts in unit hypercube. Management Sci. Res. Rep. No. 198(R). Carnegie-Mellon Univ., Pittsburgh, Pennsylvania (rev. February 1971).

Balas, E., and Padberg, M. W. (1972). On the set covering problem. *Oper. Res.* **20**, 1152–1161.

Balas, E., and Padberg, M. W. (1975). On the set-covering problem: An algorithm for set partitioning. *Oper. Res.* **23**, 74–90.

Balas, E., Bowman, V. J., Glover, F., and Sommer, D. (1971). An intersection cut from the dual of the unit hypercube. *Oper. Res.* **19**, 40–44.

Balinski, M. L. (1961). Fixed cost transportation problems. *Nav. Res. Logist. Quart.* **11**, 41–54.

Balinski, M. L. (1964). On finding integer solutions to linear programs. *Proc. IBM Sci. Comput. Symp. Combinatorial Problems, March 16–18, 1964,* pp. 225–248. White Plains, New York.

Balinski, M. L. (1965). Integer programming: Methods, uses, computation. *Management Sci.* **12**, 253–313.

Balinski, M. L. (1967). Some general methods in integer programming. *Nonlinear Programming* (J. Abadie, ed.), pp. 220–247. Wiley, New York.

Balinski, M. L. (1970a). On recent developments in integer programming. *Proc. Symp. Math. Programming, Princeton, 1967* (H. W. Kuhn, ed.), pp. 267–302. Princeton Univ. Press, Princeton, New Jersey.

Balinski, M. L. (1970b). On maximum matching, minimum covering and their connections. *Proc. Symp. Math. Programming, Princeton, 1967* (H. W. Kuhn, ed.), pp. 303–312. Princeton Univ. Press, Princeton, New Jersey.

Balinski, M. L., and Quandt, R. E. (1964). On an integer program for a delivery problem. *Oper. Res.* **12**, 300–304.

Balinski, M. L., and Spielberg, K. (1969). Methods for integer programming: Algebraic, combinatorial and enumerative. In *Progress in Operations Research—Volume III* (J. Aronofsky, ed.) pp. 195–292. Wiley, New York.

Balinski, M. L., and Wolfe, P. (1963). On Benders' decomposition and a plant location problem. *Mathematica (Cluj)* Working Paper, **ARO-27**.

Bauer, F. L. (1963). Algorithm 153, Gomory. *Commun. Assoc. Comput. Mach.* **6**, 125.

Baugh, C. R., Ibaraka, T., and Muroga, S. (1971). Results in using Gomory's all-integer algorithm to design optimum logic networks. *Oper. Res.* **19**, 1090–1096.

Baumol, W. J., and Kuhn, W. H. (1962). An approximate algorithm for the fixed-charges transportation problem. *Nav. Res. Logist. Quart.* **9**, 1–15.

Baumol, W. J., and Wolfe, P. (1958). A warehouse location problem. *Oper. Res.* **6**, 252–263.

Bazaraa, M. S., and Goode, J. J. (1975). A cutting plane algorithm for the quadratic set-covering problem. *Oper. Res.* **23**, 150–158.

Beale, E. M. L. (1954). An alternate method of linear programming. *Proc. Cambridge Phil. Soc.* **50**, 513–523.

Beale, E. M. L. (1955). Cycling in the dual simplex method. *Nav. Res. Logist. Quart.* **2**, 269–276.

Beale, E. M. L. (1958). A method for solving linear programming problems when some but not all of the variables must take integral values. Tech. Rep. No. 19. Statist. Tech. Res. Group, Princeton Univ., Princeton, New Jersey.

Beale, E. M. L. (1965). Survey of integer programming. *Operational Res. Quart.* **16**, 219–228.

Beale, E. M. L. (1968). *Mathematical Programming in Practice.* Pitman, London.

Beale, E. M. L. (1970a). Advanced algorithmic features for general mathematical programming systems. In *Integer and Nonlinear Programming* (J. Abadie, ed.). Amer. Elsevier, New York.

Beale, E. M. L. (1970b). *Applications of Mathematical Programming Techniques.* English Univ. Press, London.

Beale, E. M. L., and Small, R. E. (1965). Mixed integer programming by a branch-and-bound technique. *Proc. IFIP Congr., New York, May 1965* (W. A. Kalenich, ed.), **2**, pp. 450–451. Spartan Books, Washington, D.C.

Beale, E. M. L., and Tomlin, J. A. (1969). Special facilities in a general mathematical programming system for non-convex problems using ordered sets of variables. *Proc. Int. Conf. Operational Res., 5th, Venice, 1969* (J. Lawrence, ed.), pp. 8.21–8.28. Tavistock, London.

Beale, E. M. L., and Tomlin, J. A. (1972). An integer programming approach to a class of combinatorial problems. *Math. Programming* **3**, 339–344.

Bellman, R. E. (1956). Notes on the theory of dynamic programming IV—Maximization over discrete sets. *Nav. Res. Logist. Quart.* **3**, 67–70.

Bellman, R. E. (1957a). *Dynamic Programming.* Princeton Univ. Press, Princeton, New Jersey.

Bellman, R. E. (1957b). Comment on Dantzig's paper on discrete variable extremum problems. *Oper. Res.* **5**, 723–724.

Bellman, R. E. (1962). Dynamic programming treatment of the traveling salesman problem. *J. Assoc. Comput. Mach.* **9**, 61–63.

Bellman, R. E. (1965). An application of dynamic programming to location–allocation problems. *SIAM (Soc. Ind. App. Math.) Review,* **7**, 126–128.

Bellman, R. E., and Dreyfus, S. E. (1962). *Applied Dynamic Programming.* Princeton Univ. Press, Princeton, New Jersey.

Bellman, R. E., and Hall, M., Jr., eds. (1960). Combinatorial analysis. *Proc. Symp. Appl. Math. Amer. Math. Soc., 10th, 1960, New York.*

Bellmore, M., and Malone, J. C. (1971). Pathology of traveling salesman subtour elimination algorithms. *Oper. Res.* **19**, 278–307.

Bellmore, M., and Nemhauser, G. L. (1968). The traveling salesman problem: A survey. *Oper. Res.* **16**, 538–558.

Bellmore. M., and Ratliff, H. D. (1971a). Set covering and involutory bases. *Management Sci.* **18**, 194–206.

Bellmore, M., and Ratliff, H. D. (1971b). Optimal defense of multi-commodity networks. *Management Sci.* **18**, B174–B185.

Benders, J. F. (1962). Partitioning procedures for solving mixed-variables programming problems. *Numer. Math.* **4**, 239–252.

Benders, J. F., Catchpole, A. R., and Kuiken, C. (1959). Discrete variable optimization problems. *Rand Symp. Math. Programming, March 16–20, 1959, Santa Monica, California.*

Benichou, M., Gauthier, J. M., Girodet, P., Hentges, G., Ribiere, G., and Vincent O. (1971). Experiments in mixed-integer linear programming. *Math. Programming* **1**, 76–94.

Ben-Israel, A., and Charnes, A. (1962). On some problems in diophantine programming. *Cah. Cent. Etud. Rech. Opérationelle* **4**, 215–280.

Bennet, J. M., and Dakin, R. J. (1961). Experience with mixed linear programming problems, Mimeo. Rep. Univ. of Sydney, Sydney, Australia, October 26.

Bennet, J. M., Cooley, P. C., and Edwards, J. (1968). The performance of an integer programming algorithm with test examples. *Aust. Comput. J.* **1**, 182–185.

Bertier, P., and Roy, B. (1965). Une Procédure de Résolution pour une Classe de Problemes Pouvant avoir un Charactère Combinatoire. *ICC (Int. Comput. Cent.) Bull.* **4**, 19–28. (Transl. by W. S. Jewell, ORC Rep. 67–34. Univ. of California, Berkeley.)

Bertier, P., Nghiem, Ph. T., and Roy, B. (1965). Programmes Lineaires en Nombres Entiers. *Metra* **4**, 153–158.

Bessiere, F. (1965). Sur la Recherche du Nombre Chromatique d'un Graphe par un Programme Lineaire en Nombres Entiers. *Rev. Fr. Rech. Opérationelle* **9**, 143–148.

Blankenship, W. A. (1963). A new version of the Euclidean algorithm. *Amer. Math. Monthly* **70**, 742–745.

Bod, P. (1970). Solution of a fixed charge linear programming problem. *Proc. Symp. Math. Programming, Princeton, 1967* (H. W. Kuhn, ed.), pp. 367–375. Princeton Univ. Press, Princeton, New Jersey.

Bowman, E. H. (1956). Production scheduling by the transportation method of linear programming. *Oper. Res.* **4**, 100–103.

Bowman, E. H. (1959). The schedule-sequencing problem. *Oper. Res.* **7**, 621–624.

Bowman, E. H. (1960). Assembly line balancing by linear programming. *Oper. Res.* **8**, 385–389.

Bowman, V. J. (1969). Determining cuts in integer programming by two variable diophantine equations. Tech. Rep. No. 13. Dept. of Statist., Oregon State Univ., Corvallis.

Bowman, V. J. (1972). Constraint classification of the unit hypercube. Management Sc.. Res. Rep. No. 287. Carnegie-Mellon Univ., Pittsburgh, Pennsylvania (March, rev. July).

Bowman, V. J. (1972). A structural comparison of Gomory's fractional cutting planes and Hermitian basic solutions. *SIAM* (*Soc. Ind. App. Math.*) *J. Appl. Math.* **23**, 460–462.

Bowman, V. J. (1974). The structure of integer programs under the Hermitian normal form. *Oper. Res.* **22**, 1067–1080.

Bowman, V. J., and Glover, F. (1972). A note on zero–one integer and concave programming. *Oper. Res.* **20**, 182–183.

Bowman, V. J., and Nemhauser, G. L. (1970). A finiteness proof for modified Dantzig cuts in integer programming. *Nav. Res. Logist. Quart.* **17**, 309–313.

Bowman, V. J., and Nemhauser, G. L. (1971). Deep cuts in integer programming. *Opsearch* **8**, 89–111.

Bozoki, G., and Richard, J. (1970). A branch-and-bound algorithm for the continuous process job-shop scheduling problem. *A.I.I.E. Trans.* **2**, 246–252.

Bradley, G. H. (1969). Equivalent integer programs. *Proc. Int. Conf. Operational Res., 5th, Venice, 1969* (J. Lawrence, ed.), pp. 455–463. Tavistock, London.

Bradley, G. H. (1970). Algorithm and bound for the greatest common divisor of n integers. *Commun. Assoc. Comput. Mach.* **13**, 433–436.

Bradley, G. H. (1971a). Algorithms for Hermite and Smith normal matrices and linear diophantine equations. *Math. Comput.* **25**, 897–908.

Bradley, G. H. (1971b). Heuristic solution methods and transformed integer linear programming problems. Rep. No. 43. Dept. of Admin. Sci., Yale Univ., New Haven, Connecticut.

Bradley, G. H. (1971c). Transformation of integer programs to knapsack problems. *Discrete Math.* **1**, 29–45.

Bradley, G. H. (1971d). Equivalent integer programs and canonical problems. *Management Sci.* **17**, 354–366.

Bradley, G. H. (1973). Equivalent mixed integer programming problems. *Oper. Res.* **21**, 323–325.

Bradley, G. H., and Wahi, P. N. (1973). An algorithm for integer linear programming: A combined algebraic and enumeration approach. *Oper. Res.* **21**, 45–60.

Bradley, G. H., Hammer, P. L., and Wolsey, L. (1973). Coefficient reduction for inequalities in 0–1 variables. Res. Rep. CORR 73-6. Dept. of Combinatorics and Optimization, Univ. of Waterloo, Waterloo, Iowa.

Brooks, R. L. (1941). On colouring the nodes of a network. *Proc. Cambridge Phil. Soc.* **37**, 194–197.

Brooks, R., and Geoffrion, A. M. (1966). Finding Everett's Lagrange multipliers by linear programming. *Oper. Res.* **14**, 1149–1153.

Burdet, C. (1970). A class of cuts and related algorithms in integer programming. Management Sci. Res. Rep. No. 220. Carnegie-Mellon Univ., Pittsburgh, Pennsylvania.

Burdet, C. (1970b). Deux Modeles De Minimisation d'Une Fonction Economique Concave. *R.I.R.O.* **1**, 49–84.

Burdet, C. (1972a). Enumerative Inequalities in Integer Programming. *Math. Programming* **2**, 32–64.

Burdet, C. (1972b). Group theory or convex analysis; A combined approach. Management Sci. Res. Rep. No. 291. Carnegie-Mellon Univ., Pittsburgh, Pennsylvania.

Burdet, C. (1973a). Polaroids: A new tool in non-convex and integer programming. *Nav. Res. Logist. Quart.* **20**, 13–24.

Burdet, C. (1973b). Enumerative cuts. I. *Oper. Res.* **21**, 61–89.

Burt, O. R., and Harris, C. C., Jr. (1963). Appointment of the U.S. House of Representatives: A minimum range. integer solution, allocation problem. *Oper. Res.* **11**, 648–652.

Busacker, R. G., and Saaty, T. L. (1965). *Finite Graphs and Networks: An Introduction with Applications.* McGraw-Hill, New York.

Cabot, V. A. (1970). An enumeration algorithm for knapsack problems. *Oper. Res.* **18**, 306–311.

Cabot, V. A. (1972). Solving fixed-charge problems as nonlinear programming problems. *Bull. Oper. Res. Soc. of Amer.* (Spring), B-40.

Cabot, V. A., and Francis, R. (1970). Solving certain nonlinear quadratic minimization problems by ranking the extreme points. *Oper. Res.* **18**, 82–86.

Cabot, V. A., and Hurter, A. P. (1968). An approach to 0–1 integer programming. *Oper Res.* **16**, 1206–1211.

Camion, P. (1960). Une Methode de Resolution par l'Algebre de Boole des Problemes Combinatoires ou Interviennent de Entiers. *Cah. Cent. Etud. Rech. Opérationelle* **2**, 234–289.

Carlson, R. G., and Nemhauser, G. L. (1966). Scheduling to minimize interaction cost. *Oper. Res.* **14**, 52–58.

Castellan, J. W. (1967). Political appointment by computer. *Brown Univ. Comput. Rev.* **1**, 51–53.

Charlton, J. M., and Death, C. C. (1970). A generalized machine scheduling algorithm. *Operational Res. Quart.* **21**, 127–134.

Charnes, A. (1952). Optimality and degeneracy in linear programming. *Econometrica,* **20**, 160–170.

Clarke, G., and Wright, S. W. (1964). Scheduling of vehicles from a central depot to a number of delivery points. *Oper. Res.* **12**, 568–581.

Cobham, A., and North, J. H. (1963). Extensions of the integer programming approach to the minimization of Boolean Functions. Res. Rep. RC-915. IBM. Yorktown Heights, New York.

Cobham, A., Fridshal, R., and North, J. H. (1961). An application of linear programming to the minimization of Boolean functions. Res. Rep. RC-472. IBM. Yorktown Heights, New York.

Cobham, A., Fridshal, R., and North, J. H. (1962). A statistical study of the minimization of Boolean functions using integer programming. Res. Rep. RC-756. IBM. Yorktown Heights, New York.

Conway, R. W., Maxwell, W. L., and Miller, L. W. (1967). *Theory of Scheduling.* Addison-Wesley, Reading, Massachusetts.

Cooper, L., (1972). The transportation-location problem. *Oper. Res.* **20**, 94–108.

Cooper, L., and Drebes, C. (1967). An approximate solution method for the fixed charge problem. *Nav. Res. Logist. Quart.* **14**, 101–113.

Cooper, L., and Olson, A. M. (Unknown). Random perturbations and the MI-MII heuristics for the fixed-charge problem. Rep. No. COO-1493-7. Washington Univ., St. Louis, Missouri.

Cristofides, N., and Eilon, S. (1969). An algorithm for the vehicle dispatching problem. *Operational Res. Quart.* **20**, 309–318.

Curry, G. L., and Skeith, R. W. (1969). A dynamic programming algorithm for facility location and allocation. *A.I.I.E. Trans.* **1**, 133–138.

Cushing, B. E. (1970). The application potential of integer programming. *J. Bus.* **43**, 457–467.

Dakin, R. J. (1961). Application of mathematical programming techniques to cost optimization in power generating systems. Tech. Rep. 19. Basser Comput. Dept., Univ. of Sydney, Sydney, Australia.

Dakin, R. J. (1965). A tree-search algorithm for mixed integer programming problems. *Comput. J.* **8**, 250–255.

Dalton, R. E., and Llewellyn, R. W. (1966). An extension of the Gomory mixed-integer algorithm to mixed-discrete variables. *Management Sci.* **12**, 569–575.

Dantzig, G. B. (1955). Upper bound, secondary constraints, and block triangularity in linear programming. *Econometrica* **23**, 174–183.

Dantzig, G. B. (1957). Discrete variable extremum problems. *Oper. Res.* **5**, 266–277.

Dantzig, G. B. (1959). Notes on solving linear programs in integers. *Nav. Res. Logist. Quart.* **6**, 75–76.

Dantzig, G. B. (1960a). On the significance of solving linear programming problems with some integer variables. *Econometrica* **28**, 30–44.

Dantzig, G. B. (1960b). On the shortest route through a network. *Management Sci.* **6**, 187–190.

Dantzig, G. B. (1963). *Linear Programming and Extensions.* Princeton Univ. Press, Princeton, New Jersey.

Dantzig, G. B., and Ramser, J. H. (1959). The truck dispatching problem. *Management Sci.* **6**, 80–91.

Dantzig, G. B., Fulkerson, D. R., and Johnson, S. M. (1954). Solution of a large scale travelling salesman problem. *Oper. Res.* **2**, 393–410.

Dantzig, G. B., Fulkerson, D. R., and Johnson, S. M. (1959). On a linear programming, combinatorial approach to the travelling salesman problem. *Oper. Res.* **7**, 58–66.

Davis, P. S., and Ray, T. L. (1969). A branch-bound algorithm for the capacitated facilities location problem. *Nav. Res. Logist. Quart.* **16**, 331–344.

Davis, R. E., Kendrick, D. A., and Weitzman, M. (1971). A branch-and-bound algorithm for 0–1 mixed integer programming problems. *Oper. Res.* **19**, 1036–1044.

Day, R. H. (1965). On optimal extracting from a multiple file data storage system: An application of integer programming. *Oper. Res.* **13**, 482–494.

Denzler, D. R. (1969). An approximative algorithm for the fixed charge problem. *Nav. Res. Logist. Quart.* **16**, 411–416.

Desler, J. F., and Hakimi, S. L. (1969). A graph-theoretic approach to a class of integer programming problems. *Oper. Res.* **17**, 1017–1033.

D'Esopo, D. A., and Lefkowitz, B. (1963). Certification of algorithm 153. *Commun. Assoc. Comput. Mach.* **6**, 223–226.

D'Esopo, D. A., and Lefkowitz, B. (1964). Note on an integer linear programming model for determining a minimum embarkation fleet. *Nav. Res. Logist. Quart.* **11**, 79–82.

Dijkstra, E. W. (1959). A note on two problems in connection with graphs. *Numer. Math.* **1**, 269–271.

Dragan, I. (1968). Un Algorithme Lexicographique pour la Résolution des Programmes Polynomiaux en Variables Entières. *R.I.R.O.* **2**, 81–90.

Dragan, I. (1969). Un Algorithme Lexicographique pour la Resolution des Programmes Lineanes en Variables Binaires. *Management Sci.* **16**, 246–252.

Driebeek, N. J. (1966). An algorithm for the solution of mixed integer programming problems. *Management Sci.* **12**, 576–587.

Drysdale, J. K., and Sandiford, P. J. (1969). Heuristic warehouse location—A case history using a new method. *Can. Operational Res. Soc. J.* **7**, 45–61.

Eastman, W. L. (1958). Linear programming with pattern constraints. Ph.D. Dissertation, Harvard Univ., Cambridge, Massachusetts.

Eastman, W. L. (1959). A note on the multi-commodity warehouse problem. *Management Sci.* **5**, 327–331.

Echols, R. E., and Cooper, L. (1968). Solution of integer linear programming problems by direct search. *J. Assoc. Comput. Mach.* **15**, 75–84.

Efroymson, M. A., and Ray, T. L. (1966). A branch-and-bound algorithm for plant location. *Oper. Res.* **14**, 361–368.

Eilon, S., and Cristofides, N. (1971). The loading problem. *Management Sci.* **17**, 259–268.

Ellwein, L. B., and Gray, P. (1971). Solving fixed charge allocation problems with capacity and configuration constraints. *A.I.I.E. Trans.* **3**, 290–299.

Elmaghraby, S. E. (1966). *The Design of Production Systems*, Reinhold, New York.

Elmaghraby, S. E. (1970). The theory of networks and management science, Pt. I. *Management Sci.* **17**, 1–34.

Elmaghraby, S. E., ed. (1973). *Symp. Theory Scheduling and Appl. Raleigh, North Carolina, May 1972.* Springer-Verlag, Berlin and New York.

Elmaghraby, S. E., and Wig, M. K. (1970). On the treatment of cutting stock problems as diophantine programs. Dept. of Ind. Eng., North Carolina State Univ., Raleigh.

Elshafei, A. N. (1972). Facilities location: Formulations, methods of solution, applications and some computational experience. Memo. No. 276. Inst. of Nat. Planning, Cairo, Egypt.

Elson, D. G. (1972). Site location via mixed integer programming. *Operational Res. Quart.* **23**, 31–44.

Everett, H. (1963). Generalized Lagrange multiplier method for solving problems of optimum allocation of resources. *Oper. Res.* **11**, 399–417.

Faaland, B. (1970). Generalized equivalent integer programs and canonical problems. Tech. Rep. No. 21. Dept. of Oper. Res., Stanford Univ., Stanford, California.

Faaland, B. (1971). Solution of the value independent knapsack problem by partitioning. Tech. Rep. No. 22. Dept. of Oper. Res., Stanford Univ., Stanford, California.

Faaland, B. (1972a). On the number of solutions to a diophantine equation. *J. Combinatorial Theory* **13**, 170–175.

Faaland, B. (1972b). Estimates and bounds on computational effort in the accelerated bound-and-scan algorithm. Tech. Rep. No. 10. Dept. of Oper. Res., Stanford Univ., Stanford, California.

Faaland, B. (1973). Solution of value independent knapsack problem by partitioning. *Oper. Res.* **21**, 332–33.

Faaland, B., and Hillier, F. S. (1972a). An accelerated bound-and-scan algorithm for integer programming. Tech. Rep. No. 9. Dept. of Oper. Res., Stanford Univ., Stanford, California.

Faaland, B., and Hillier, F. S. (1972b). A constructive theory and applications for the solution of mixed integer systems of linear equations. Tech. Rep. No. 11. Dept. of Oper. Res., Stanford Univ., Stanford, California.

Feldman, E., Lehrer, F. A., and Ray, T. L. (1966). Warehouse location under continuous economies of scale. *Management Sci.* **12**, 670–684.

Fiorot, J. Ch. (1972). Generation of all integer points for given sets of linear inequalities. *Math. Programming* **3**, 276–295.

Fisher, M. L. (1973). Optimal solution of scheduling problems using Lagrange multipliers: Pt. I. *Oper. Res.* **21**, 1114–1127.

Fisher, M. L. (1972). Optimal solutions of scheduling using Lagrange multipliers: Pt. II. *Symp. Theory Scheduling and Appl., 1972* (S. E. Elmaghraby, ed.). Springer-Verlag, Berlin and New York.

Fleischmann, B. (1967). Computational experience with the algorithm of Balas. *Oper. Res.* **15**, 153–155.

Florian, M., and Robillard, P. (1971). Programmation hyperbolique en variables bivalents. *R.I.R.O.* **1**, 3–9.

Florian, M., Trepant, P., and McMahon, G. B. (1971). An implicit enumeration algorithm for the machine sequencing problem. *Management Sci.* **17**, B782–B792.

Ford, L. R., Jr., and Fulkerson, D. R. (1962). *Flows in Networks.* Princeton Univ. Press, Princeton, New Jersey.

Forrest, J. J. H., Hirst, J. P. H., and Tomlin, J. A. (1974). Practical solution of large mixed integer programming problems with UMPIRE. *Management Sci.* **20**, 736–773.

Fortet, R. (1959). L'Algebre de Boole et se Application en Recherche Operationelle. *Cah. Cent. Etud. Rech. Operationelle* **4**, 215–280.

Francis, R. L., and White, J. A. (1974). *Facility Layout and Location: An Analytic Approach.* Prentice-Hall, Englewood Cliffs, New Jersey.

Freeman, R. J. (1966). Computational experience with a "balasian" integer programming algorithm. *Oper. Res.* **14**, 935–941.

Garfinkel, R. S. (1971). An improved algorithm for the bottleneck assignment problem. *Oper. Res.* **19**, 1747–1751.

Garfinkel, R. S. (1973). On partitioning the feasible set in a branch-and-bound algorithm for the asymmetric traveling salesman problem. *Oper. Res.* **21**, 340–342.

Garfinkel, R. S., and Nemhauser, G. L. (1969). The set partitioning problem: Set covering with equality constraints. *Oper. Res.* **17**, 848–856.

Garfinkel, R. S., and Nemhauser, G. L. (1970). Optimal political districting by implicit enumeration techniques. *Management Sci.* **16**, B495–B508.

Garfinkel, R. S., and Nemhauser, G. L. (1972a). Optimal set covering: A survey. In *Perspectives on Optimization: A Collection of Expository Articles* (A. M. Geoffrion, ed.), pp. 164–183. Addison-Wesley, Reading, Massachusetts.

Garfinkel, R. S., and Nemhauser, G. L. (1972b). *Integer Programming.* Wiley, New York.

Garfinkel, R. S., and Rao, M. R. (1971). The bottleneck transportation problem. *Nav. Res. Logist. Quart.* **18**, 465–472.

Gass, S. (1969). *Linear Programming Methods and Applications*, 3rd ed. McGraw-Hill, New York.

Gavett, J. W., and Plyter, N. V. (1966). The optimal assignment of facilities to locations by branch and bound. *Oper. Res.* **14**, 210–232.

Geoffrion, A. M. (1967). Integer programming by implicit enumeration and Balas' method. *SIAM (Soc. Ind. Appl. Math.) Rev.* **7**, 178–190.

Geoffrion, A. M. (1969). An improved implicit enumeration approach for integer programming. *Oper. Res.* **17**, 437–454.

Geoffrion, A. M., ed. (1972a). *Perspectives on Optimization: A Collection of Expository Articles.* Addison-Wesley, Reading, Massachusetts.

Geoffrion, A. M. (1972b). Lagrangean relaxation for integer programming. Working Paper No. 195. Univ. of California, Los Angeles (December, rev. December 1973).

Geoffrion, A. M. (1972c). Generalized Benders' decomposition. *J. Optimization Theory Appl.* **10**, 237–260.

Geoffrion, A. M., and Marsten, R. E. (1972). Integer programming: A framework and state-of-the-art survey. *Management Sci.* **18**, 465–491.

Giglio, R. J., and Wagner, H. M. (1964). Approximate solutions to the three-machine scheduling problem. *Oper. Res.* **12**, 306–324.

Gilmore, P. C. (1962). Optimal and sub-optimal algorithms for the quadratic assignment problem. *SIAM (Soc. Ind. Appl. Math.) J. Appl. Math.* **10**, 305–313.

Gilmore, P. C., and Gomory, R. E. (1961). A linear programming approach to the cutting stock problem. *Oper. Res.* **9**, 849–859.

Gilmore, P. C., and Gomory, R. E. (1963). A linear programming approach to the cutting stock problem—Part II. *Oper. Res.* **11**, 863–888.

Gilmore, P. C., and Gomory, R. E. (1965). Many stage cutting stock problems of two or more dimensions. *Oper. Res.* **13**, 94–120.

Gilmore, P. C., and Gomory, R. E. (1966). The theory of computation of knapsack functions. *Oper. Res.* **14**, 1045–1074.

Glover, F. (1965a). A bound escalation method for the solution of integer linear programs. *Cah. Cent. Etud. Rech. Opérationelle* **6**, 131–168.

Glover, F. (1965b). A hybrid-dual integer programming algorithm. *Cah. Cent. Etud. Rech. Opérationelle* **7**, 5–23.

Glover, F. (1965c). A multiphase-dual algorithm for the zero–one integer programming problem. *Oper. Res.* **13**, 879–919.

Glover, F. (1966). Generalized cuts in diophantine programming. *Management Sci.* **13**, 254–268.

Glover, F. (1967a). A pseudo primal-dual integer programming algorithm. *J. Res. Nat. Bur. Stand. Sect. B*, **71**, 187–195.

Glover, F. (1967b). Maximum matching in a convex bipartite graph. *Nav. Res. Logist. Quart.* **14**, 313–316.

Glover, F. (1967c). Stronger cuts in integer programming. *Oper. Res.* **15**, 1174–1176.

Glover, F. (1968a). A new foundation for a simplified primal integer programming algorithm. *Oper. Res.* **16**, 727–740.

Glover, F. (1968b). Surrogate constraints. *Oper. Res.* **16**, 741–749.

Glover, F. (1968c). A note on linear programming and integer feasibility. *Oper. Res.* **16**, 1212–1216.

Glover, F. (1968d). Faces on an integer polyhedron for an additive group. Publ. AMM-11. Graduate School of Bus., Univ. of Texas, Austin.

Glover, F. (1969). Integer programming over a finite additive group. *SIAM (Soc. Ind. Appl. Math.) Contr.* **7**, 213–231.

Glover, F. (1971a). Faces of the Gomory polyhedron for cyclic groups. *J. Math. Anal. Appl.* **35**, 195–208.

Glover, F. (1971b). A note on extreme point solutions and a paper by Lemke, Salkin and Spielberg. *Oper. Res.* **19**, 1023–1026.

Glover, F. (1972a). New results for reducing integer linear programs to knapsack problems. *Management Sci. Rep.* Ser. No. 72-7. Univ. of Colorado, Boulder.

Glover, F. (1972b). Cut search methods in integer programming. *Math. Programming* **3**, 86–100.

Glover, F. (1973). Convexity cut and cut search. *Oper. Res.* **21**, 123–134.

Glover, F., and Klingman, D. (1973a). Concave programming applied to a special class of 0–1 integer problems. *Oper. Res.* **21**, 135–140.

Glover, F., and Klingman, D. (1973b). The generalized lattice point problem. *Oper. Res.* **21**, 141–155.

Glover, F., and Litzler, L. (1969). Extension of an asymptotic integer programming algorithm to the general integer programming problem. Graduate School of Bus., Univ. of Texas, Austin.

Glover, F., and Sommer, D. (1972). Pitfalls of rounding in discrete management decision problems. Management Sci. Rep. Ser. No. 72-2. Univ. of Colorado, Boulder.

Glover, F., and Woolsey, R. E. (1970). Aggregating diophantine equations. Rep. No. 70-4. Univ. of Colorado, Boulder.

Glover, F., and Woolsey, R. E. (1973). Further reduction of zero–one polynomial programming problems to zero–one linear programming problems. *Oper. Res.* **21**, 141–161.

Glover, F., and Woolsey, R. E. (1974). Converting the 0–1 polynomial programming problem to a 0–1 linear program. *Oper. Res.* **22**, 180–182.

Glover, F., and Zionts, S. (1965). A note on the additive algorithm of Balas. *Oper. Res.* **13**, 546–549.

Golomb, S. W., and Baumert, L. O. (1965). Backtrack programming. *J. Assoc. Comput. Mach.* **12**, 516–524.

Gomory, R. E. (1958). Outline of an algorithm for integer solutions to linear programs. *Bull. Amer. Math. Soc.* **64**, 275–278.

Gomory, R. E. (1960a). All-integer integer programming algorithm. RC-189. IBM, Yorktown Heights, New York; also in *Industrial Scheduling* (J. F. Muth and G. L. Thompson, eds.), pp. 193–206. Prentice-Hall, Englewood Cliffs, New Jersey, 1963.

Gomory, R. E. (1960b). An algorithm for the mixed integer problem. RM-2597. RAND Corp. Santa Monica, California.

Gomory, R. E. (1960c). Solving linear programming problems in integers. *Proc. Symp. Appl. Math. Amer. Math. Soc., 10th, 1960* (R. E. Bellman and M. Hall, Jr., eds.), pp. 211–216.

Gomory, R. E. (1963). An algorithm for integer solutions to linear programs. In *Recent Advances in Mathematical Programming* (R. L. Graves, and P. Wolfe, eds.), pp. 269–302. McGraw-Hill, New York.

Gomory, R. E. (1965). On the relation between integer and non-integer solutions to linear programs. *Proc. Nat. Acad. Sci. U.S.* **53**, 260–265.

Gomory, R. E. (1967). Faces on an integer polyhedron. *Proc. Nat. Acad. Sci. U.S.* **57**, 16–18.

Gomory, R. E. (1969). Some polyhedra related to combinatorial problems. *Linear Algebra Appl.* **2**, 451–558.

Gomory, R. E., and Baumol, W. J. (1960). Integer programming and pricing. *Econometrica* **28**, 521–550.

Gomory, R. E., and Hoffman, A. J. (1963). On the convergence of an integer programming process. *Nav. Res. Logist. Quart.* **10**, 121–123.

Gomory, R. E., and Johnson, E. L. (1972). Some continuous functions related to corner polyhedra. *Math. Programming* **3**, 23–85.

Gondran, M., and Lauriere, J. (1974). Une algorithme pour le problème de partitionnement. *R.I.R.O.* **1**, 27–40.

Gorry, G. A., and Shapiro, J. F. (1971). An adaptive group theoretic algorithm for integer programming. *Management Sci.* **17**, 285–306.

Gorry, G. A., Nemhauser, G. L., Northrup, W. D., and Shapiro, J. F. (1970). An improved branching rule for the group theoretic branch-and-bound integer programming algorithm. Oper. Res. Cent., MIT, Cambridge, Massachusetts.

Gorry, G. A., Shapiro, J. F., and Wolsey, L. A. (1972). Relaxation methods for pure and mixed integer programming problems. *Management Sci.* **18**, 229–239.

Gould, F. J., and Rubin, D. S. (1973). Rationalizing discrete programs. *Oper. Res.* **21**, 343–345.

Graves, G. W., and Whinston, A. B. (1968). A new approach to discrete mathematical programming. *Management Sci.* **15**, 177–190.

Graves, G. W., and Whinston, A. B. (1970). An algorithm for the quadratic assignment problem. *Management Sci.* **16**, 692–707.

Graves, R. L., and Wolfe, P., eds. (1963). *Recent Advances in Mathematical Programming.* McGraw-Hill, New York.

Gray, P. (1967). Mixed integer programming algorithms for site selection and other fixed charge problems having capacity constraints. Tech. Rep. No. 101. Dept. of Oper. Res., Stanford Univ., Stanford, California.

Gray, P. (1971). Exact solution of the fixed-charge problem. *Oper. Res.* **19**, 1529–1537.

Greenberg, H. (1968). A branch bound solution to the general scheduling problem. *Oper. Res.* **16**, 353–361.

Greenberg, H. (1969a). An algorithm for the computation of knapsack functions. *J. Math. Anal. Appl.* **26**, 159–162.

Greenberg, H. (1969b). A dynamic programming solution to linear integer programs. *J. Math. Anal. Appl.* **26**, 454–459.

Greenberg, H. (1969c). A quadratic assignment problem without column constraints. *Nav. Res. Logist. Quart.* **16**, 417–422.

Greenberg, H. (1971). *Integer Programming.* Academic Press, New York.

Greenberg, H., and Hegerich, R. L. (1970). A branch search algorithm for the knapsack problem. *Management Sci.* **16**, 327–332.

Grunspan, M., and Thomas, M. E. (1973). Hyperbolic integer programming. *Nav. Res. Logist. Quart.* **20**, 341–356.

Guha, D. (1973). The set covering problem with equality constraints. *Oper. Res.* **21**, 348–350.

Guignard, M., and Spielberg, K. (1972). Mixed integer algorithm for zero-one knapsack problems. *IBM J. Res. Develop.* **16**, 424–430.

Hadley, G. (1962). *Linear Programming.* Addison-Wesley, Reading, Massachusetts.

Hadley, G. (1964). *Nonlinear and Dynamic Programming.* Addison-Wesley, Reading, Massachusetts.

Haldi, J. (1964). 25 integer programming test problems. Working Paper No. 43. Graduate School of Bus., Stanford Univ., Stanford, California.

Haldi, J., and Isaacson, L. M. (1965). A computer code for integer solutions to linear problems. *Oper. Res.* **13**, 946–959.

Hammer, P. L. (1971a). A B-B-B method for linear and nonlinear bivalent programming. *Developments in Operations Research, 1971* (B. Avi-Itzhak, ed.). Gordon and Breach, New York.

Hammer, P. L. (1971b). BABO—A Boolean approach for bivalent optimization. Cent. des Rech. Math., Univ. of Montreal, Montreal, Canada.

Hammer, P. L. (1974). A note on the monotonicity of pseudo-boolean functions. *Z. Oper. Res.* **18**, 47–50.

Hammer, P. L., and Rudeanu, S. (1968). *Boolean Methods in Operations Research and Related Areas.* Springer-Verlag, Berlin and New York.

Hammer, P. L., and Rudeanu, S. (1969). Pseudo-Boolean programming. *Oper. Res.* **17**, 233–261.

Harris, P. M. J. (1964). The solution of mixed integer linear programs. *Operational Res. Quart.* **15**, 117–133.

Healey, W. C., Jr. (1964). Multiple choice programming. *Oper. Res.* **12**, 122–138.

Held, M., and Karp, R. M. (1970). The traveling salesman problem and minimum spanning trees. *Oper. Res.* **18**, 1138–1162.

Held, M., and Karp, R. M. (1971). The traveling salesman problem and minimum spanning trees: Pt. II. *Math. Programming* **1**, 6–25.

Heller, I. (1957). On linear systems with integral valued solution. *Pac. J. Math.* **7**, 1351–1364.

Heller, I., and Tompkins, C. B. (1958). An extension of a theorem of Dantzig. In *Linear Inequalities and Related Systems* (H. W. Kuhn and A. W. Tucker, eds.), pp. 247–254. Princeton Univ. Press, Princeton, New Jersey.

Hess, S., Weaver, J., Siegfeldt, H., Whelan, J., and Zitlau, P. (1965). Nonpartisan political redistricting by computer. *Oper. Res.* **13**, 993–1006.

Hillier, F. S. (1963). Derivation of probalistic information for the evaluation of risky investments. *Management Sci.* **10**, 443–457.

Hillier, F. S. (1967). Chance-constrained programming with 0–1 or bounded continuous decision variables. *Management Sci.* **14**, 34–57.

Hillier, F. S. (1969a). Efficient heuristic procedures for integer linear programming with an interior. *Oper. Res.* **17**, 600–637.

Hillier, F. S. (1969b). A bound-and-scan algorithm for pure integer linear programming with general variables. *Oper. Res.* **17**, 638–679.

Hillier, F. S., and Connors M. M. (1967). Quadratic assignment problem algorithms and the location of indivisible facilities. *Management Sci.* **13**, 42–57.

Hirsch, W. M., and Dantzig, G. B. (1954). The fixed charge problem. Paper P-648, The RAND Corp., Santa Monica, California. Also in *Nav. Res. Logist. Quart.* **15**, 413–424.

Hirsch, W. M., and Dantzig, G. B. (1968). The fixed charge problem. *Nav. Res. Logist. Quart.* **15**, 413–424.

Hoffman, A. J. (1953). Cycling in the Simplex algorithm. *Nat. Bur. Stand.* (*U.S.*) *Rep. No.* 2974 (December).

Hoffman, A. J., and Kruskal, J. B. (1958). Integral boundary points of convex polyhedra. In *Linear Inequalities and Related Systems* (H. W. Kuhn and A. W. Tucker, eds.), pp. 223–246. Princeton Univ. Press, Princeton, New Jersey.

Hohn, F. (1955). Some mathematical aspects of switching. *Amer. Math. Monthly* **62**, 75–90.

House, R. W., Nelson, L. D., and Rado, J. (1966). Computer studies of a certain class of linear integer problems. In *Recent Advances in Optimization Techniques* (A. Lavi and T. Vogl, eds.), pp. 241–280. Wiley, New York.

Hu, T. C. (1963). Multi-commodity network flows. *Oper. Res.* **11**, 344–360.

Hu, T. C. (1966). Decomposition on traveling salesman type problem. *Proc. IFORS* A32-A44.

Hu, T. C. (1968). Decomposition algorithm for shortest paths in a network. *Oper. Res.* **16**, 91–102.

Hu, T. C. (1969). *Integer Programming and Network Flows.* Addison-Wesley, Reading, Massachusetts.

Hu, T. C. (1970). On the asymptotic integer algorithm. *Linear Algebra Appl.* **3**, 279–294.

Hu, T. C. (1971). Some problems in discrete optimization. *Math. Programming* **1**, 102–112.

Huard, P. (1967). Resolutions of mathematical programming with nonlinear constraints by the method of centers. In *Nonlinear Programming* (J. Abadie, ed.), pp. 207–219. Wiley, New York.

Huard, P. (1970). Programmes Mathématiques Nonlineares à Variables Bivalentes. *Proc. Symp. Math. Programming, Princeton, 1970* (H. W. Kuhn, ed.), pp. 313–322. Princeton Univ. Press, Princeton, New Jersey.

Ignal, E., and Schrage, L. (1965). Applications of the branch-and-bound technique to some flow-shop scheduling problems. *Oper. Res.* **13**, 400–412.

Ignizio, J. P. (1971). A heuristic solution to generalized covering problems. Unpublished Ph.D. Dissertation, Virginia Polytech. Inst. Blacksburg.

Iri, M. (1960). A new method for solving transportation network problems. *J. Oper. Res. Soc. Jap.* **3**, 27–87.

Ivanescu, P. L. (1966). *Pseudo-Boolean Programming Methods for Bivalent Programming.* Springer-Verlag, Berlin and New York.

Jeroslow, R. G. (1971). Comments on integer hulls of two linear constraints. *Oper. Res.* **19**, 1061–1069.

Jeroslow, R. G. (1973). There cannot be any algorithm for integer programming with quadratic constraints. *Oper. Res.* **21**, 221–224.

Jeroslow, R. G., and Kortanek, K. O. (1969). Dense sets of two variable integer programs requiring arbitrarily many cuts by fractional algorithms. Management Sci. Res. Rep. No. 174. Carnegie-Mellon Univ., Pittsburgh, Pennsylvania.

Jeroslow, R. G., and Kortanek, K. O. (1971). On an algorithm of Gomory. *SIAM* (*Soc. Ind. Appl. Math.*) *J. Appl. Math.* **21**, 55–60.

Johnson, E. L. (1966). Networks on basic solutions. *Oper. Res.* **14**, 619–624.

Johnson, S. (1954). Optimal two and three stage production schedules with setup times included. *Nav. Res. Logist. Quart.* **1**, 15–21.

Johnson, S. (1959). Discussion: Sequencing in jobs on two machines with arbitrary time lags. *Management Sci.* **5**, 299–303.

Jones, A. P., and Soland, R. M. (1969). A branch-and-bound algorithm for multilevel fixed charge problems. *Management Sci.* **16**, 67–76.

Kalenich, W. A., ed. (1965). *Proc. IFIP Congr. New York, 1965*, **2**, Spartan Books, Washington, D.C.

Kalymon, B. A. (1971). Note regarding "A new approach to discrete mathematical programming." *Management Sci.* **17**, 777–778.

Kaplan, S. (1966). Solution of the Lorie-Savage problem and similar integer programming problems. *Oper. Res.* **14**, 1130–1136.

Karg, R. L. and Thompson, G. L. (1964). A heuristic approach to solving traveling salesman problem. *Management Sci.* **10**, 225–248.

Karp, R. M. (1972). Reducibility among combinatorial problems. Tech. Rep. 3. Comput. Sci., Univ. of California, Berkeley.

Kelley, J. E., Jr. (1960), The cutting plane method for solving convex programs. *SIAM (Soc. Ind. Appl. Math.)* **8**, 703–712.

Kianfar, F. (1971). Stronger inequalities for 0, 1 integer programming using knapsack functions. *Oper. Res.* **19**, 1374–1392.

Kirby, M. J. L., and Scobey, P. F. (1970). Production scheduling on N identical machines. *Can. Oper. Res. Soc. J.* **8**, 14–27.

Kolesar, P. J. (1967). A branch and bound algorithm for the knapsack problem. *Management Sci.* **13**, 723–735.

Koopmans, T. C., and Beckmann, M. J. (1957). Assignment problems and the location of economic activities. *Econometrica* **25**, 53–76.

Krolak, P. (1969). Computational results of an integer programming algorithm. *Oper. Res.* **17**, 743–749.

Krolak, P., Felts, W., and Marble, G. (1971). A man-machine approach toward solving the traveling salesman problem. *Commun. Assoc. Comput. Mach.* **14**, 327–334.

Kuehn, A. A., and Hamburger, M. J. (1963). A heuristic program for locating warehouses. *Management Sci.* **9**, 643–666.

Kuhn, H. W. (1955). The Hungarian method for the assignment problem. *Nav. Res. Logist. Quart.* **2**, 83–97.

Kuhn, H. W. (1956). Solvability and consistency of linear equations and inequalities. *Amer. Math. Monthly* **63**, 217–232.

Kuhn, H. W., ed. (1970). *Proc. Symp. Math. Programming, Princeton, 1967*. Princeton Univ. Press, Princeton, New Jersey.

Kuhn, H. W., and Baumol, W. J. (1962). An approximate algorithm for the fixed charge transportation problem. *Nav. Res. Logist. Quart.* **9**, 1–15.

Kuhn, H. W., and Tucker, A. W., eds. (1958). *Linear Inequalities and Related Systems*. Princeton Univ. Press, Princeton, New Jersey.

Kunzi, H. P., and Oettli, W. (1963). Integer quadratic programming. In *Recent Advances in Mathematical Programming* (R. L. Graves and P. Wolfe, eds.), pp. 303–308. McGraw-Hill, New York.

Lambert, F. (1960). Programmes Lineaires Mixtes. *Cah. Cent. Etud. Rech. Opérationnelle* **2**, 47–126.

Lambert, F. (1962). Programmes en Nombero Entiers et Programmes Mixtes. *Metra* **1**, 11–15.

Land, A. H., and Doig, A. G. (1960). An automatic method for solving discrete programming problems. *Econometrica* **28**, 497–520.

Land, A. H., and Doig, A. G. (1965). A problem of assignment with interrelated costs. *Operational Res. Quart.* **14**, 185–199.

Laughhunn, D. J. (1970). Quadratic binary programming with applications to capital-budgeting problems. *Oper. Res.* **18**, 454–461.

Lavi, A., and Vogl, T., eds. (1966). *Recent Advances in Optimization Techniques*. Wiley, New York.

Lawler, E. L. (1963). The quadratic assignment problem. *Management Sci.* **9**, 586–599.

Lawler, E. L. (1966). Covering problem: Duality relations and a new method of solution. *SIAM (Soc. Ind. Appl. Math.) J. Appl. Math.* **14**, 1115–1132.

Lawler, E. L., and Bell, M. D. (1966). A method for solving discrete optimization problems. *Oper. Res.* **14**, 1098–1112.

Lawler, E. L., and Wood, D. E. (1966). Branch-and-bound methods: A survey. *Oper. Res.* **14**, 699–719.

Lawrence, J., ed. (1969). *Proc. Int. Conf. Operational Res. 5th, Venice, 1969.* Tavistock, London.

Lemke, C. E. (1954). The dual method for Solving the linear programming problem. *Nav. Res. Log. Quart.* **1**, 36–47.

Lemke, C. E., and Spielberg, K. (1967). Direct search zero-one and mixed integer programming. *Oper. Res.* **15**, 892–914.

Lemke, C. E., Salkin, H. M., and Spielberg, K. (1971). Set covering by single branch enumeration with linear programming subproblems. *Oper. Res.* **19**, 998–1022.

Lin, S. (1965). Computer solutions of the traveling salesman problem. *Bell Syst. Tech. J.* **44**, 2245–2269.

Little, J. D. C. (1966). The synchronization of traffic signals by mixed integer linear programming. *Oper. Res.* **14**, 568–594.

Little, J. D. C., Murty, K. G., Sweeney, D. W., and Karel, C. (1963). An algorithm for the traveling salesman problem. *Oper. Res.* **11**, 979–989.

Lorie, J., and Savage, L. J. (1955). Three problems in capital rationing. *J. Bus.* **28**, 229–239.

McCluskey, E. J. (1956). Minimization of Boolean functions. *Bell Syst. Tech. J.* **35**, 1417–1444.

McMahon, G. B., and Burton, P. G. (1967). Flow shop scheduling with the branch-and-bound method. *Oper. Res.* **15**, 473–481.

Maio, A. D., and Roveda, C. (1971). An all zero-one algorithm for certain class of transportation problems. *Oper. Res.* **19**, 1406–1418.

Mallette, A., and Francis, R. (1972). A generalized assignment approach to optimal facility layout. *A.I.I.E. Trans.* **4**, 144–147.

Manne, A. S. (1960). On the job-shop scheduling problem. *Oper. Res.* **8**, 219–223.

Manne, A. S. (1964). Plant location under economies of scale-decentralization and computation. *Management Sci.* **11**, 213–235.

Mao, J. C. T., and Wallingford, B. A. (1968). An extension of Lawler and Bell's method of discrete optimization with examples from capital budgeting. *Management Sci.* **15**, B51–B60; corrections and comments. *Management Sci.* **15**, 481 (1969).

Markowitz, H. M., and Manne, A. S. (1957). On the solution of discrete programming problems. *Econometrica* **25**, 84–110.

Marsten, R. E. (1971). An implicit enumeration algorithm for the set partitioning problem with side constraints. Ph.D. Dissertation, Univ. of California, Los Angeles.

Marsten, R. E. (1974). An algorithm for large set partitioning problems. *Management Sci.* **20**, 774–787.

Martin, G. T. (1963). An accelerated Euclidean algorithm for integer linear programming. In *Recent Advances in Mathematical Programming* (R. L. Graves and P. Wolfe, eds.), pp. 311–318. McGraw-Hill, New York.

Martin, G. T. (1966). Solving the traveling salesman problem by integer linear programming. CEIR, New York.

Martin-Lof, A. (1970). A branch-and-bound algorithm for determining the minimal fleet size of a transportation system. *Transportation Sci.* **4**, 159–163.

Mason, A. T., and Moodie, C. L. (1971). A branch-and-bound algorithm for minimizing cost in project scheduling. *Management Sci.* **18**, B158–B173.

Mathews, G. (1897). On the partition of numbers. *Proc. London Math. Soc.* **28**, 486–490.

Mathis, S. J., Jr. (1971). A counterexample to the rudimentary primal integer programming algorithm. *Op. Res.* **19**, 1518–1522.

Miercort, R. A., and Soland, R. M. (1971). Optimal allocation of missiles against area and point defenses. *Oper. Res.* **11**, 605–617.

Miller, B. L. (1971). On minimizing non-separable functions defined on the integers with an inventory application. *SIAM (Soc. Ind. Appl. Math.) J. Appl. Math.* **21**, 166–185.

Miller, C. E., Tucker, A. W., and Zemlin, R. A. (1960). Integer programming formulation of travelling salesman problems. *J. Assoc. Comput. Mach.* **7**, 326–329.

Mitra, G., Richards, B., and Wolfenden, K. (1970). An improved algorithm for the solution of linear programs by the solution of associated diophantine equations. *R.I.R.O.* **1**, 47–66.

Mitten, L. G. (1970). Branch-and-bound methods: General formulation and properties. *Oper. Res.* **18**, 24–34.

Mizukami, K. (1968). Optimum redundancy for maximum system reliability by the method of convex and integer programming. *Oper. Res.* **16**, 392–406.

Moore, J. E. (1974). *An Improved Branch and Bound Algorithm for Integer Linear Problems.* Ph.D. Dissertation, University of Arkansas, Fayetteville.

Moore, J. M. (1968). An *n*-job one machine sequencing algorithm for minimizing the number of late jobs. *Management Sci.* **15**, 102–109.

Morrison, D. R. (1969). Matching algorithms. *J. Combinatorial Theory* **6**, 20–32.

Murty, K. G. (1968). Solving the fixed charge problem by ranking the extreme points. *Oper. Res.* **16**, 268–279.

Murty, K. G. (1972). A fundamental problem in linear inequalities with applications to the traveling salesman problem. *Math. Programming* **2**, 296–308.

Murty, K. G. (1973). On the set representation and set covering problems. *Symp. Theory Scheduling and Appl., Raleigh, North Carolina, 1972* (S. E. Elmaghraby, ed.), pp. 143–162. Springer-Verlag, Berlin and New York.

Muth, J. F., and Thompson, G. L., eds. (1963). *Industrial Scheduling.* Prentice-Hall, Englewood Cliffs, New Jersey.

Nemhauser, G. L. (1966). *Introduction to Dynamic Programming.* Wiley, New York.

Nemhauser, G. L., and Ullman, Z. (1968). A note on the generalized Lagrange multiplier solution to an integer programming problem. *Oper. Res.* **16**, 450–452.

Nemhauser, G. L., and Ullman, Z. (1969). Discrete dynamic programming and capital allocations. *Management Sci.* **15**, 494–505.

Orchard-Hays, W. (1968). *Advanced Linear Programming Computing Techniques.* McGraw-Hill, New York.

Orden, A. (1956). The Transshipment Problem. *Management Sci.* **2**, 276–285.

Padberg, M. (1970). Equivalent knapsack-type formulations of bounded integer linear programs. Management Sci. Res. Rep. No. 227. Carnegie-Mellon Univ., Pittsburgh, Pennsylvania.

Padberg, M. (1971). Simple zero-one problems: Set covering, matchings and coverings in graphs. Management Sci. Res. Rep. No. 235. Carnegie-Mellon Univ., Pittsburgh, Pennsylvania.

Padberg, M. (1973). On the facial structure of set packing polyhedra. *Math. Programming* **5**, 199–215.

Padberg, M. (1974). Perfect zero-one matrices. *Math. Programming* **6**, 180–196.

Pandit, S. N. N. (1962). The loading problem. *Oper. Res.* **10**, 639–646.

Petersen, C. C. (1967). Computational experience with variants of the Balas algorithm applied to the selection of R and D projects. *Man. Sci.* **13**, 736–750.

Peterson, D. E. and Laughhunn, D. (1971). Capital expenditure programming and some alternative approaches to risk. *Management Sci.* **17**, 320–336.

Picard, J.-C., and Ratliff, H. (1973). A graph-theoretic equivalence for integer programs. *Oper. Res.* **21**, 261–269.

Pierce, J. F. (1968). Application of combinatorial programming to a class of all-zero-one integer programming problems. *Management Sci.* **15**, 191–209.

Pierce, J. F., and Crowston, W. B. (1971). Tree search algorithms in quadratic assignment problems. *Nav. Res. Logist. Quart.* **18**, 1–36.

Pierce, J. F., and Lasky, J. S. (1970). Improved combinatorial programming algorithms for a class of all-zero-one integer programming problems. IBM Cambridge Sci. Cent. Rep. Cambridge, Massachusetts.

Plane, D. R., and McMillan, C. (1971). *Discrete Optimization: Integer Programming and Network Analysis for Management Decisions.* Prentice-Hall, Englewood Cliffs, New Jersey.

Pnueli, A. (1971). An improved starting point for integer linear programming algorithms. *Developments in Operations Research, 1971* (B. Avi-Itzhak, ed.), pp. 83–93. Gordon and Breach, New York.

Pritsker, A. A., Watters, L. J., and Wolfe, P. M. (1969). Multiproject scheduling with limited resources: A 0–1 programming approach. *Management Sci.* **16**, 93–108.

Raghavachari, M. (1969). On connections between 0–1 integer programming and concave programming under linear constraints. *Oper. Res.* **17**, 680–684; Suppl. *Oper. Res.* **15**, 546 (1970).

Rebelein, P. R. (1968). An extension of the algorithm of Driebeek for solving mixed integer programming problems. *Oper. Res.* **16**, 193–197.

Reiter, S., and Rice, D. B. (1966). Discrete optimizing solution procedures for linear and nonlinear integer programming problems. *Management Sci.* **12**, 829–850.

Reiter, S., and Sherman, G. (1965). Discrete optimizing. *SIAM (Soc. Ind. Appl. Math.) J. Appl. Math.* **13**, 864–899.

Revelle, C., Marks, D., and Liebman, J. C. (1970). An analysis of private and public sector location models. *Management Sci.* **11**, 692–707.

Richmond, T. R., and Ravindran, A. (1974). A generalized Euclidean procedure for integer linear programming. *Nav. Res. Logist. Quart.* **21**, 125–144.

Roberts, S. D., and Villa, C. D. (1970). On a multiproduct assembly line balancing problem. *A.I.I.E. Trans.* **2**, 361–364.

Robillard, P. (1971). (0, 1) hyperbolic programming. *Nav. Res. Logist. Quart.* **18**, 47–58.

Root, J. G. (1964). An application of symbolic logic to a selection problem. *Oper. Res.* **12**, 519–526.

Roth, J. P. (1958). Algebraic topological methods for the synthesis of switching systems—I. *Trans. Amer. Math. Soc.* **88**, 301–326.

Roth, J. P., and Karp, R. M. (1962). Minimization over Boolean graphs. *IBM J. Res. Develop.* **6**, 227–238.

Roth, J. P., and Wagner, E. G. (1960). Minimization over Boolean trees. *IBM J. Res. Develop.* **4**, 543–558.

Roth, R. H. (1969). Computer solutions to minimum cover problems. *Oper. Res.* **17**, 455–466.

Roth, R. H. (1970). An approach to solving linear discrete optimization problems. *J. Assoc. Comput. Mach.* **17**, 303–313.

Roveda, C., and Schmid, R. (1971). Two algorithms for a plant storehouse location problem. *Unternehmensforsch.* **15**, 30–44.

Roy, B., Benayoun, R., and Tergny, J. (1970). From S. E. P. procedure to the mixed OPHELIE program. In *Integer and Nonlinear Programming* (J. Abadie, ed.), pp. 419–436. Amer. Elsevier, New York.

Rubin, A. A., and Hammer, P. L. (1969). Quadratic programming with 0–1 variables. Rep. No. 53. Faculty of Ind. and Management Eng., Technion, Israel.

Rubin, D. S. (1970). On the unlimited number of faces in integer hulls of linear programs with a single constraint. *Oper. Res.* **18**, 940–946.

Rubin, D. S. (1972). Redundant constraints and extraneous variables in integer programs. *Management Sci.* **18**, 423–427.

Rubin, D. S., and Graves, R. L. (1972). Strengthened Dantzig cuts for integer programming. *Oper. Res.* **20**, 173–177.

Rudeanu, S. (1969). Irredundant optimization of a pseudo-Boolean function. *J. Optimization Theory Appl.* **4**, 253–259.

Rutledge, R. W. (1967). A simplex method for 0–1 mixed integer linear programs. *J. Math. Anal. Appl.* **18**, 377–390.

Sà, G. (1969). Branch-and-bound and approximate solutions to the capacitated plant location problem. *Oper. Res.* **17**, 1005–1016.

Saaty, T. L. (1970). *Optimization in Integers and Related Extremal Problems.* McGraw-Hill, New York.

Salkin, H. M. (1970). On the merit of the generalized origin and restarts in implicit enumeration. *Oper. Res.* **18**, 549–554.

Salkin, H. M. (1971). A note on Gomory's fractional cut. *Oper. Res.* **19**, 1538–1541.

Salkin, H. (1972). A note on comparing Glover's and Young's simplified primal algorithms. *Nav. Res. Logist. Quart.* **19**, 399–402.

Salkin, H. (1975). *Integer Programming.* Addison-Wesley, Reading, Massachusetts.

Salkin, H. M., and Koncal, R. D. (1973). Set covering by an all integer algorithm: Computational experience. *J. Assoc. Comput. Mach.* **20**, 180–193.

Salveson, M. E. (1955). The assembly line balancing problem. *J. Ind. Eng.* **6**, 18–25.

Scheurmann, A. C. (1971). *Heuristic Integer Linear Programming.* Ph.D. Dissertation, Univ. of Arkansas, Fayetteville.

Schrage, L. (1970). Solving resource-constrained network problems by implicit enumeration—Nonpreventive case. *Oper. Res.* **18**, 263–278.

Scientific Control Systems Ltd. (1970). UMPIRE User's Guide, London (June).

Senju, S., and Toyoda, Y. (1968). An approach to linear programming with 0–1 variables. *Management Sci.* **15**, B196–B207.

Shaftel, T. (1971). An integer approach to modular design. *Oper. Res.* **19**, 130–134.

Shannon, R. E., and Ignizio, J. P. (1970). A heuristic programming algorithm for warehouse location. *A.I.I.E. Trans.* **2**, 361–364.

Shapiro, D. (1966). Algorithms for the solution of the optimal cost traveling salesman problem. Sc.D. Dissertation, Washington Univ., St. Louis, Missouri.

Shapiro, J. F. (1968a). Dynamic programming algorithms for the integer programming problem-I: The integer programming problem viewed as a knapsack type problem. *Oper. Res.* **16**, 103–121.

Shapiro, J. F. (1968b). Group theoretic algorithms for the integer programming problem-II: Extension to a general algorithm. *Oper. Res.* **16**, 928–947.

Shapiro, J. F. (1968c). Shortest route methods for finite state space deterministic dynamic programming problems. *SIAM (Soc. Ind. Appl. Math.) J. Appl. Math.* **16**, 1232–1250.

Shapiro, J. F. (1970). Turnpike theorems for integer programming problems. *Oper. Res.* **18**, 432–440.

Shapiro, J. F. (1971). Generalized Lagrange multipliers in integer programming. *Oper. Res.* **19**, 68–76.

Shapiro, J. F., and H. M. Wagner, (1967). A finite renewal algorithm for the knapsack and turnpike models. *Oper. Res.* **15**, 319–341.

Shareshian, R. (1966). A modification of the mixed integer algorithm of N. Driebeek. Rep. No. 939007. IBM.

Shareshian, R., and Spielberg, K. (1966). The mixed integer algorithm of N. Driebeek. Rep. No. 939013. IBM.

Shaw, M. (1970). Review of computational experience in solving large mixed integer programming problems. In *Applications of Mathematical Programming Techniques* (E. M. L. Beale, ed.), pp. 406–412. English Univ. Press, London.

Simonnard, M. (1966). *Linear Programming.* Prentice-Hall, Englewood Cliffs, New Jersey. (Transl. by W. S. Jewell from original French edition, Dunod, Paris.)

Spielberg, K. (1969a). Algorithms for the simple plant-location problem with some side conditions. *Oper. Res.* **17**, 85–111.

Spielberg, K. (1969b). Plant location with generalized search origin. *Management Sci.* **16**, 165–178.

Srinivasan, A. V. (1965). An investigation of some computational aspects of integer programming. *J. Assoc. Comput. Mach.* **12**, 525–535.

Steinberg, D. I. (1970). The fixed charge problem. *Nav. Res. Logist. Quart.* **17**, 217–236.

Steinmann, H., and Schwinn, R. (1969). Computational experience with a 0–1 programming problem. *Oper. Res.* **17**, 917–920.

Svestka, J., and Huckfeldt, V. (1973). Computational experience with an M-salesman traveling salesman algorithm. *Management Sci.* **19**, 790–799.

Swarc, W. (1971). Some remarks on the time transportation problem. *Nav. Res. Logist. Quart.* **18**, 473–487.

Taha, H. A. (1971a). Sequencing by implicit ranking and zero-one polynomial programming. *A.I.I.E. Trans.* **2**, 157–162.

Taha, H. A. (1971b). On the solution of zero-one linear programs by ranking the extreme points. Tech. Rep. No. 71-2. Dept. of Ind. Eng., Univ. of Arkansas, Fayetteville.

Taha, H. A. (1971c). Solution of integer linear programs using cuts and imbedded zero-one problems. Tech. Rep. No. 71-5. Dept. of Ind. Eng., Univ. of Arkansas, Fayetteville.

Taha, H. A. (1971d). Hyperbolic programming with bivalent variables. Tech. Rep. No. 71-7. Dept. of Ind. Eng., Univ. of Arkansas, Fayetteville.

Taha, H. A. (1971e). *Operations Research: An Introduction.* Macmillan, New York.

Taha, H. A. (1972a). A Balasian-based algorithm for zero-one polynomial programming. *Management Sci.* **18**, B328–B343.

Taha, H. A. (1972b). Further improvements in the polynomial zero-one algorithm. *Management Sci.* **19**, B226–B227.

Taha, H. A. (1973). Concave minimization over a convex polyhedron. *Nav. Res. Logist. Quart.* **20**, 533–548.

Thangavelu, S. R., and Shetty, C. M. (1971). Assembly line balancing by 0–1 integer programming. *A.I.I.E. Trans.* **3**, 64–69.

Thompson, G. L. (1964). The stopped simplex method: Basic theory for mixed integer programming. *Rev. Fr. Rech. Opérationelle* **8**, 159–182.

Tomlin, J. A. (1970). Branch and bound methods for integer and non-convex programming. In *Integer and Nonlinear Programming* (J. Abadie, ed.), pp. 437–450. Amer. Elsevier, New York.

Tomlin, J. A. (1971). An improved branch-and-bound method for integer programming. *Oper. Res.* **19**, 1070–1074.

Toregas, C., Swain, R., Revelle, C., and Bergman, L. (1971). The location of emergency service facilities. *Oper. Res.* **19**, 1363–1373.

Trauth, C. A., and Woolsey, R. E. (1968). MESA; A heuristic integer linear programming Technique. Res. Rep. SC-RR-68-299. Sandia Labs., Albuquerque, New Mexico.

Trauth, C. A., and Woolsey, R. E. (1969). Integer linear programming: A study in computational efficiency. *Management Sci.* **15**, 481–493.

Trubin, V. A. (1969). On a method of solution of integer programming problems of a special kind. *Soviet Math.* **10**, 1544–1546.

Tuan, N. P. (1971). A flexible tree search method for integer programming problems. *Oper. Res.* **19**, 115–119.

Tuy, H. (1964). Concave programming under linear constraints. *Sov. Math.* **5**, 1437–1440.

Unger, V. E. (1970). Capital budgeting and mixed 0–1 integer programming. *A.I.I.E. Trans.* **2**, 28–36.

Uskup, E., and Smith, S. B. (1975). A branch-and-bound algorithm for two-stage production-sequencing problem. *Oper. Res.* **23**, 118-136.

Valenta, J. R. (1969). Capital equipment decisions: A model for optimal system interfacing. M.S. Thesis, MIT, Cambridge, Massachusetts.

Veinott, A. F., Jr. (1968). Extreme points of Leontief substitution systems. *Linear Algebra Appl.* **1**, 181–194.

Veinott, A. F., Jr., and Dantzig, G. B. (1968). Integral Extreme Points. *SIAM (Soc. Ind. Appl. Math.) Rev.* **10**, 371–372.

Wagner, H. M. (1957). The dual simplex algorithm for bounded variables. *Nav. Res. Logist. Quart.* **4**, 257–261.

Wagner, H. M. (1959). An integer linear programming model for machine scheduling. *Nav. Res. Logist. Quart.* **6**, 131–140.

Wagner, H. M. (1969). *Principles of Operations Research.* Prentice-Hall, Englewood Cliffs, New Jersey.

Wagner, H. M., Giglio, R. J., and Glaser, R. G. (1964). Preventive maintenance scheduling by mathematical programming. *Management Sci.* **10**, 316–334.

Wagner, W. H. (1968). An application of integer programming to legislative redistricting. *Nat. Meeting of ORSA, 34th, 1968.*

Wahi, P. N., and Bradley, G. H. (1969). Integer programming test problems. Rep. No. 28. Dept. of Administrative Sci., Yale Univ., New Haven, Connecticut.

Walker, R. J. (1960). An enumerative technique for a class of combinational problems. *Proc. Symp. Appl. Math. Amer. Math. Soc., 10th, 1960* (R. E. Bellman and M. Hall, Jr., eds.), pp. 90–94.

Warszawki, A. (1974). Pseudo-boolean solutions to multidimensional location problems. *Oper. Res.* **22**, 1081–1085.

Watters, L. J. (1967). Reduction of integer polynomial problems to zero–one linear programming problems. *Oper. Res.* **15**, 1171–1174.

Webb, M. H. J. (1971). Some methods of producing approximate solutions to travelling salesmen problems with hundreds or thousands of cities. *Opernational Res. Quart.* **22**, 49–66.

Weingartner, H. M. (1963). *Mathematical Programming and the Analysis of Capital Budgeting Problems.* Prentice-Hall, Englewood Cliffs, New Jersey.

Weingartner, H. M. (1966). Capital budgeting of interrelated projects: Survey and synthesis. *Management Sci.,* **12**, 485–516.

Weingartner, H. M., and Ness, D. N. (1967). Methods for the solution of multi-dimensional 0/1 knapsack problems. *Oper. Res.* **15**, 83–103.

White, J. A., and Francis, R. L. (1971). Solving a segregated storage problem using branch-and-bound and extreme point ranking methods. *A.I.I.E. Trans.* **3**, 37–44.

White, W. W. (1961). On Gomory's mixed integer algorithm. Senior Thesis, Princeton Univ., Princeton, New Jersey.

White, W. W. (1966). On a group theoretic approach to linear integer programming. ORC 66-27. Oper. Res. Cent., Univ. of California, Berkeley.

Wilson, R. B. (1967). Stronger cuts in Gomory's all-integer programming algorithm. *Oper. Res.* **15**, 155–157.

Wilson, R. B. (1970). Integer programming via modular representations. *Management Sci.* **16**, 289–294.

Witzgall, C. (1963). An all-integer programming algorithm with parabolic constraints. *SIAM* (*Soc. Ind. Appl. Math.*) *J. Appl. Math.* **11**, 855–871.

Witzgall, C., and Zahn, C. T., Jr. (1965). Modification of Edmonds' maximum matching algorithm. *J. Res. Nat. Bur. Stand. Sect. B* **69**, 91–98.

Wolsey, L. A. (1970). Mixed integer programming: Discretization and the group-theoretic approach. Ph.D. Dissertation, MIT, Cambridge, Massachusetts.

Wolsey, L. A. (1971a). Extensions of the group theoretic approach in integer programming. *Management Sci.* **18**, 74–83.

Wolsey, L. A. (1971b). Group-theoretic results in mixed integer programming. *Oper. Res.* **19**, 1691–1697.

Woolsey, R. E. (1972). A candle to Saint Jude, or four real world applications of integer programming. *Interfaces* **2**, 20–27.

Young, R. D. (1965). A primal (all-integer) integer programming algorithm. *J. Res. Nat. Bur. Stand. Sect. B* **69**, 213–250.

Young, R. D. (1968). A simplified primal (all-integer) integer programming algorithm. *Oper. Res.* **16**, 750–782.

Young, R. D. (1971). Hypercylinderically deduced cuts in 0–1 integer programming. *Oper. Res.* **19**, 1393–1405.

Zimmermann, H.-J., and Pollatsckek, M. A. (1975). The probability distribution function of the optimum of a 0–1 linear program with randomly distributed coefficients of the objective function and the right-hand side. *Oper. Res.* **23**, 137–149.

Zionts, S. (1968). On an algorithm for the solution for mixed integer programming problems. *Management Sci.* **15**, 113–116.

Zionts, S. (1969). Toward a unifying theory of integer linear programming. *Oper. Res.* **17**, 359–367.

Zionts, S. (1973). *Linear and Integer Programming.* Prentice-Hall, Englewood Cliffs, New Jersey.

Zoutendijk, G. (1970). Enumeration algorithms for the pure and mixed integer programming problem. *Proc. Symp. Math. Programming, Princeton, 1967* (H. W. Kuhn, ed.), pp. 232–338. Princeton Univ. Press, Princeton, New Jersey.

Zwart, P. (1973). Nonlinear programming: Counter examples to two global optimization algorithms by Ritter and Tui. *Oper. Res.* **21**, 1260–1266.

Index

A

Accelerated fractional cut
  Gomory's, 190
  Martin's, 226
Additional constraint, 62
Additive algorithm of Balas, 94
  interpreted as a branch-and-bound method, 144
  relationship to linear programming, 98–99
Aggregation of constraints as a knapsack model, 266–269
Airline crew scheduling problem, 21
All-integer cut
  dual, 192–197, *see also*, Bound-escalation cut
  primal, 216–225
Almost dual methods, 28, 213, *see also* branch-and-bound
Alternative optima in linear programming, 49
Applications of integer programming, 8–27, 30–33
Approximate integer algorithms, 347, *see also* Heuristics
Arc, 30, 31, 242
Artificial constraint, 60

B

Artificial variable, 38, *see also* Two-phase method
Assignment problem, 134
  quadratic, 30
  solution by branch-and-bound, 173
  traveling salesman algorithms, 308–316
  traveling salesman formulation, 305
Asymptotic integer algorithm, 230
  sufficient condition for feasibility of basic variables, 239, 261, 262

Backtracking, *see* Backward move
Backward move, 90–91
Basic solutions, 35–36
  relationship to extreme points, 36
  relationship to solution of zero–one problems, 86, 98–99
Benders' cut, 127
Benders' partitioning approach, 67
  applied to mixed zero–one problem, 126–132
  relationship to implicit enumeration, 130
Bound-escalation cut, 202

Bounded variables, 74
  dual algorithm, 77
  primal algorithm, 75
  use in asymptotic algorithm, 256
  use in branch-and-bound, 152
Bounding, 140, 143, *see also* Branch-and-bound
Branch-and-bound, *see also* Zero–one algorithm
  branching variable selection, 147, 161, 169
  concept, 139
  fixed-charge, 289–292
  general integer algorithms, 144–164
  group problem, 255–257
  mixed zero–one problem, 119–126, 131–132
  node selection, 167
  node swapping, 167
  nonlinear problems, 171
  plant location, 300–303
  set covering, 323–326
  traveling salesman, 308–316
Branching, 140, 143, *see also* Branch-and-bound
Branching variable, rules for, 147, 161, 169

C

Capital budgeting, 8
  chance constraints, 10
  risk aversion, 10
Cargo loading problem, *see* Knapsack problem
Coloring problem, 2, 30
Column simplex method, 63
Commercial integer codes, 166, 346
Complementary slackness, 58
Completion, 90
Composite algorithm, 346–347
Composite cut, *see* Accelerated fractional cut
Computations in integer programming, *see also* Formulating integer models
  asymptotic problem, 260
  branch-and-bound, 164–171
  cutting methods, 190, 224
  fixed-charge, 288, 297–298
  mixed zero–one problem, 126
  nonlinear zero–one problem, 119
  plant location, 303
  polynomial zero–one problem, 115

set covering, 326, 331
traveling salesman, 308, 311, 316
zero–one problem, 100, 107, 115, 138
Concave function, minimization of, 32
  applied to fixed-charge problem, 292
Concave program representation of zero–one problem, 133
Cone, 232
Congruence, 258
Convex hull, 232
Convexity cut, 205
  use in fixed-charge problem, 294
Corner polyhedron, 232
  faces of, 259
Covering problem, *see* Set covering problem
Cut
  all-integer, 192, 216
  Benders', 127
  bound-escalation, 202
  convexity or intersection, 205, **294**
  Dantzig's, 178
  fractional, 179
  mixed, 198
  primal, 216
Cut search method, 213
Cutting methods, special
  fixed-charge, 292
  set covering, 326
  traveling salesman, 306
Cutting plane, *see* Cut
Cutting stock problem, 264–266
Cycling in simplex method, 48, 65

D

Dantzig's cut, 178
Degeneracy in simplex method, 48
Dichotomies, 26
Diophantine equations, aggregation of, 266
Direct search methods, 347–350
Driebeek's penalties, 120
Dual cutting-plane methods, 28, 179
  all-integer, 192
  bound-escalation, 202
  convexity or intersection, 205
  fractional, 179
  mixed, 198
Dual simplex algorithm, 59
  bounded variables, 77

column tableau, 63
 lexicographic, 65
Duality in linear programming, 52
 relationships between primal and dual solutions, 54
 use in Benders' partitioning approach, 67

E

Either-or constraint, 26, 30
Enumeration, *see* Implicit enumeration, concept of
Extreme point, 36
 determination of adjacent extreme point, 281
 relationship to basic solution, 36
 relationship to solution of zero–one problems, 86–87, 98–99
Extreme point ranking, 276, 293, 339

F

Fathoming, 88
 of partial solutions, 90
Fathoming tests, 94
 linear zero–one problem, 94
 nonlinear zero–one problem, 116
 polynomial zero–one problem, 107
 quadratic binary problem, 137
 surrogate constraints, 100
Fixed-charge problem, 22, 285, *see also* Plant location problem
 approximate algorithm, 287
 exact algorithms, 289, 292
 solution as a linear program, 286
Fixed-charge transportation problem, 339
Fixed-order tree search, 116–117
Flexible tree search, 93, 98
Flyaway kit problem, *see* Knapsack problem
Formulating integer models, 343, *see also* Applications of integer programming
Forward move, 90–91
Four color problem, 30
Fourier-Motzkin elimination, 274
 for solving knapsack problem, 272–276
Fractional cut, 179
 acceleration, 190, 226
 convergence, 188

properties, 181
 selection of source row, 184
 strength, 183
Fractional zero–one programming, *see* Hyperbolic zero–one programming
Free variable, 90

G

Generalized origin, 138
Glover's enumeration scheme, 90
Group
 cyclic, 238
 finite, 238
Group problem in asymptotic algorithm, 241
 dynamic programming algorithm, 247
 shortest-route algorithm, 241
 solution of integer problem, 253–258

H

Hermite normal form, 259
Heuristics
 branch-and-bound, 165–171
 fixed-charge, 287
 knapsack, 279
 plant location, 299
 set covering, 322
 traveling salesman, 316
Hyperbolic zero–one programming, 136
Hypercylindrically deduced cut, 205

I

Implicit enumeration, concept of, 87
 zero–one problem, 85
Information retrieval, 33
Integer forms, method of, *see* Fractional cut
Integer programming
 mathematical definition of, 2
 mixed problem, 2
 pure problem, 2
Integer programming methods, 27
 asymptotic or group, 230–262
 branch-and-bound, 139–176
 cutting-plane, 177–229
 implicit enumeration, 85–138
 specialized, 263–341
Intersection cut, 208, *see also* Convexity cut
Involuntary basis in set covering, 319

K

Knapsack problem, 18, 263, 336, 338, *see also* Group problem in asymptotic algorithm algorithms, 269–284
application to cutting stock problem, 264–266
approximate solution, 279
reduction of integer problem to, 266–269

L

Land–Doig's algorithm, 145
Lexicographic dual simplex method, 65
in cutting-plane method, 185
LIFO rule, 92, 313
Line balancing, application of set covering to, 316
Linear equivalence of polynomial terms, 109
Linear programming cone, 232
Linear programming, definition of, 34
additional constraint, 62
alternative optima, 49
artificial constraint, 60
artificial variable, 38
basic solution, 35
bounded variables, 74–79
column simplex method, 63
complementary slackness, 58
cycling, 48
degeneracy, 48
dual problem, 52–54
dual simplex method, 59
extreme point, 36
feasibility condition, 40
lexicographic simplex method, 65
matrix representation, 44
optimality condition, 42
relationships between primal and dual, 54–58
revised simplex method, 49–52
simplex method, 37–44
two-phase method, 39–40, 80
Linear zero–one problem, 94
Linearization of a curve, *see* Separable programming

M

Martin's cut, 226
applied to traveling salesman problem, 308
Matrix representation of simplex tableau, 44
Minimal cuts in network flows, 31
Mixed asymptotic integer problem, 258
Mixed cut, 198
viewed as a convexity cut, 212
Mixed zero–one problem, 119–132

N

Network flows, 31
Node selection in branch-and-bound methods, 167
best projection criterion, 167
percentage-error criterion, 169
pseudocosts criterion, 168
Node swapping in branch-and-bound methods, 167
Nonlinear integer problem, 116, 161, *see also* Fixed-charge problem, Traveling salesman problem
Normal forms, 259

P

Partial enumeration, *see* Implicit enumeration, concept of
Partial solution
completion of, 90
definition, 90
Partitioning approach, 67–74
Partitioning in zero–one mixed problems, 126, 131
Penalties
computational experience, 166
failure of penalty approach, 166
fractional cut, 184
general integer problem, 154, 166
zero–one problem, 121
Periodicity of knapsack problem, 337
Piecewise linearization, *see* Separable programming
Plant location problem, 298, 339
approximate algorithm, 299
exact algorithm, 300

Political districting, 31
Polynomial zero–one problem, 107, *see also* Zero–one algorithm, nonlinear
linear equivalence of, 109
Primal cut, 216
Primal methods, 28
Pure integer problem
    asymptotic algorithm, 230
    branch-and-bound, 144
    cutting plane, 179, 192, 202, 213, 216
    implicit enumeration, 85

Q

Quadratic assignment problem, 30
Quadratic zero–one problem, 116, 137, *see also* Capital budgeting

R

Ratio problem, 32
Relaxed problem in asymptotic algorithm, *see* Group problem in asymptotic algorithm
Revised simplex method, 49
Rounding continuous optimum, 4, *see also* Asymptotic integer algorithm
    limitations of, 5

S

Scheduling, 10, *see also* Sequencing
Separable programming, 24, 32
Separation, *see* Branching
Sequencing, 12, 26, 30
Set covering problem, 316
    approximate algorithm, 322
    conversion of set partitioning, 333
    exact algorithms, 323–331
    properties, 317
    size reductions, 321
Set partitioning problem, 332
    equivalence to set covering, 333
    size reduction, 332, 340, 341
Shortest route algorithm
    group problem, 241
    knapsack problem, 270
    network representation, 242

Simplex method, 37
    for bounded variables, 75
Source row, 180
Subtour elimination algorithm, 309, *see also* Traveling salesman problem
Surrogate constraints, 100, 134, 135
Switching theory, application of set covering, 316

T

Tour building algorithm, 311, *see also* Traveling salesman problem
Traveling salesman problem, 19, 304
    branch-and-bound, 308–316
    cutting plane, 306–308
    relationship to assignment problem, 305
    tours and subtours, 305
    triangular inequality, 304
Tree search, order of
    flexible, 93, 98
    fixed, 116–117
    LIFO, 92, 313
Triangular inequality, 304
Trim loss, *see* Cutting stock problem
Two-phase method, 39, 80

U

UMPIRE, 166
Unbounded linear programming solution, 49
Unimodular property, 5

W

Wilson's all-integer cut, 194–195

Z

Zero–one algorithm
    linear, 94
    mixed, 119
    nonlinear, 116
    polynomial, 107
    quadratic, 116

Zero–one equivalence of integer problems, 86
Zero–one implicit enumeration, 85
  enumeration scheme, 89
  generalized origin, 138

Zero–one problem
  concave programming representation, 133
  relationship to extreme point solution,
    86–87, 98–99

A
B 5
C 6
D 7
E 8
F 9
G 0
H 1
I 2
J 3